T. W. Schultz

ANIMAL AGRICULTURE

SPONSORS OF THE CONFERENCE

American Association for the Advancement of Science
American Dairy Science Association
American Meat Science Association
American Society of Animal Science
Association of American Veterinary Colleges
College of Veterinary Medicine, Michigan State University
Michigan State Agricultural Experiment Station
Poultry Science Association
U.S. Department of Agriculture
Winrock International Livestock Research and Training Center

Agricultural experiment stations at the following universities supported the conference by partial financial support of participants:

University of Arizona
University of California
Colorado State University
Cornell University
University of Florida
University of Georgia
University of Illinois
Iowa State University
Kansas State University
University of Kentucky
Michigan State University
University of Minnesota
University of Missouri
University of Nebraska
New Mexico State University
North Carolina State University
Ohio State University
Oklahoma State University
Pennsylvania State University
Purdue University
Texas A & M University
Virginia Polytechnic Institute and State University
Washington State University
West Virginia University
University of Wisconsin
University of Wyoming

Colleges of veterinary medicine at the following universities supported the conference by partial financial support of participants:

University of California
Cornell University
University of Georgia
University of Illinois
Iowa State University
Michigan State University
University of Minnesota
University of Pennsylvania
Purdue University
Virginia Polytechnic Institute and State University

The following organizations also supported the conference by partial financial support of participants:

American Meat Institute
Animal Welfare Institute
Michigan Farm Bureau
Michigan Milk Producers Association
National Academy of Science
National Cattlemen's Association
National Dairy Council
National Dairy Herd Improvement Association
National Food Processors Association
National Pork Producers' Council
Select Sires, Inc.
United States Animal Health Association
U.S. Department of Health, Education, and Welfare

STEERING COMMITTEE

Ned Bayley
**Robert Bray
David M. Burns
Larry J. Connor
**Hank Fitzhugh
*Harold D. Hafs
*Virgil W. Hays
Gilbert A. Leveille
William Stadelman
Sylvan Wittwer
John R. Welser

*Steering Committee Co-Chairman
**Program Committee Co-Chairman

KEYNOTE SPEAKERS

Philip H. Abelson
Editor, *Science* Magazine

Carol Tucker Foreman
Assistant Secretary, Food and Consumer Services
U.S. Department of Agriculture

Anson R. Bertrand
Director, Science and Education Administration
U.S. Department of Agriculture

Glenn W. Salisbury
Director Emeritus, Agricultural Experiment Station
University of Illinois

Sylvan H. Wittwer
Director, Agricultural Experiment Station
Michigan State University

These proceedings represent the best collective judgment on research priorities by a group of informed and dedicated people concerned with the future role of animal agriculture in meeting human needs

ANIMAL AGRICULTURE

RESEARCH TO MEET HUMAN NEEDS IN THE 21ST CENTURY

edited by

Wilson G. Pond, Robert A. Merkel,
Lon D. McGilliard, and V. James Rhodes

Westview Press / Boulder, Colorado

Any opinions, findings, conclusions, or recommendations expressed in this publication are those of the contributors and do not necessarily reflect the views of the sponsors.

Published for the Conference on Animal Agriculture in 1980 in the United States of America by:

Westview Press, Inc.
5500 Central Avenue
Boulder, Colorado 80301
Frederick A. Praeger, Publisher

Library of Congress Catalog Card Number: 80-52211
ISBN: 0-86531-032-7

Printed and bound in the United States of America

CONTENTS

ORGANIZERS' PREFACE

Animal agriculture serves human needs. Three-fourths of the protein, one-third of the energy, most of the calcium and phosphorus and substantial amounts of essential vitamins and other minerals in the American diet are from animal products. Productivity of animal agriculture rapidly increased during a period of abundant national resources and low-cost energy, primarily as a consequence of applying technology developed through research.

This era has passed and the future for animal agriculture holds obvious challenges. National resources will be increasingly scarce and much more expensive. Efficiency of resource utilization as well as productivity of food will be essential to meeting needs of an expanding human population. The reservoir of yet unapplied technology is nearly exhausted.

Recognizing that innovative research holds the key to meeting these challenges, 210 concerned individuals convened at Boyne Mountain, Michigan, May 4 to 9, 1980. Their goal was to identify priorities for future research to enable animal agriculture to efficiently and effectively serve human needs in the 21st century.

This Conference differed from others largely in the focus on animal agriculture as a complex system, serving the interests of consumers as well as of producers of animal products. Joining with animal scientists, veterinarians, producers and industry representatives were agronomists, ecologists, nutritionists, chemists, sociologists, anthropologists, economists, engineers and political scientists. The unifying ties among this broad and talented array of specialists were their understanding and appreciation of research and their common concern for human welfare.

The structure of the Conference was designed to promote productive discussion among individuals with differing training, experience and attitudes. First, Chairpersons and Rapporteurs - all respected, experienced professionals - were carefully selected. Then, although well over 2,500 persons were nominated, the number per working group was strictly limited to enable effective dialogue. Ten working groups were organized to address the critical challenges facing animal agriculture. Of these ten groups - Human Nutrition; Food Processing and Acceptability; Food Safety; Animal Nutrition; Animal Genetics and Reproduction; Animal Health; Feed Production; Production, Marketing and Distribution; Resources and Environment; and Public Policy - seven were outside the boundaries of the traditional animal sciences.

Interactions between subject matters were identified and dialogue between working groups was encouraged. Keynote speakers were selected because they had a special message; they were asked to stimulate, not placate; they succeeded.

We, the organizers of the Conference, felt that this breadth of approach was essential to the development of innovative, effective research strategies for the future. It was a gamble. This was not to be a comfortable gathering of people accustomed to talking together. And, in fact, the initial attitudes of participants ranged from skepticism to apprehension and pessimism that agreement could ever be reached on the diverse and complex issues with which they were challenged. Even communication was difficult at first; the jargon of one scientific discipline was not known by other scientists, much less by producers in the working group. But confrontation, conflict and even chaos gradually diminished until, by the third day, there was a noticeable development of esprit de corps, a uniting of individuals for the technical expertise and practical perspective of their cohorts. True to the organizers' expectations, professionalism and common concern for human needs united the conferees.

The gamble was successful! The comprehensive program developed by the Conference can meet the challenges of the future.

In this moment of pride and pleasure with the success of the Conference, it is fitting to reflect on its evolution and recognize the support that made it possible. Two years have passed since the officers of four scientific societies - American Dairy Science Association, American Meat Science Association, American Society of Animal Science and Poultry Science Association - activated an Intersociety Research Committee to develop a comprehensive research plan for the animal sciences.

As this group worked with scientists, producers and industry representatives, they became aware that an even more broadly based approach was necessary to design a highly effective research program. Special concerns surfaced. The impact of animal agriculture on human nutrition, food safety and resource conservation could not be adequately addressed by animal scientists alone.

As the dimensions of the challenges unfolded, it became increasingly apparent that a holistic approach to developing a research program would be necessary. So the Steering Committee was expanded. Necessary financial support was provided by newly added organizations, all concerned with the future role of animals in meeting human needs.

The scope of the Conference was restricted in two ways: principal attention was devoted to animal agriculture in the U.S.; and emphasis was placed on the major domesticated food-producing species (poultry, swine, sheep and cattle). Notwithstanding these restraints, the products of the Conference should have applicability to all segments of animal agriculture, both national and international.

The Steering Committee disbanded at the close of the Conference and a Task Force was established to enhance implementation of the research priorities identified at the Conference.

With optimism and enthusiasm for the future, the sponsors, organizers and participants anticipate that realization of the research goals developed in the Conference will enable animal agriculture to serve human needs in the 21st century.

"...number one on the list of human needs is a dependable food supply."

Anson Bertrand

"There will be a worldwide shift from...a natural-resource-based to a more science-based agriculture."

Sylvan Wittwer

EDITORS' PREFACE

The Editors were given the responsibility to edit intermediate and final copies of chapters prepared by each of 10 working groups of the Conference on Animal Agriculture and to edit and arrange the format of the text of five keynote addresses of the Conference. The Executive Summary synthesizes the overall state of the art and science of each subject of the 10 working groups, indicates constraints and identifies broad areas of imperatives for research in animal agriculture to meet human needs in the 21st century.

Co-Chairpersons and Rapporteurs recorded the consensus within each working group and presented the summary of their efforts to Hank Fitzhugh and Harold Hafs of the Conference Steering Committee who prepared an Executive Summary which includes Conference Imperatives, published as Chapter 1 of the Conference Proceedings. The Executive Summary is published as a separate document for readers who have an interest in the overall conclusions and recommendations arising from the Conference. All working groups reached unanimity in their final reports; this simplified the work of the Editors and reflected the cooperative and mission-oriented spirit that seemed to pervade the atmosphere throughout the Conference. The busy reader should find the Executive Summary useful for selective reading of other chapters.

The Conference participants, listed with their institutional affiliation at the end of the complete Conference Proceedings, were selected by the Conference Steering Committee from a broad range of disciplines bearing directly and indirectly on animal agriculture. Chapters 2 through 11 of the Conference Proceedings were written by each of 10 subject matter groups of 15-20 participants literally during the 5 days of the Conference. The group dynamics, evident in identifying constraints on animal research and in enunciating research imperatives that resulted in this book, were a remarkable phenomenon requiring exemplary cooperation and coordination among the Co-Chairpersons, Rapporteurs and other participants and the typists, all of whom worked long, hectic hours during the entire week of the

Conference to complete the task. The Editors are grateful for this team spirit and effort.

The Editors express special gratitude to Naomi Revzin, who handled many of the arrangements before, during and after the Conference to make the work of the Editors easier and more pleasant. The Editors are deeply indepted to Theresa Fillwock, Rose Hannewald, Sylvia Jupin, Libby LaGoe and Linda Smith for typing all drafts of the Proceedings under exceedingly high time pressures. Bernice Wilson also is acknowledged for her efficient editorial assistance, and Limin Kung, Kristen Sejrsen and Leon Spicer are recognized gratefully for performing a variety of duties that were of value to the Editors. Lynne Rienner, of Westview Press, was helpful in providing guidance to the Editors in preparing final copy for the printer.

Finally, we are indebted to Harold Hafs and Virgil Hays, Co-Chairmen of the Conference, for their invitation to edit the Proceedings and for their patience and support in the final hours of the process.

We hope the finished product proves to be an effective vehicle for transferring the important information and challenges to the readers who will influence the future course of animal agriculture research in meeting the human needs of the 21st century.

Wilson G. Pond
Robert A. Merkel
Lon D. McGilliard
V. James Rhodes

"...the base of fundamental knowledge from which most of us in science are working is not very great... Unless we begin now a long-term commitment to restock our storehouse of fundamental knowledge, it will have been picked bare long before the year 2000. I do not believe that the general public, whose human needs are in the forefront in setting our priorities, has the slightest idea of how precarious our position is."

Anson Bertrand

ACKNOWLEDGMENTS

The work of this Conference was shared by many persons. The contributions of the Keynote Speakers, Chairpersons, Rapporteurs and Participants obviously deserve principal recognition. However, there are others whose work behind the scene was critical to the success of the Conference.

Naomi Revzin shouldered responsibility for coordinating local arrangements, supervising manuscript preparations and uncounted other tasks for months before, during and after the Conference. Her professionalism, good humor and untiring efforts set an example for all.

Wallace Barr (Ohio State University), Ernest Briskey (Oregon State University), Bart Cardon (Arizona Feeds), Raymond Loehr (Cornell University) and Sanford Miller (Bureau of Foods, FDA) contributed during the preparative stages but were unable to participate during the Conference.

Our fellow Steering Committee members were outstanding in their enthusiasm and willingness to contribute their time and expertise. Particular recognition is due to Sylvan Wittwer for his leadership in conceptualizing the Conference and for his consistent support through all phases of Conference development.

Finally the four of us wish to express personal appreciation to our secretaries - Veryl Kreigh, Libby Fowler, Ruth Shell and Alice Mann - for their efficient efforts in behalf of the Conference.

Robert Bray   Harold Hafs
Hank Fitzhugh   Virgil Hays

The Conference Participants

ANIMAL AGRICULTURE

1

H. A. Fitzhugh and H. D. Hafs

CONFERENCE IMPERATIVES

"Every great advance in science has issued from a new audacity of imagination."
--Keynote speaker Carol Tucker Foreman quoting John Dewey, Boyne Mountain, Michigan, May 1980.

It was no trivial challenge.

The charge was bold; the challenge was exciting: Using science as the foundation, fashion research goals for the next 100 years to ensure that food and other products from animals will nurture and sustain a better quality of human life, for consumers and producers alike.

This was the call that went out to 210 scientists, consumerists, producers, economists, industry people, engineers, human nutrition and medical research workers and plant scientists gathered at the May 1980 conference at Boyne Mountain, Michigan, on "Animal Agriculture: Meeting Human Needs in the 21st Century." And hearing it, they answered. A summary of their responses follows.

The Conference focused on research. However, improving animal agriculture so that the well-being of people continues to be served suggests imperatives beyond specific scientific studies.

First, enough scientists must be found to do the research recommended by the conference participants. In some cases, this will require training scientists in new fields; in other cases, it will dictate that individuals refocus and redirect their research efforts. Scientists must be trained to work in the multidisciplinary teams needed to research complex problems and questions. Society must provide researchers with signals of funding and moral support to assure that sufficient human and financial resources are spent on appropriate research.

Second, a broad range of people has a vested interest in the successful realization of the recommended research priorities. In the words of Carol Tucker Foreman ". . . producers and consumers need each other. They are natural allies, not born antagonists." An imperative of this Conference is to promote collaboration among all people involved with and benefiting from animal agriculture.

Third, categorical distinctions should not be made among types of research which should have priority. The conflict between basic and applied research need not exist if the focus remains on research to meet human needs. The tremendous advancements already made in animal agriculture came from a foundation of basic research. But, as Keynoter Anson Bertrand noted,". . . the base of fundamental knowledge from which most of us in science are working today is not very great. . .the general public (does not have) the slightest idea how precarious our position is." We require basic knowledge to feed applicational research.

Some participants were concerned at first that the parochial interests within each Conference working group might limit the scope of their respective recommendations on research priorities, but "common concerns" surfaced among all of the working groups. These concerns sharply influenced research recommendations, so we list the three major ones here before enumerating the research imperatives.

SECURITY OF NATURAL RESOURCES

The public is aware that our supplies of fossil fuel energy are rapidly dwindling. The resultant shocks to our economy make this a primary concern. Keynoter Anson Bertrand noted, "Innovation is needed. . . throughout our food system. Much of the current technology was developed in an era of cheap energy." As mismanagement, erosion, pollution and urban encroachment have taken their toll on our once abundant agricultural resources - land and water - concern has deepened about the future of food production in the U.S. The still growing human population must feed itself in the future from less abundant resources. Our options are limited. Keynoter Sylvan Wittwer predicted, "There will be a worldwide shift from. . .a natural resource based-agriculture to a more science based-agriculture."

CONFIDENCE IN AND CONCERN FOR THE ROLE OF SCIENCE

A public accustomed to scientific miracles in communication, transportation and computation expects that science will resolve the problems of population and scarce resources in the future. Space age technology has already had major impact on animal agriculture. Genetic engineering is a reality. More such advances are confidently expected. But the public is anxious about science as an unquestioned positive force in the U.S. way of life. Societal ambivalency about science had its effect on the imperatives. The conferees recognized that clear communication of the values of scientific animal agriculture to all segments of society is essential to a public mandate for research.

HUMAN WELFARE

Human welfare was the most important of the common concerns which influenced the Conference. With the focus on food from animal agriculture, Keynoter Carol Tucker Foreman made the leading key point: ". . . American people are concerned about the food they eat. . ." Concern for human welfare stimulated numerous recommendations for research on dietary habits and needs and for regulation of production, processing and marketing practices. Later, Keynoter Glenn Salisbury emphasized that ". . . the ultimate beneficiary (of agricultural research) is the consuming public." In this Conference, concern for welfare of humans - both consumers and producers - was the focal point.

Keeping these important concerns in mind, it is not surprising that a well-informed, thoughtful group of experienced professionals would identify a few central issues for special emphasis. These central issues were the organizing basis for what evolved into the Conference Imperatives. Although these imperatives are broadly stated, they are all researchable. Details of specific research objectives are provided in each of the following chapters.

Conference Imperatives are six categories of high priority research:

1. Resource Conservation
2. Human Health
3. Systems Research
4. Communications
5. Biological Engineering
6. Analytical Methods

RESOURCE CONSERVATION

Concern for this critical issue was manifest in the research to reduce fat in animal products recommended by all groups. Animal scientists likewise recommended research in physiology, nutrition and genetics to increase efficiency of protein synthesis and to increase the capability to control fat deposition.

Efficient use of resources was the target of research proposed by Resources and Environment, Feed Production and Animal Nutrition groups. Research was recommended to reduce erosion and pollution, harvest nutrients in animal waste, decrease requirements for cultural energy, promote use of noncompetitive feedstuffs such as ligno-cellulosic (plant fiber) materials, and exploit ability of legumes to fix nitrogen (conserve fertilizer).

Livestock management and health specialists recommended research to improve the health, welfare and environment of livestock in order to improve efficiency of production. This concern for biological efficiency followed through in the recommendations of the Genetics and Reproduction group for germ plasm improvement and utilization.

Losses of animal products due to disease, contamination or poor packaging were of particular concern. Food processing and safety specialists emphasized the need to preserve both quantity and quality of food by basic research on mechanisms of deterioration, leading to new ways to conserve nutrients in animal food products.

Somewhat less obvious, but just as real, are the losses of resources due to inefficiencies in production, marketing and distribution. Research was recommended to reduce these "slippages." Similarly, recommended research leading to enlightened public policies can improve the effectiveness of animal agriculture as a conserving force.

HUMAN HEALTH

The Human Nutrition group urged research to determine requirements for animal product-supplied nutrients and to determine the possible relationships between diet and human health problems such as cardiovascular disease and cancer. Food Processing, Food Safety and Animal Health groups recommended research to detect and prevent contamination by pathogens and toxic substances. Transfer of resistance factors between pathogenic species was priority research recommended by the Animal Health group. Less direct, but similarly critical to future human welfare, were the expressed needs for research to preserve environmental quality and natural resources for future generations.

SYSTEMS RESEARCH

Conferees agreed on the need to evaluate and deal with the sensitive relationships and linkages among the components of animal agriculture. All groups recognized the importance of a systems approach to research on animal agriculture and its integral components. Development of tools such as quantitative models was urged. Three groups: Genetics and Reproduction; Resources and Environment; and Production, Marketing and Distribution specifically suggested quantitative modeling to evaluate strategies for resource allocation and utilization.

The Human Nutrition and Animal Health groups recommended modeling to study epidemiology of diverse health risks to humans and animals alike. Research on integrated health care delivery systems was given top priority by the Animal Health group. Benefit/risk or benefit/cost analyses were urged by several groups as a preferred technique for objective decision making.

COMMUNICATIONS

This priority research area follows from the emphasis on the systems approach. Most groups recognized that, to be effective, research results must be clearly communicated to consumers, public officials and regulatory agencies. Many of current restraints on effective and efficient operation of the animal agriculture system may be traced to poor communications.

Methods to improve communication were considered researchable. For example, the Human Nutrition and Processing and Acceptability groups recognized a need for more precise statements of human nutrient requirements and for better communication through labeling. The Food Safety and Animal Health groups proposed research to improve quality of information on hazards due to toxic substances in feed and food products. Research is needed to provide this precision and quality of information.

The group dealing with Production, Marketing and Distribution recommended research to develop information bases on supply and demand factors for contingency planning and for more effective and accurate communication of market signals throughout the production, marketing and distribution chain. Finally, the Public Policy group recommended that

that research be conducted and results expressed in ways to improve communication and level of understanding between the body politic and policy-makers.

BIOLOGICAL ENGINEERING

Research was recommended with the objective of genetic and environmental engineering. These recommendations emanated primarily from the biological disciplines; however, emphases expressed by other groups were compatible. Recommended research included that intended to regulate body protein and fat deposition, improve gut microbial populations, understand factors controlling appetite, improve disease resistance and immunity and provide basic knowledge for interventions such as gene splicing, sexed semen and embryo transfer. Interdisciplinary research to improve feed production on marginal lands (e.g., high saline, drought prone) also was given high priority. These research approaches generally involve basic genetic or physiological research leading to means of manipulating animal and plant resources, including unlocking cellulose, the most abundant organic substance in the world, for use by animals in producing human food.

ANALYTICAL METHODS

Research to develop more precise and efficient analytical tools is essential. Emphasis on better analytical tools was stimulated by concern for feed and food quality and safety, and the resulting regulation of production and processing practices. Efforts to develop low cost, noncompetitive diets adequate for profitable levels of animal production necessitate rapid and accurate analyses of feedstuffs.

Concern for both human and animal health stimulated recommendations for improved diagnostic tools. The costs of inaccurate diagnoses affect producer and consumer alike and inhibit the development of effective regulatory policies for animal disease and toxic substances. Research to develop accurate computer models for use in systems analysis was another frequently cited need.

****

EPILOGUE

The tremendous effort which went into this Conference must not end with these Proceedings. A continuing, cooperative effort will be instituted to build on the foundation of this Conference. This effort will include an annual updating of research priorities in response to new developments and new human needs, and it must be combined with continuing improvement in communication among all segments of the system: producers, industry people, consumers, policy makers and scientists.

2

A. E. Harper (Co-Chairman), J. Dwyer (Co-Chairman),
M. L. Brown (Rapporteur)

C. E. Allen, W. G. Bergen, R. Dierks, H. A. Guthrie,
M. Hegsted, L. M. Henderson, N. L. Jacobson,
M. Kellough, G. A. Leveille, M. C. Nesheim,
A. Simopoulos, and E. Speckman

HUMAN NUTRITION

SUMMARY

The nutritional needs of the U.S. population are met, to a large extent, by foods of animal origin. Such foods are good to excellent sources of most essential nutrients. Nevertheless, with a high incidence of obesity in the population and with concern about relationships between high fat and sodium intakes and the incidence of certain chronic degenerative diseases, questions have been raised about the appropriateness of the U.S. diet and food supply. Some of the assumptions about the relationships between diet and the incidence of chronic, degenerative diseases are controversial.

In view of the questions that have been raised about relationships between diet, health and disease in the U.S., there are several lines of research pertaining to animal agriculture that should be pursued at the present time. Some are included in the recommendations listed below. It should be recognized that many questions about diet-disease relationships are far from being resolved, that the validity of assumptions about them is controversial, and that many aspects of them are important topics for human nutrition research. A strong, continuing program of human nutrition research will be needed to provide the information required to resolve current controversies. When information about currently controversial topics becomes more definitive, some of the assumed constraints for animal agriculture may be removed. At the same time, others may arise either as the result of new information about human nutrition or as the result of the development of new products, presently unforeseen. It is, therefore, important to encourage interaction among researchers in animal agriculture, human nutrition and health sciences.

The working group on Human Nutrition identified four priority research areas:

ALTERING COMPOSITION OF ANIMAL PRODUCTS IN RELATION TO HUMAN HEALTH

Foods of animal origin are recognized as important sources of many essential nutrients. There is concern, however, about the fat content and resulting caloric intake from diets containing an abundance of these foods. Another concern is the postulated relationship of dietary saturated fat and cholesterol to the incidence of circulatory disorders.

A substantial portion of the U.S. population is affected by hypertension or at risk of hypertension. These individuals are encouraged to restrict sodium intake. The phosphate content of U.S. diets is increasing and is currently in excess of recommended levels.

Priority research:

- Investigation of the potential for reducing the fat content of fresh and processed meat and milk products by breeding, feeding and processing methods.
- Investigation of the potential for reducing sodium and phosphate in processed animal products.

NUTRITIONAL VALUE OF FOODS OF ANIMAL ORIGIN

Assessment of the nutrient contribution of fresh and processed foods of animal origin to human nutrition requires knowledge of their nutrient composition. This is particularly important for modified products, processed foods and substitute products.

Priority research:

- Development of economical and rapid methods for analysis of nutrients and other food constituents.
- Analysis of components of fresh and processed foods of animal origin and the effect of storage and preparation.
- Determination of bioavailability of nutrients in foods of animal origin.
- Studies of the nutritional adequacy and comparability of meat, dairy and egg substitutes.

DIET AND HEALTH

Several human health problems are associated directly or indirectly with diet. Perhaps the most widely publicized is the coronary heart disease problem where the roles of fat and protein have not been satisfactorily delineated. Relationships between diet and severe chronic degenerative diseases (e.g., osteoporosis, atherosclerosis, cancer) have been postulated.

Priority research:

- Expanded research on diet-cardiovascular relationships including the metabolism of lipoproteins and cholesterol oxidation products, and thrombogenesis.
- Expanded research on diet-cancer relationships.
- Studies of availability of nutrients as affected by constituents of animal food products.
- Studies of the relationships between animal-food products and bone and oral health.

CONSUMER INFORMATION

Foods of animal origin have received a great deal of publicity in relation to the substances they contain, their place in the diet and their role in relation to health and disease. The present nutrition labeling format does not communicate sufficient information effectively. Research on evaluating the effectiveness of various approaches on consumer knowledge, attitude and food consumption practices is also required. A system of monitoring trends in food consumption changes is important to the evaluation of areas needing emphasis.

Priority research:

- Identification of consumer information needs.
- Development of effective means of providing information, including labeling, in an understandable and usable manner.
- Studies on effective evaluation of consumer information programs, including changes in knowledge and food-related behavior.
- Development of food consumption monitoring systems which permit evaluation of dietary changes, including changes in consumption patterns of specific food items and the relationship between such changes and health and disease status.

INTRODUCTION

During this century, nutrients that are essential for man have been identified and needs for many of them have been quantified. This knowledge has brought understanding of dietary deficiency diseases and, subsequently, with the exception of iron-deficiency anemia, there has been success in eliminating them as public health problems. Similar successes have been achieved in the control of infectious diseases. As dietary deficiency and infectious diseases have been brought under control, life expectancy has increased. At the same time, obesity has become more prevalent and chronic degenerative diseases, which become more prevalent with advancing age, are the major medical problems of the U.S. This has resulted in attention being shifted away from essential nutrient needs and toward possible relationships between diet and chronic degenerative diseases and obesity. Questions have been raised about the roles of total food consumption, consumption of specific foods and nutrients and various environmental factors in relation to the changing pattern of disease. With a high proportion of food energy and fat being obtained from animal products in the U.S., it is not surprising that attention has been focused on the possibility of a link between the high consumption of these products and their relationships to health and disease.

It is the objective of this paper to assess the nutritional contributions of animal products to human nutrition, to identify current issues in human nutrition which may represent constraints to the further development of animal agriculture and to recommend directions in research that will provide answers to help resolve these issues. It is important to note that the assurance of an adequate diet requires consideration of the total foods consumed and cannot focus on animal products alone.

FOOD FROM ANIMALS

The newborn infant begins life with food of animal origin, usually human milk or a formula based on milk products. Milk is one of the most nearly complete foods available to the human race. Foods derived from milk include butter, ice cream, cheeses and yogurt (fermented milks). The components of milk include lactose, fat, casein, lactalbumin, vitamins and minerals.

Eggs represent another class of nearly complete foods available to man. Eggs are food that serves in nature as the sole source of nutriment during embryonic development of the chick. Eggs are eaten as such, but like milk, may be a component of many complex foods. Eggs may also be processed, for example, as whole dried eggs and egg white.

Both muscles and organs from beef, pork, lamb and poultry represent foods from the major domesticated animal species. Meats are marketed as minimally processed foods. They are also marketed as processed products that have undergone some form of treatment to minimize microbial spoilage and thereby extend their keeping qualities. These products differ in composition from fresh meat as a result of the addition of salt and water, as with ham, or may be more extensively modified as with many luncheon meats and sausages.

Processed meat products vary considerably in composition. They tend to have less protein and more fat than unprocessed meats, but may vary over a wide range as do the different cuts of meat. The most prominent change is an increase in sodium content from about 65 mg/ 100 g to 750-1,300 mg/100 g (Watt and Merrill, 1963). They also frequently have some added carbohydrate.

Fish are consumed to a limited extent in contrast to meats; for example, in 1979 the per capita disappearance of meat was 67 kg (147 lb) and of poultry, 28 kg (62 lb), whereas disappearance of fish amounted to only 6 kg (13 lb) (Moore, 1980).

NUTRIENTS PROVIDED BY ANIMAL FOODS

Foods derived from animal flesh are composed mainly of protein and fat, whereas foods of plant origin, with the exception of nuts and seeds, are composed primarily of carbohydrate and protein. Animal products also do not provide "fiber", whereas many plant products are sources of this complex of nonnutrient substances that provide "bulk" in the diet.

Table 2.1 illustrates the composition of some representative animal products before cooking. The composition of milk, some dairy products and eggs are given in table 2.2 for comparison. The content of vitamins and minerals in some foods of animal origin is presented in table 2.3.

With the exceptions of some of the processed products, meats, eggs and cheeses are high protein, high fat, low carbohydrate foods. Most dairy products are high protein foods but they also contain the sugar, lactose. Dairy products may be high or low in fat, depending upon the fat content of the milk from which they are prepared.

The proteins of milk and eggs are of high nutritional quality. Meat proteins are utilized somewhat less efficiently than milk and egg proteins. With the exception of soybean proteins and possibly those of some other legumes, individual plant proteins are of lower quality than animal proteins. Among proteins of animal origin, only certain derived proteins, such as gelatin, are of poor quality.

The human infant grows rapidly during the first year of life. Human milk, in which 50% of the calories are provided by fat, has a sufficiently high caloric density to meet the high energy requirement for rapid growth. After 4-6 months of age other foods must be added to provide the additional energy and nutrients needed for normal growth (Fomon, 1974). Animal foods are particularly valuable as weaning foods and as foods for the active, growing youngster whose energy needs remain high for several years. Between the ages of 1 and 10, energy needs fall gradually from 100 kcal/kg of body weight to about 75 kcal/kg of body weight. This compares to a value of between 35 and 40 kcal/kg body weight for the much less active adult.

Animal products are unique providers of several essential nutrients. With the exception of products obtained by bacterial fermentation they are the only reliable food source of vitamin B_{12} (table 2.4). Beef, eggs and milk are particularly rich sources of this vitamin, with pork and chicken providing lesser quantities.

Dairy products are the major source of calcium in the U.S. diet, providing approximately 75% of the total dietary calcium and only about 11% of the energy. Meat and eggs are poor sources of calcium.

Meat is a major source of iron in the diet of most industralized nations. The heme iron, a major form of iron in animal tissues, is particularly well absorbed (on the order of 20%), whereas nonheme iron is absorbed only to the extent of 3-5%. The presence of meat in a meal improves the absorption of nonheme iron. Increasing meat intake from 30-90 g in a meal can improve the absorbability of nonheme iron from 3 up to 8%, thereby increasing the utilization of iron from plant foods and the nonheme iron from animal foods (Monsen et al., 1978).

Dairy products are major sources of vitamin B_{12} and riboflavin, providing approximately 21% and 40%, respectively, of these vitamins in the U.S. diet. Tryptophan is an efficient precursor of niacin. Animal proteins are good sources of tryptophan and, therefore, the amount of niacin obtained from animal-food products is much greater than would be indicated from measurements of niacin content alone. For example, milk, which is a poor source of niacin was used to cure niacin deficiency, an effect attributable to its high tryptophan content. Milk is a major source of niacin equivalents because of its content of preformed niacin and of tryptophan.

Animal products are also good sources of thiamin, vitamin B_6 and vitamin A. Pork is particularly rich in thiamin. Animal products do not, however, provide significant quantities of ascorbic acid. Milk is an excellent source of vitamin D since in the U.S. it is usually fortified with this vitamin. Milk and butter are also good sources of vitamin A.

Animal products are also excellent sources of most of the trace elements. Meats are particularly rich in zinc, and dairy products provide significant amounts of magnesium. The available evidence suggests that zinc from plant sources is poorly absorbed.

All in all, animal products are excellent sources of all but a few of the essential nutrients. They contain most essential nutrients in appropriate proportion to the calories.

Recent dietary surveys (DHEW, 1973; 1978; Pennington, 1976; King et al., 1978) indicate that several nutrients are consumed in marginal amounts relative to the Recommended Dietary Allowances. The nutrients of concern include iron, calcium, zinc, magnesium, vitamin B_6 and vitamin A. With the exception of iron, however, there is little clinical evidence of deficiencies. Animal food products are good to excellent sources of all of these nutrients. Consequently, any major reduction in the consumption of foods from animals would require careful dietary planning to avoid potential inadequacies.

COMPETITION BETWEEN ANIMALS AND MAN

One issue receiving considerable public attention is the competition between man and animals for grain resources. This is related to the concerns about hunger and malnutrition, particularly in the developing world. Decisions regarding exports of cereal grains and legumes, and the market price for these commodities will clearly influence the proportion diverted to animal production. However, many other moral, ethical and economic factors will influence the final decision on the relative use of feed grains and, consequently, the production and availability of animal-food products. This topic has been reviewed recently (CAST, 1980).

Feed grains are in high demand for nonruminants, whereas for ruminants the extent of feed grain use depends on the desired production level and degree of finish. Grains produced in the U.S. may be used domestically or exported for 1) animal production, 2) direct use as food by humans, or 3) as raw material for other products (e.g., fuels, ethanol). Major factors modulating grain supply are weather, disease, yield, prices and government policies.

As world population increases, competition for basic grain resources used directly as human food, compared to conversion to animal-food products by animal production, may become an increasingly important determinant of the use of animal products for human food.

During the 1970s some major fluctuations in grain supplies affected domestic meat production in the U.S. Decreases in grain production in the middle 1970s because of drought, disease and increased grain exports caused a major adjustment in cattle, swine and poultry production and the degree of finish in beef cattle. Thus, market and external production forces played a key role in overall animal production.

Proposals have been advanced that animal production be reduced in developed countries and that the grain resources so saved be used to feed people in less developed countries. Currently, both the proponents and critics of the so-called - eating down on the food scale - concept have accepted the view that the answer to world hunger and malnutrition does not lie solely in reduced animal food consumption in developed countries. It has become clear that the long-range solution to world hunger and malnutrition is dependent upon increasing food production and improving food preservation in the developing countries.

Even if man-animal competition for plant products increases markedly, animal production will still play an important role in the conversion of feedstuffs to human food. The relative importance of ruminant species as compared to swine and poultry as human food may increase, but animal products will continue to have a substantial role as sources of human food, even if extreme competition for food grain resources takes place between animals and man. This unique role of ruminant animals results from the usefulness of large areas of land unsuitable for grain production but suitable for production of forages utilized solely by ruminants.

If substantial changes in availability of grain for animal feeding were to occur, there could be significant changes in nutrient sources in diets of the U.S. population. This would require careful analysis of the nutritional consequences of any substantial shifts in the types of food produced and consumed to ensure that health of the U.S. population would not deteriorate.

The socio-economic forces affecting utilization of animal sources of food will not be considered in this report. Rather, this report will address the health issues that may be constraints in animal agriculture.

CONTROVERSIES IN RELATION TO ROLE OF ANIMAL PRODUCTS IN HEALTH AND DISEASE

The issues relating to diet and disease focus mainly on the potential significance of: 1) the type and quantity of fat and the quantity of cholesterol in the diet as factors contributing to the development of cardiovascular diseases and, more recently, certain types of cancers; 2) sodium and potassium intakes as factors contributing to the development and treatment of hypertension; 3) modification of fat, carbohydrate and fiber intake in relation to management of diabetes, the occurrence of cardiovascular diseases, certain types of cancer and control of obesity; 4) phosphate and protein intake in relation to the utilization and retention of calcium, and bone diseases such as osteoporosis.

Issues of current interest, especially in relation to public policy pertaining to food and nutrition, fall for the most part into the category of diet-disease relationships. The fact that heart disease and cancer are the two leading causes of morbidity and mortality in the U.S. and that most deaths are the result of chronic degenerative diseases has focused attention on the possibility of improving health status and delaying the onset of such diseases. The results of a variety of epidemiologic studies, animal experiments and some clinical and intervention studies have suggested that relationships exist between diet composition and intake, and the time of onset and severity of chronic degenerative diseases. This has led to acceptance of the view that lifestyles, including dietary practices, are in some way associated with morbidity and mortality from chronic degenerative diseases.

There is no controversy about the role of animal products in providing valuable sources of energy, protein and a variety of other essential nutrients. The value of dairy products as sources of calcium and riboflavin and the value of meat products as sources of niacin, iron and trace minerals is undisputed. The importance of animal products as sources of vitamin B_{12} and other B vitamins is clearly and unequivocally established. The controversies over the effects of animal products on health do not involve their contributions to meeting essential nutrient needs. Rather, these controversies relate to the type and quantity of fat animal products provide, their content of cholesterol, their lack of fiber and, for processed products, their salt content, as factors that may contribute to the development of chronic degenerative diseases.

DIETARY FAT, CHRONIC DEGENERATIVE DISEASES AND OBESITY

The health status of the U.S. population has been assessed clearly in the Surgeon General's report of 1979 entitled "Healthy People" (DHEW, 1979). The report begins, "The health of the American people has never been better." It includes tables and charts showing that infant mortality has reached an all time low and that death rates at every age have fallen throughout the century so that life expectancy at birth has increased from 47 years in 1900 to 73 years at present. For those who survive to age 45, life expectancy for males is between 71 and 74 and for females, between 77 and 79. The end result of this, which is attributable to improvements in nutrition, sanitation, housing, immunization and medical care, is that those over 65 now make up about 11% of the population instead of 4% as was true in 1900.

During the 20th century the major causes of death in industrialized countries with high standards of living and medical technology shifted from gastrointestinal and respiratory diseases (including tuberculosis) to acute and chronic cardiovascular disease, cancer and other degenerative diseases. This changing pattern of diseases that are major causes of death and disability has stimulated great interest in the role of environmental factors and their effects on health. Substantial research efforts have been directed toward discovering the underlying causes of these major diseases. These efforts have led to the concept of risk factors in the development of coronary artery disease.

The etiology of chronic degenerative diseases is accepted as being complex (NRC, 1980a). Few specific causes of such diseases have been identified. Interactions among variables that are considered to contribute to the development of these diseases are many and are poorly understood. This situation has led to development of the "risk factor" concept in which the strength of associations between variables and the incidence of various diseases is assessed by correlation measurements. Because of the complexity of these interactions it is important to discuss them in their totality rather than as individual variables.

Two approaches have been used in attempts to identify possible causal factors of these diseases. One has been to examine relationships between the incidence of the various diseases in different countries and the environmental characteristics of the countries; the other has been to study relationships between the characteristics of individuals in a population and their susceptibility to these diseases. From the first type of study, epidemiological studies, associations between certain environmental factors, particularly dietary factors, and death rates from chronic degenerative diseases have been identified; from the second, associations between characteristics and susceptibility to the diseases have been identified. These studies have given rise to the "risk factor" concept (Kannel, 1978). This is based on the assumption that the degree of correlation between characteristics of individuals, and the probability that persons so characterized will develop chronic degenerative diseases permits assessment, from analysis of risk factors, of the probability that a particular individual will be susceptible to these diseases. The "risk factor" concept does not identify cause-effect relationships, only associations that have some predictive value.

One of the features associated with industrialization of countries and a rising standard of living is a shift in dietary patterns. The proportion of fat in the diet increases, the proportion of carbohydrate decreases, and the proportion of simple sugars increases. These changes are in large measure the result of decreased consumption of cereal grains and root crops and increased consumption of animal products, refined sugar and separated oils and fats. The association between these changes and increased incidence of chronic degenerative diseases that accompanies them has focused attention on high consumption of total fat, particularly of the more saturated fat from animal products, as factors possibly contributing to the development of cardiovascular disease and cancer. It is important to note that these dietary changes are usually accompanied by other changes in the lifestyle, such as reduced physical activity. Diet-disease relationships have received major emphasis because clinical studies and studies in experimental animals have provided evidence of relationships between diet modification and metabolism. Alterations in susceptibility to the action of carcinogens related to diet also have been demonstrated in experimental animals, and through epidemiologic studies in man in a limited number of studies for some types of cancer.

Studies of associations between the characteristics of individuals and susceptibility to chronic degenerative diseases - risk factor-analysis - also have focused attention on diet-disease relationships. Although many potential risk factors have been investigated, only a few are considered "major risk factors". Among these are elevated serum total cholesterol concentration, hypertension, abnormal electrocardiogram, diabetes, cigarette smoking, obesity, low physical activity, and psychosocial behavior that is characterized by an excessively competitive and aggressive attitude (Kannel, 1978). Relationships between diet and serum cholesterol, blood pressure and obesity, established in experimental studies with animals and human subjects, suggest a potential for reducing these risk factors by diet modification.

The subject of diet and chronic degenerative diseases, particularly diet and cardiovascular disease, has been reviewed frequently (Levy et al., 1979; Ahrens and Connor, 1979). Kritchevsky (1979; 1979a) has emphasized that in studies with animals, many dietary factors have been shown to influence both serum total cholesterol concentration and carcinogenesis. However, many of the dietary effects of individual food components or nutrients are modified by changes in the quantity or type of other dietary components. Nevertheless, major attention has been given to observations from epidemiologic studies which show a relationship between total fat intake and incidence of certain cancers, especially cancer of the breast and of the large bowel. Both clinical trials and animal studies in which elevations of serum cholesterol have been demonstrated as the result of feeding high quantitites of saturated fats, and in which reductions in serum cholesterol concentration have been observed after feeding large quantities of polyunsaturated fats, have added evidence to the significance of these epidemiologic studies (Ahrens and Connor, 1979).

Despite acceptance of the results of the epidemiologic and experimental studies as providing evidence of links between dietary fat and chronic degenerative diseases by some investigators, the general conclusion that modification of the diets of populations is an appropriate preventive health measure is viewed with skepticism by many investigators for several reasons (Ahrens and Connor, 1979). It has been pointed out that cause and effect relationships between dietary fat and arteriosclerosis have not been established (Mann, 1973; Reiser, 1978; Ahrens and Connor, 1979; Olson, 1979). In the epidemiologic studies, despite the positive correlation observed within countries between saturated fat intake and the incidence of coronary heart disease and that of breast and colon cancer, there are exceptions that indicate the inadequacy of the hypothesis. Intervention trials in which diet modification and drug treatment have been used in an effort to reduce the incidence of coronary heart disease in middle-aged men have not been successful. In seven large scale trials involving 3,000 men for 2-10 years, decreases in serum total cholesterol of from 7-16% occurred and a marginal decrease in new events of coronary artery disease was observed, but no reduction was detected in overall mortality (Ahrens, 1976; 1979; Report from the Committee of Principal Investigators, 1978). There are currently underway studies of simultaneous reduction of three risk factors: smoking, hypertension and serum cholesterol.

There is also some question as to the range of serum cholesterol that constitutes a risk for greater incidence of coronary heart disease. Some investigators have interpreted the evidence as indicating a risk from elevated serum cholesterol over the full range of serum cholesterol observed. Others have suggested that only levels above a certain threshold constitute a risk (Reiser, 1978; Blackburn, 1979).

Obesity is a widespread condition in the U.S. that is associated with increased mortality. It is accompanied by increased risk of diabetes and hypertension, two identified risk factors for coronary heart disease (Van Itallie and Hirsch, 1979). There is also suggestive evidence that some cancers are more prevalent among the obese (Van Itallie and Hirsch, 1979). Obesity is a complex condition that results from consumption of food sources of energy in excess of the amount of energy expended. Understanding of the underlying causes of obesity is limited. The success rate in long-term maintenance of weight reduction is low.

Relationships between diet composition and the occurrence of obesity have not been established. Nevertheless, foods that provide a large quantity of energy per unit of weight or volume can be consumed in excessive quantities more easily than those providing fewer calories per unit of weight. Such foods tend to be highly palatable because of their high fat content. The probability that weight control is more easily achieved by reducing the intake of such foods has not been established. Nevertheless, reduction in total fat intake is obviously an appropriate approach to body weight control.

All in all, despite the controversy over the dietary lipid-heart disease hypothesis, emphasis on a high fat intake as a risk factor for health has become established and is widely accepted. The high fat content of many animal products and especially their high content of saturated fatty acids must be considered in deciding future research directions for animal agriculture.

CHOLESTEROL AND CARDIOVASCULAR DISEASE

In both epidemiologic studies of relationships between diet and cardiovascular disease and in most experimental studies of relationships between diet and lipid metabolism, cholesterol and saturated fats are closely linked. This is because cholesterol is found only in animal products (table 2.5) and, where saturated fat intake is high, animal products usually are important diet constituents. As with saturated fats, increasing dietary cholesterol has been shown in epidemiological studies of populations to be associated with increasing incidence of cardiovascular disease. This type of association has not been observed regularly or consistently in studies of individuals within populations.

Animals vary greatly in their serum cholesterol response to changes in cholesterol intake but in susceptible species, cholesterol usually must be included together with a high saturated fat content of the diet to produce atherosclerotic lesions. Humans are relatively resistant to dietary cholesterol as a factor that affects serum cholesterol. In studies with human subjects consuming high fat diets, increasing cholesterol intake over the range from zero to about 600 mg/day is associated with gradually increasing serum cholesterol concentration, but intakes greatly in excess of this amount are associated with a limited further response. When large numbers of individuals are examined, there is little evidence of a relationship between cholesterol intake and the occurrence of arteriosclerotic disease (McGill, 1979).

The dietary guidelines of USDA-HEW (1980) recommend reduction of cholesterol intake by the U.S. population. The Canadian Health

Protection Branch (Health and Welfare Canada, 1979) did not recommend reduction of cholesterol intake by the Canadian population as a whole as a measure for reducing the incidence of cardiovascular disease. It did, nevertheless, recommend reduced intake for individuals who were identified as being at risk for coronary heart disease.

In the minds of the public, cholesterol has become associated with cardiovascular disease due in large measure to advertising of low cholesterol products, the inability to distinguish between dietary cholesterol and serum cholesterol, and the educational campaigns of organizations such as the American Heart Association. There is no doubt, therefore, that the cholesterol content of animal products represents another probable constraint that will influence directions in animal agriculture. Eggs are the richest source of cholesterol among major food products. Decreased egg consumption during the past several years is assumed to be the result of acceptance of the idea that a high cholesterol intake is a health hazard.

Two additional developments in relation to diet and cardiovascular disease deserve special attention by the animal industry. The first of these has to do with observations indicating that milk and some milk products have a hypocholesterolemic effect in experimental animals (Kritchevsky et al., 1979) and in human subjects (Mann, 1977). The hypothesis has been put forward that this is attributable to a unique unidentified factor in milk that is carried over in other dairy products, but not into butterfat. This observation emphasizes that it is inappropriate to indict individual components of complex foods and food mixtures without regard to other components of the diet.

The second development that deserves prompt and serious consideration is the observation that some oxidation products of cholesterol may have specific toxic effects on the cardiovascular system. Recognition of this possibility arose from studies in which the effects of old, stored cholesterol and freshly prepared cholesterol were examined in rabbits and in aortic cell suspensions in vitro (Taylor et al., 1979). Oxidized cholesterol, in the diet or in the medium, was found to be associated with an increase in the number of degenerative aortic cells. Some specific oxidation products that have this effect have been identified.

SODIUM AND HYPERTENSION

Epidemiologic studies suggest a relationship between sodium intake and onset of hypertension. A proportion of the U.S. population, estimated at between 10 and 20%, is susceptible to development of hypertension, particularly with advancing age. Hypertension is rare in nonindustrialized populations, whose salt intake is only about 2 g per day, whereas the incidence of hypertension is high in populations with salt intakes exceeding 20-25 g per day (Tobian, 1979). However, hypertension is not simply a reflection of salt toxicity. Hypertension in man is associated with genetic as well as environmental factors: race, family history, variations in endocrine and kidney function, obesity and lifestyle.

The role of sodium in the cause and prevention of hypertension is not known and is under investigation. Data from epidemiologic studies indicate that there may be racial differences in ability to handle sodium excretion, and recent evidence that blood pressure can be raised in some hypertensive individuals by incremental increases in dietary sodium, makes concern about excess sodium appropriate. There is insufficient information to make recommendations of a specific level of sodium intake for normal individuals. The NRC (1980), however, has published estimated safe and adequate daily dietary intake levels for sodium; for healthy adults, the range is 1.1-3.3 g sodium per day, or 2.75-8.25 g sodium chloride (salt) per day.

Individuals with a family history of hypertension are advised to avoid obesity and to limit their salt/sodium intake. In individuals susceptible to hypertension, sodium chloride intake of 2-10 g per day may elicit hypertension. Although control of hypertension with drugs is considered to be the treatment of choice, sodium restriction and weight control are often advised. A daily sodium intake below .5 g sodium chloride (.2 g sodium) is clinically beneficial, but is not practical in free-living populations (Meneely and Batterbee, 1976).

For man, the requirement for sodium as sodium chloride is considerably less than 1 g per day (NRC, 1980). The average daily intake of sodium chloride from all sources is 10-12 g, of which about 45% is from processing, 17% is of indigenous origin, and 38% is discretionary addition by the user. Salt preference is individual and, to some extend, is culturally determined. The sodium content of raw meat is within the range of that found in vegetables, 50-60 mg/100 g for beef and pork as compared with 47 mg/100 g for carrots, 60 mg for beets and 70 mg for spinach. Most fresh vegetables and fruits contain less than 15 mg of sodium/100 g. Whole milk has a sodium content of 50 mg/ 100 ml and eggs 122 mg/100 g (Watt and Merrill, 1963). In contrast, processed meats and cheeses contain as much as 1,000 mg or more of sodium per 100 g (table 2.6).

It is clear that sodium is an essential nutrient that is consumed in great excess in relation to need in the U.S. On the principle that consumption of great excesses of individual nutrients is undesirable and is often associated with adverse effects, a recommendation for reduction of salt intake can be supported. At present, public health programs are directed toward identifying individuals with hypertension. At the same time, education programs are making the public aware of the excessive dietary use of salt in the U.S. As these programs gain momentum it is most likely that a trend toward reduced salt consumption will develop, if it has not already begun to do so. This should not represent a constraint for animal production generally, as most unprocessed animal products do not contain excessive amounts of salt. For the branch of the food industry that produces processed animal products which have high salt content, it could be a serious constraint.

OSTEOPOROSIS AND CALCIUM METABOLISM

Osteoporosis is a rather common condition in postmenopausal women and in older men. It is characterized by loss of bone mass, leading to thinning of bones and bone fractures. Factors affecting bone metabolism have been extensively studied, but the metabolic abnormality which causes osteoporosis is not known. Lack of exercise, hormonal changes, low calcium and high phosphorus intakes have been implicated by various investigators (Albanese et al., 1975; Whedon, 1980). Dairy products provide as much as 75% of the calcium intake. Meats, especially some processed products, are major sources of dietary phosphate. These foods, and others to which phosphate is added during processing, may have an important influence on the dietary intake of these two elements. More research is needed on the effects of the levels and the ratio of calcium and phosphorus as well as other dietary components, e.g., protein, in those segments of the population who are likely to experience osteoporosis.

Mechanical deboning of meats gives more complete recovery of muscle. Mechanically deboned meat can be added to processed meats. It has only minor implications for dietary calcium and phosphorus since only small quantities of bone minerals are extracted by this process. Findings concerning the safety of this product have been reassuring, but further research on composition and biological implications is needed.

PROTEIN INTAKE

As previously indicated, animal products are among the better food sources of some of the nutrients that may be marginal in the diets of Americans. Animal products also provide about 70% of the dietary protein, a nutrient consumed in the U.S. at a level nearly double the recommended dietary allowance. It has been suggested that consumption of high levels of some proteins may result in greater calcium excretion (Anand and Linkswiler, 1974), contribute to the development of hypercholesterolemia (Kritchevsky, 1979) and show epidemiological correlations with incidence of some types of cancers (Gori, 1979). These effects are not clearly the result of excessive proten intake, however, and further research must be done to define the limits of safe protein intake for adult populations. Excessive protein intake has been shown to be clearly detrimental to young infants (Fomon, 1974) and to individuals with kidney disease.

LACTASE DEFICIENCY-LACTOSE INTOLERANCE

Dairy foods are unique in that they contain lactose. Lactose appears to promote a favorable gastrointestinal flora and to facilitate the absorption of nutrients such as calcium (National Dairy Council, 1974).

Lactase deficiency or lactose intolerance is prevalent in the adult population, particularly in nonwhites. Lactose is normally digested or hydrolyzed in the intestine into its two component sugars. When this does not occur, bacteria in the intestine utilize the unabsorbed lactose to produce acid and hydrogen gas which leads to the symptoms of lactose intolerance. Some individuals who have a clinical lactase deficiency can tolerate small servings of milk. Thus, clinical lactase deficiency is not necessarily synonymous with milk intolerance. (Simopoulos, 1979; Torun et al., 1979).

Most individuals who are milk intolerant can still derive the nutritional benefits of dairy foods by: 1) consuming smaller servings more frequently, 2) selecting dairy foods low in lactose such as cheese and certain fermented products, or 3) by using commercial lactase preparations to hydrolyze the milk lactose.

FOOD ALLERGY

Allergic reactions to many foods, including meats, milk and milk products, and eggs, occur in people of all ages, but the identification of the allergen remains a serious obstacle in the management of many of the food allergies (Taylor, 1980). Therefore, a number of elimination diets have been developed to deal with the problem. Because of the difficulty of identifying the antigen, it is at present difficult to recommend specific research activities bearing on animal agriculture.

The problem of food allergy during infancy has diminished with the return to breast-feeding and the avoidance of milk-based formulas.

Research on the antibody transfer from mother to the infant via breast milk indicates that the breast fed infant can tolerate milk and meat products if they had been part of the mother's diet, since the baby has received passive immunity against these foreign proteins (Hanson et al., 1979). This type of research needs to be expanded. It is quite possible that meat and dairy products can be more easily assimilated by the breast-fed infant if they have been part of the mother's diet.

CONTAMINANTS AND RESIDUES

During recent years, a good deal of concern has developed about the effect of antibiotics, toxins, chemicals and microbial contaminants in minimally processed animal foods on human health (CAST, 1980). Monitoring systems exist for some, but not all, of the contaminants and residues. There is no system currently, for example, for monitoring inadvertent contaminants.

Growth stimulants such as diethylstilbestrol (DES) that produce hormonal changes have been especially controversial (CAST, 1977). Newer products are being developed as effective substitutes; however, the mode of action and other related information must be elucidated for each of these products. Research on hormonal compounds such as estradiol also should be continued to assure that they are effective substitutes for DES that provide growth promotion without creating long-term tissue residues that may affect human health.

The widespread use of herbicides and pesticides in cereal grain and forage production should be accompanied by the development of more effective monitoring systems to assure the continued production of animal food products that are free of those chemicals. The strengthening of diagnostic laboratory-field epidemiology services to assist the grain and forage milling industries and farmers in preventing the contamination of the food chain or food animal tissues is badly needed.

The episomal transfer of drug resistance to enteric microorganisms of the human is a theoretical concern that is currently controversial. Antibiotics are being incorporated into animal diets to promote more efficient growth as well as the prevention of specific animal diseases. The effect of these practices on human health requires continuing evaluation.

Natural toxins such as mycotoxins and other microbial products may have both carcinogenic and toxic effects on the animals consuming them directly as well as secondarily, on man if consumed in contaminated foods of animal origin. Research is needed on conditions required for the production of toxins, methods of detection and methods of elimination from the food chain. The effects of contaminants on the utilization of nutrients by the host and their possible effects on the reaction of the host to specific contaminants also are areas of concern.

Attention must be addressed to the need for better mechanisms for the effective interaction of local, state and federal agencies in the prevention and elimination of many contamination and residue problems.

NUTRITION LABELING AND PRODUCT GRADING

NUTRITION LABELING

Regulatory

Food labeling requirements under the Food, Drug, and Cosmetic Act were relatively simple and remained largely unchanged from 1938 until 1970. In 1975, new Food and Drug Administration (FDA) regulations for labeling placed increased emphasis on nutrition. The format chosen was based on consumer testing of alternative methods of labeling. The reason for promulgating these regulations was the consumers' "right to know" what was in their food. This information can help consumers make more informed food choices and may provide basic information on which a nutrition education program could be based. USDA labeling format is similar to that of the FDA. Currently, nutrition labeling is voluntary, except on products that make a nutrition claim or contain added nutrients.

In 1978, FDA, USDA and the Federal Trade Commission (FTC) held hearings throughout the country to reconsider current regulatory policy governing food labeling and nutrition. Oral and written comments were received, and a survey was conducted of 1,374 primary food shoppers, to determine the future direction of labeling. FDA also solicited marketing research on alternative labeling formats to provide guidance for revision of labeling regulations.

Use of Labeling by Industry

An FDA nutrition labeling survey (Schucker, 1978) of the extent of nutrition labeling 1 year after the regulation became effective showed that 40% of leading national brands were labeled. This indicates that sectors of industry will label if the labeling constraints are reasonable. The existing compliance constraints are not deemed reasonable by most commodity producers, including the meat industry.

Industry has been, and continues to be, cooperative in developing reasonable, cost-effective nutrition labeling. A constructive step toward increasing the use of labeling would be to permit the use of values from a national data bank. The nutrient data bank, "average nutrient value", approach is appropriate and reasonable for commodities for which 1) the existing analytical compliance costs are unduly burdensome and costly, and 2) the nutrient variability may be considerable. A data base should be compiled to provide representative nutritional values for specific categories of all food products. Nutrition labeling could then be based on these average values.

A second constructive step is to approach nutrition labeling as a consumer information tool, rather than as a regulatory compliance problem or a clinical nutrition exercise. First, some of the analytical methods for micronutrients do not have the precision the label implies. Second, since nutritional deficiencies are rare in the general U.S. population (with the possible exception of iron), there is no nutritional need for the degree of precision currently required on labels. Third, nutrient needs vary from person to person, and even from day to day. Nutrition labeling does not need to be more precise than the methods used to determine the nutrients in foods or the nutrient needs of humans.

Those industries that have used nutrition labeling have found the costs to be considerable on a yearly or cumulative basis. It is, therefore, important that 1) nutrition labeling remain voluntary, 2) it be based on averages or a data bank, if desired, and 3) an economic impact statement be prepared by the responsible agency before any changes are made in the labeling regulations.

Consumer Use of Nutrition Labeling

Food selection is based on four key factors: cost, taste, nutrition/health and convenience (Roper Reports, 1979a, General Mills, unpublished data). About one-half of the consumers look at the nutrition information on labels, especially when purchasing a food for the first time (Roper Reports, 1977) and to compare nutrient content of competing brands (NRC, 1976). Fewer than 10% of consumers look at the label for information for meal preparation (Roper Reports, 1979b). The 1978 FDA consumer food labeling survey showed that only 8% of consumers interviewed had specific problems with current food labeling; 59% said they were "completely satisfied," and approximately 33% expressed some need for improvement in food labeling. The majority of these wanted improved ingredient labeling, simplifications of language (e.g., vitamin B_{12}, not cyanocobalamin) and improved format (Consumer Food Res., 1979).

The information on the labeling panel needs to be better organized and updated to meet current consumer needs. Consumers' nutrition concerns center around having information that will enable them to avoid certain food ingredients; hence, this is the major use consumers make of labels and should be the major focus of the information presented. Consumers would like to see calories (55%), protein (about 36%), fat (32%) and vitamins/minerals (48%) (Congressional Federal Register, 1979). Protein, fat and carbohydrate in animal products should be listed in grams per ready-to-eat serving; percentage labeling of protein, fat and carbohydrate is not meaningful to consumers and may be misleading.

An increasing number of consumers are concerned with cholesterol and sodium content of foods, but voluntary listing of these food components should be permitted unless a claim is made. Interestingly, in the 1978 FDA Food Labeling Survey, only 1% of consumers indicated a concern for sodium labeling.

The data for type and amount of fat, sodium and cholesterol are scanty for some animal foods and well developed for others. The accuracy of the older data for some of the minerals such as iron and some of the trace elements may be questionable in view of the great changes that have been made in analytical methodology and in formulation of processed animal food products.

Whatever nutrients are listed, the format and information should be understandable to consumers and, where possible, should build on information consumers already know and understand. The current inconsistency between the labeling approaches of the regulatory agencies, the cost of regulatory compliance, and lack of knowledge about how best to communicate nutrition information to consumers are key constraints to expanding nutritional labeling. No changes should be made in nutrition labels without extensive consumer research to verify that the change is necessary and that the information is understood.

Ingredient labeling furnishes another type of useful information for consumers in helping them to avoid specific food ingredients. However, it is not a substitute for nutrition labeling. Both are useful and should provide meaningful information for consumers.

In summary, voluntary nutrition labeling (based, where appropriate, on average nutrient values) would seem to be desirable . Calories, and perhaps proximate analysis, are likely to be most useful to consumers. The label format and information probably need to be simplified. Research should be conducted with consumers before any changes are made on labels.

MEAT GRADING

Although many consumers do not fully understand the method used for today's meat grading system, they do associate their individual preference with one of the retailer's grades or traditional USDA Quality Grades for beef and lamb (Prime, Choice, Good, etc). USDA beef quality grades are currently based on carcass maturity, color and texture of the lean and marbling or fat deposited within the muscle. All of these factors are used as indicators of meat palatability. Marbling is the most critical factor in determining the quality grade of a carcass. As marbling increases, the USDA quality grade increases. USDA Prime is the one with the most marbling and is, therefore, the highest grade. However, research has demonstrated that the marbling levels required for USDA Prime, and the upper portion of the USDA Choice grade, are not necessary to achieve optimum palatability.

Lowering the fat content of human diets through reduction of fat in animal products will require 1) research to establish that the products will be palatable and continue to provide the necessary nutrients in the diet; 2) certain changes in government grades and policies; and 3) an extensive consumer education program which relates to human health.

For some foods of animal origin, the reduction in fat content could be accomplished rather easily. Consumer acceptance of lower fat content in foods of animal origin is necessary to improve the economic feasibility of providing these products. Available information suggests that such products indeed are desired by consumers.

SCIENTIFIC KNOWLEDGE AND PUBLIC POLICY

Knowledge about the physical world, including biological systems, is obtained through application of the scientific method. The starting point of this method is that unanswered questions stimulate speculation about possible explanations for natural phenomena. The essence of the method is that information bearing on these phenomena is evaluated critically so that contradictions and inconsistencies can be identified. Once this has been done, the approach is to design and carry out experiments that have the potential for disproving assumptions or hypotheses.

In cases where it is not possible, or is extremely difficult, to carry out experiments to test hypotheses directly (e.g., with problems of diet-chronic disease relationships), theories can be developed based upon several lines of experimentation related to the hypotheses. Theories developed in this way may be modified by additional evidence without direct experimentation. The strength of the evidence supporting such hypotheses is subject to judgment by individual scientists and may change with time.

The method of science does not prove that an hypothesis is true, but by continuous, rigorous testing and by elimination of assumptions that are shown to be false, the probability increases that those that survive represent increasingly reliable approximations of truth. Dissemination of information obtained by this method exposes the conclusions to wide criticism and evaluation by others and further increases the probability of detecting error. Ideally, scientific controversies are resolved in this way but, in practice, it is often difficult to separate issues that deal with evidence from those that involve judgment. As long as inconsistencies remain, the questions cannot be considered to be resolved.

Public policy involves decisions about the application of established knowledge for human welfare. Policy issues currently of concern in relation to animal products pertain to recommendations for changes in lifestyle which include diet modification as an intervention for improving health. Decisions about public policy are not determined by the scientific approach alone. Public policy decisions may be based on scientific information but they involve political, social and fiscal judgments which may or may not be influenced by the information provided by scientists.

Thus, two steps are involved in public policy actions. The first is in establishing the extent and reliability of the information on which an action may be based. This can be assessed accurately only through use of the scientific method. The reliability of scientific information is not established by consensus. The second step which follows after acceptance of the adequacy of information is a public decision. This is judgmental and, in democratic societies, consensus is often an appropriate way of reaching a decision.

RESEARCH IMPERATIVES

The nutritional needs of the U.S. population are met to a large extent by foods of animal origin. Such foods are good to excellent sources of most essential nutrients. Nevertheless, with a high incidence of obesity in the population and with concern about relationships between high fat and sodium intakes and the incidence of certain chronic degenerative diseases, questions have been raised about the appropriateness of the U.S. diet and food supply. Some of the assumptions about the relationships between diet and the incidence of chronic, degenerative diseases are controversial. Because public policy decisions are being made on the basis of the information currently available, it is incumbent upon all of those concerned to support research in an effort to resolve the scientific controversies and, thereby, to provide a sound, factual basis for public policy decisions. Resistance to consideration of the implications of potential problems usually aggravates, rather than alleviates the problem, and can lead to policy decisions being made before sufficient scientific information has been acquired on which to base such decisions.

It is important that research to provide definitive solutions for potential problems be supported and undertaken as soon as possible after the possibility of problems arising is recognized.

The responsibility for undertaking the necessary research is not a unique responsibility for any one group. Obviously the Federal Government has a responsibility for assuring the safety and adequacy of the food supply. It should, therefore, devote funds from the research resources available to it to ensure that any threat to human well-being from food be thoroughly investigated. The responsibility is not solely that of the Federal Government. It is also a responsibility of the producers and processors of products about which questions of safety and adequacy are raised. When such questions arise, research support to investigate the validity of the assumptions arising from the questions should be accepted as an obligation by the industry involved. Universities also have a responsibility to investigate questions that are of public concern.

In view of the questions that have been raised about relationships between diet, health and disease in the U.S., there are several lines of research pertaining to animal agriculture that should be pursued at the present time. Some are included in the recommendations listed below. It should be recognized that many questions about diet-disease relationships are far from being resolved; that the validity of assumptions about them is controversial; and that many aspects of them are important topics for human nutrition research. A strong continuing program of human nutrition research will be needed to provide the information needed to resolve current controversies and provide needed information. When information about currently controversial topics becomes more definitive, some of the assumed constraints for animal agriculture may be removed. At the same time, others may arise either as the result of new information about human nutrition or as the result of the development of new products, presently unforeseen. It is, therefore, important to encourage interaction among researchers in animal agriculture, human nutrition and the health sciences.

The introduction of new products or technologies into animal production and processing without concomitant understanding or appreciation of potential implications for the consuming public can result in controversy. Many such controversies can most readily be resolved by securing, through research, fundamental knowledge of the overall implications of adoption of new technologies. Some industries producing foods of animal origin have appreciated this and have supported research on a continuing basis to answer questions arising from the use of new technologies. A partnership to conduct research to deal even more effectively with these issues should be encouraged between the private sector, university and government researchers.

ALTERING COMPOSITION OF ANIMAL PRODUCTS IN RELATION TO HUMAN HEALTH

Foods of animal origin are recognized as important sources of many essential nutrients. There is concern, however, about the fat content and the resulting large contribution to the caloric intake from diets containing an abundance of these foods. Another concern is the postulated relationship of dietary saturated fat and cholesterol to the incidence of circulatory disorders.

The addition of salt and sodium-based flavoring agents results in sodium intakes far in excess of the requirement. A substantial portion of the U.S. population is affected by hypertension, and many others are at risk of hypertension. These individuals are encouraged to restrict sodium intake and, therefore, often avoid eating many foods of animal origin which have been processed by current methods involving the addition of salt or other sodium-containing compounds. The phosphate content of U.S. diets is increasing and is currently in excess of recommended levels. This is due in part to the use of phosphate-containing compounds in processed animal products and other foods and beverages. Since the effect of high phosphate intake on human health is not known, this question and the means of reducing phosphate in processed animal products should be investigated.

Priority Research Areas

- Investigation of the potential for reducing the fat content of fresh and processed meat and milk products by breeding, feeding and processing methods.
- Investigation of the potential for reducing sodium and phosphate in processed animal products.

NUTRITIONAL VALUE OF FOODS OF ANIMAL ORIGIN

Assessment of the nutrient contribution of fresh and processed foods of animal origin to human nutrition requires knowledge of their nutrient composition. This is particularly important for modified products, processed foods and substitute products. Information relative to the bioavailability of these nutrients is also essential. Finally, the nutritional consequence of changes that may occur in animal products during processing, storage and preparation should be studied.

Priority Research Areas

- Development of economical and rapid methods for analysis of nutrients and other food constituents.
- Analysis of components of fresh and processed foods of animal origin and effects of storage and preparation on these components.
- Determination of bioavailability of nutrients in foods of animal origin.
- Studies of the nutritional adequacy and comparability of meat, dairy and egg substitutes.

DIET AND HEALTH

Several human health problems are associated directly or indirectly with diet. Perhaps the most widely publicized is coronary heart disease, where the roles of fat, cholesterol and protein in the development of this disease have not been satisfactorily delineated. Foods of animal origin are major dietary sources of these substances.

Relationships between diet and severe chronic degenerative diseases (e.g., osteoporosis, atherosclerosis, cancer) have been postulated. To understand these diet-disease relationships, continued research is needed on the biological events which lead to the manifestation of disease. Recent advances in nutrition, cell biology and biochemistry provide techniques and concepts that should lead to an understanding of the disease process. These advances will become significant biological accomplishments in the 21st century.

Priority Research Areas

- Expanded research on diet-cardiovascular relationships including the metabolism of lipoproteins and cholesterol oxidation products, and thrombogenesis.
- Expanded research on diet-cancer relationships.
- Studies of availability of nutrients as affected by constituents of animal food products.
- Studies of the relationships between animal food products and bone and oral health.

CONSUMER INFORMATION

Foods of animal origin have received a great deal of publicity in relation to the substances they contain, their place in the diet and their role in relation to health and disease. The present nutrition labeling format does not communicate sufficient information effectively.

Research on labeling and other forms of information needed by consumers, and how best to present, is necessary for both fresh and processed foods of animal origin. Cooperative efforts of producers, processors, government agencies, educators and consumers are necessary to assure the development of effective labeling and educational efforts for these products. Calories, fat, protein, sodium and cholesterol deserve particular attention. Research to evaluate the effectiveness of various approaches related to consumer knowledge, attitudes and food consumption practices is also required. Consumer interest and issues related to health and disease must be addressed as part of such research. A system of monitoring trends in food consumption changes is important to the evaluation of areas needing emphasis in any program aimed at providing consumer information and education relative to nutrition and the contributions of various foods including foods of animal origin.

Priority Research Areas

- Identification of consumer information needs.
- Development of effective means of providing information, including labeling, in an understandable and usable manner.
- Studies on effective evaluation of consumer information programs including changes in knowledge and food-related behavior.
- Development of food consumption monitoring systems which permit evaluation of dietary changes, including changes in consumption patterns of specific food items, and the relationship between such changes and health and disease status.

REFERENCES

Ahrens, E.H., Jr. 1976. The management of hyperlipidemia. Whether, rather than how. Ann. Internat.. Med. 85:87.

Ahrens, E.H., Jr. 1979. Dietary fats and coronary heart disease: unfinished business. Lancet 2:1345.

Ahrens, E.H., Jr. and W.E. Connor. 1979. Symposium: Report on dietary factors relating to the national health. Amer. J. Clin. Nutr. 32(Suppl):2627.

Albanese, A.A., A.H. Edelson, E. Lorenze, Jr., M.L. Woodhull and E. H. Wein. 1975. Problems of bone health in the elderly - A ten year study. NY State J. Med. 75:326.

Anand, C.R. and H.M. Linkswiler. 1974. Effect of protein intake on calcium balance on young men given 500 mg calcium daily. J. Nutr. 104:695.

Blackburn, H. 1979. Diet and mass hyperlipidemia; A public health view. In: R. Levy, B. Rifkind, B. Dennis and N. Ernst (Ed.) Nutrition, Lipids and Coronary Heart Disease. Raven Press, NY.

Better Homes and Gardens. 1976. Consumer Rep. 2.

CAST.1977. Hormonally active substances in foods: a safety evaluation. Council Agr. Sci. Technol. Rep. No. 66, March 1977, Ames, IA.

CAST.1980. Foods from animals: Quantity, quality and safety. Council Agr. Sci. Tech. Rep. No. 82. Ames, IA.

Committee of Principal Investigators Report. 1978. A co-operative trial in the primary prevention of ischaemic heart disease using colfibrate. Brit. Heart J. 40:1069.

Congress Fed. Register. 1979. Vol. 44:75990, Dec.

FDA. 1979. The 1978 Consumer Food Labeling Study. Consumer Res. Staff, Div. Consumer Studies, Bureau of Foods, Washington, DC.

DHEW.1978. Selected findings: Food consumption profiles of white and black persons 1-74 years of age in the United Sattes. Adv. Data No. 21 USDHEW Pub. Hlth. Serv.

DHEW.1979. Healthy people: the Surgeon General's report on health promotion and disease prevention. DHEW (PHS) Pub. No. 79-55071. Washington, DC, 228 p.

Fomon, S.J. 1974. Infant nutrition, 2nd Ed. W.B. Saunders, Philadelphia, PA.

Glueck, C.J. 1979. Appraisal of dietary fat as a causative factor in atherogenesis. Amer. J. Clin. Nutr. 32:2637.

Gori,J.B. 1979. Food as a factor in the etiology of certain human cancers. Food Tech. 33:48.

Hanson, L., B. Carlson, J.R. Cruz, B. Garcia, J. Holmgren, S. Khan, B.S. Lindblastz, A.M. Svennerholm, B. Svennerholm and J. Urrutia. 1979. Immune response in the mammary gland. In: P.L. Ogra and D.H. Dayton (Ed.). Immunology of the Breast Milk. Raven Press, NY.

Health and Welfare Canada. 1979. Canada's food guide. Dept. Nat. Hlth. Welfare, Ottawa, Can.

Highlights: Ten State Nutrition Survey. 1968-1970. U.S. Department of Health, Education, and Welfare. Hlth Serv. Ment. Hlth Adm. Center Dis. Control. Atlanta, GA

Kannal, W.B. 1978. Status of coronary heart disease risk factors. J. Nutr. Educ. 10:10.

King,J.C., S.H. Cohenour, C.G. Corruccini and P. Schneeman. 1978. Evaluation and modification of the Basic Four Guide. J. Nutr. Educ. 10:27.

Kritchevsky, D. 1979. Dietary interventions, p. 229-246. In: R.J. Levy, B.M. Rifkind, B.H. Dennis and N.D. Ernst (Ed.) Nutrition, Lipids and Coronary Heart Disease. Raven Press, NY.

Kritchevsky, D. 1979. Nutrition and heart disease. Food Technol. 33:39.

Kritchevsky, D. S.A. Tepper, R.B. Morrissay, S.K. Czarnecki and D.M. Klurfeld. 1979. Influence of whole or skim milk on cholesterol metabolism in rats. Amer. J. Clin. Nutr. 32:597.

Levy,R.I. B.M. Rifkind, B.H. Dennis and N.D. Ernst (Ed.). 1979. Nutrition, Lipids and Coronary Heart Disease: Nutrition in Health and Disease, Vol. I. Raven Press, NY.

Mann,G.V. 1972. The saturated vs unsaturated fat controversy. Proc. Meat Indust. Res. Conf., Amer. Meat. Inst.Fd, p. 73.

Mann,G.V. 1977. A factor in yogurt which lowers cholesterolemia in man. Artherosclerosis 26:335. Natl. Food Rev. 1977. P. 49.

Marston, R.M. and B.B. Paterkin. 1980. Nutrient content of the national food supply. Natl. Food Rev., Winter, p. 21.

Moore, R.A. 1980. Food consumption in 1979. Natl. Food Rev., Winter, p.26.

McGill, H.C., Jr. 1979. Cholesterol: Appraisal of cholesterol as a causation factor in atherogenesis. Amer. J. Clin. Nutr. 32(Suppl.):2632.

Meneely, G.R. and H.D. Batterbee. 1976. Sodium and potassium. Nutr. Rev. 34:255.

Monsen, E.R., L Hallberg, M. Layrisse, D.M. Hegsted, J.D. Cook, W. Mertz and C.A. Finch. 1978. Estimation of available dietary iron. Amer. J. Clin. Nutr. 31:314.

National Center for Health Statistics. 1979. Plan and operation of the Health and Nutrition Examination Survey, United States 1971-1973. H.W. Miller, Vital and Health Statistics, Series 1. No. 10a and 10b DHEW Pub. No (PHS) 79-1318. Public Health Service, Washington, DC. U.S. Government Printing Office.

National Center for Health Statistics. 1979. Overweight adults in the United States - Advanced data: No. 51, Vital Health Statistics.

National Center for Health Statistics. 1979. Fat cholesterol and sodium intake in the diet of persons 1-74 years: United States Advanced Data. No. 54. Vital and Health Statistics.

National Center for Health Statistics. 1980. Consumption patterns in the United States 1971-1974 Vital and Health Statistics, Series 11. Public Health Service, DHEW, Hyattsville, MD (In press).

National Dairy Council. 1974. The role of alctose in the diet. Dairy Council Digest 45, No. 5. Rosemont, IL.

National Research Council. 1976. Consumer Nutrition Study. Washington, DC.

National Research Council (Committee on Dietary Allowances, Food and Nutrition Board). 1980. Recommended Dietary Allowances. 9th Ed. Natl. Acad. Sci., Washington, DC. 185 p.

National Research Council (Food and Nutrition Board). 1980. Toward Healthful Diets. Natl. Acad. Sci. Washington, DC. 47 p.

Olson, R. 1979. Is there an optimum diet for the prevention of coronary heart disease? P. 349-364. In: R.J. Levy, B.M. Rifkind, B.H. Dennis and ND.D Ernst (Ed.). Nutrition, Lipids and Coronary Heart Disease. Raven Press, NY.

Pennington, J. 1976. Dietary Nutrient Guide, AVI Pub. Co., Inc., Westport, CT.

Reiser, R. 1978. Oversimplification of diet: Coronary heart disease relationships and exaggerated diet recommendations. Amer. J. Clin. Nutr. 31:865.

Roper Rep. 1977. The Roper Organization, New York.

Roper Rep. 1979a. The Roper Organization, New York.

Roper Rep. 1979b. The Roper Organization, New York.

Schucker, R.E. 1978. A surveillance of nutrition labeling in the retail packaged food supply. Bureau of Foods, FDA, Washington, DC.

Simopoulos, A. (Ed.) 1979. The Biomedical and Behavioral Basis of Clinical Nutrition: a projection for the 1980's. U.S. DHEW. NIH Pub. No. 80-1970. Nov. 87 p.

Taylor, C. B., S. K. Peng, N.T. Werthessen, P. Tham and K. T. Lee. 1979. Spontaneously occurring angiotoxic derivatives of cholesterol. Amer. J. Clin. Nutr. 32:40.

Taylor, S. 1980. Food allergy. The enigma and some potential solutions. J. Food Protection. 43:300.

Tobian, L., Jr. (Consultants: E.D. Greis, H. Langford, L. Page). 1979. The relationship of salt to hypertension. Amer. J. Clin. Nutr. 32:2739.

Torun, B., N.W. Solomons and F.E. Viteri. 1979. Lactose malabsorption and lactose intolerance: Implications for general milk consumption. Arch. Latinoamericanas de Nutricion 29:445.

USDA-HEW. 1980. Nutrient content of the National Food Supply. NFR-9, Winter, 1980. Washington, DC.

Van Itallie, T.B. and J. Hirsch. 1979. Calories: Appraisal of excess calories as a factor in the causation of disease. Amer. J. Clin. Nutr. 32(Suppl.):2621.

Watt, B.K. and A.L. Merrill. 1963. Composition of foods - raw, processed, prepared. USDA Handbook 8(rev.):190 p.

Whedon, G.D. 1980. Recent advances in management of osteoporosis. 4th International Workshop on Phosphate and Other Minerals (In press).

"Animal agriculture is obviously a critical part of our agricultural resource base. It plays a vital role in providing quality nutrients in proportions needed by humans. Currently, animal products provide U.S. consumers with one-third of their energy, two-thirds of their protein, two-thirds of their fat, four-fifths of their calcium, two-thirds of their phosphorus, 38% of their iron, 42% of their vitamin A, 37% of their thiamine, 79% of their riboflavin, 47% of their niacin, 60% of their vitamin B_6 and virtually all of their vitamin B_{12}."

Anson Bertrand

"Food is first among the needs of people...it is our most important renewable resource."

Sylvan Wittwer

TABLE 2.1. COMPOSITION OF SOME REPRESENTATIVE MEAT PRODUCTS[a]

	Beef Good Grade Total Edible	Pork Medium Fat Composite	Lamb Choice Composite	Chicken Roasting Total Edible	Liver Calf	Sausage Pork	Bologna
Kcal/100 g	263	308	263	239	140	498	304
Water %	60	56	61	63	71	38	56
Protein %	18.5	15.7	16.5	18.2	19.2	9.5	12.5
Fat %	20.4	26.7	21.3	17.9	4.7	50.8	27.0
Carbohydrate %	0	0	0	0	4.7	trace	1.8
Protein % kcal	28	20.4	25	30.5	54.8	7.6	16.4
Fat % kcal	70	78	73	67.4	30.2	91.8	81.7
Carbohydrate % kcal	-	-	-	-	11.7	-	-
Kcal/g dry weight	6.6	7.0	6.7	6.5	4.8	8.0	6.9

[a]Watt and Merrill (1963).

TABLE 2.2. COMPOSITION OF DAIRY PRODUCTS AND EGGS[a]

	Milk Whole Cow	Milk Skim Cow	Milk Human	Ice Cream Fat, 12%	Cheese Cheddar	Yogurt Whole Milk	Eggs Whole Fresh	Eggs Whole Dried
Kcal/100 g	65	36	77	207	398	62	163	592
Water %	87	90.5	85.2	62	37	88	73.7	4.1
Protein %	3.5	3.6	1.1	4	25	3	12.9	47
Fat %	3.5	.1	4.0	12.5	32.2	3.4	11.5	41.2
Carbohydrate %	4.9	5.1	9.5	20.6	2.1	4.9	.9	4.1
Protein % kcal	21.5	40	5.7	7.7	25.1	19.3	31.9	31.8
Fat % kcal	48.5	2.5	46.8	54.3	72.8	49.3	63.5	62.6
Carbohydrate % kcal	30.1	56.7	49.4	39.8	2.1	31.6	2.2	2.8
Kcal/g dry weight	5	3.8	5.2	5.4	6.3	5.2	6.0	6.2

[a]Watt and Merrill (1963).

TABLE 2.3. ESSENTIAL NUTRIENT CONTENT OF SOME FOODS OF ANIMAL ORIGIN (UNITS/KG)[a]

	Milk		Eggs	Beef	Pork	Chicken
	Human	Cow				
Vitamin A (IU)	1898	1024	11,800	200	(0)	9200
Vitamin D (IU)	22	14				
Vitamin E (mg)	1.8	.4				
Ascorbic acid (mg)	43	11	----	----	----	----
Thiamin (mg)	.16	.44	1.1	.09	7.6	.8
Riboflavin (mg)	.36	1.75	3.0	1.8	1.8	1.9
Niacin (mg)	1.47	.94	1	48	41	67
Vitamin B_6 (mg)	.10	.64	1.1	4.3	3.5	5.0
Folacin (μg)	52	55		105		
Vitamin B_{12} (μg)	.3	4	20	18	5.5	4.0
Calcium (mg)	340	1170	540	120	90	100
Phosphorus (mg)	140	920	2050	2030	1750	1760
Magnesium (mg)	40	120	110	250	180	230
Iron (mg)	.5	.5	23	30	23	16
Zinc (mg)	4	4	10	34	27	12
Iodine (μg)	30	47				

[a]Watt and Merrill (1963).

TABLE 2.4. NUTRIENT CONTRIBUTION OF ANIMAL FOODS TO THE NATIONAL FOOD SUPPLY, PERCENT OF TOTAL INTAKE FOR EACH NUTRIENT[a]

	Energy	Prot.	Fat	Chol.	Ca	P	Fe	Mg	A	B_1	B_2	Niacin	B_6	B_{12}	C
Meat[b]	20.6	42.0	35.2	.1	3.9	27.4	20.0	13.5	22.5	27.1	22.2	45.0	40.0	69.2	2.0
Eggs	1.8	4.9	2.7	.1	2.3	5.2	4.9	1.2	5.8	2.0	4.9	.1	2.1	8.5	0
Dairy[c]	10.7	21.7	11.8	6.3	73.8	34.7	2.5	21.2	13.2	8.1	39.1	1.4	11.6	20.7	3.7
Total animal sources	33.1	68.6	49.7	6.5	80.0	67.3	36.4	35.9	41.5	37.2	66.2	46.5	53.7	98.4	5.7
Fats and oils	18.2		42.7	--											
Sugars and other sweeteners	17.1		--	38.6											
Miscellaneous	.6		1.1	.5											
Total	35.9		43.8	39.1											

[a]USDA (1980, R.M. Marston and B.B. Paterkin).
[b]Includes meat (pork fat cuts), poultry and fish.
[c]Includes butter.

TABLE 2.5. CHOLESTEROL CONTENT OF SOME ANIMAL PRODUCTS[a]

	mg/100 g of edible product
Beef, raw	70
Butter	250
Cheese - cheddar	100
- cottage	15
Chicken	60
Egg, whole	550
Fish	70
Kidney	375
Lard	95
Liver	300
Lobster	200
Milk, fluid, whole	11
Shrimp	125

[a]Watt and Merrill (1963).

TABLE 2.6. SODIUM AND POTASSIUM CONTENT OF SELECTED FOODS[a]

	Sodium mg/100 g	Potassium mg/100 g
Beef	65	335
Lamb	75	295
Pork	70	285
Eggs (whole)	122	129
Milk (whole)	50	144
Luncheon meats	1234	222
Bologna	1300	230
Peas (raw)	35	1005
Potatoes (raw)	3	407
Bread (whole wheat)	527	273

[a]Watt and Merrill (1963).

3

Z. L. Carpenter (Co-Chairman), J. R. Kirk (Co-Chairman),
P. Kifer (Rapporteur)

J. Birdsall, R. Bray, J. Doull, T. W. Downes,
W. Dunkley, D. Farkas, D. E. Farris, R. A. Field,
R. Hagen, N. Jerome, J. Judy, R. V. Lechowich,
D. Lund, W. W. Marion, R. A. Merkel, J. Richert,
W. M. Roberts, and W. Stadelman

FOOD PROCESSING AND ACCEPTABILITY

SUMMARY

Food conservation research is imperative if we are to meet the human needs of the 21st century. Food conservation technologies could ensure that the raw materials produced by animal agriculture would be effectively converted to usable food and food ingredients. Basic research should be carried out to develop appropriate food conservation systems. Such research should be aimed at eliminating losses of edible products which now occur in current processing systems. The overall goal of food conservation research must be to develop processing methods which will ensure that animal foods consistent with consumers' health needs, economic resources and lifestyles will be available.

The principal constraints impeding the development and adoption of new and improved food conservation systems include resource availability, health hazards, dietary needs, regulatory limitations, economic influences, biological and technological limitations imposed by limitations of the present knowledge base relating to food preservation, and consumer attitudes and ideology.

To overcome these constraints and provide for continued and expanded availability of animal foods to meet human needs in the 21st century, research funding and activities must be focused on key areas of food conservation research and provide for the availability of trained scientists and technologists.

It is imperative to develop alternative processing technologies, those currently needed by food processors to cope with the unprecedented rise in the cost of energy and other resources.

Research efforts also must be centered on the conservation of animal products by development of more effective and efficient processing and delivery strategies which will considerably reduce losses that occur within the existing system.

Consumer concerns about quality and health risks are now greater than at any time in the past, and they require response. New systems must assure the consumer that animal foods are nutritious, uniform in quality and pose no significant health risk.

Other research to improve food conservation in the 21st century includes:

1) Conservation of foods of animal origin from collection to consumers.
2) Conservation and recycling of resources.
3) Modification of regulatory activities affecting conservation.
4) Food acceptability and quality assessment.
5) Rapid, accurate and economical methods for assessing animal food quality.
6) Effects of processing, storage, distribution and preparation techniques on chemicals added directly or indirectly to animal food products.
7) Packaging systems for animal food products.

It is imperative that the animal food processing sector establish priorities for basic and applied research, and that these priorities be coordinated with those of the production sector. These priorities also should be communicated to the public sector and appropriate funding agencies.

INTRODUCTION

FOOD CONSERVATION

Food conservation is defined as the optimum utilization of all potential food materials within the animal agriculture system from production through consumption. It includes the need for new or modified technologies in preservation, packaging, food formulation and quality assessment to convert raw agricultural commodities into acceptable, nutritious foods which pose no significant risk to humans. These technologies must be developed to optimize the utilization of associated resources (energy, water, metals, etc.) at minimal environmental impact and cost.

Because world food supplies are projected to be limited in the 21st century, research to improve technologies for food conservation becomes imperative to extend and improve available food supplies.

Basic and applied science with foods of animal origin has significantly upgraded human life in the 20th century. Therefore, as we move toward the 21st century, the need for more basic information that can be translated into technologies for food conservation will become paramount in view of projected food supplies, population and resource availability.

The primary purpose of animal food conservation is to provide a wide variety of foods that are excellent sources of food nutrients and have sensory qualities desired by the general population. The principal categories of animal foods are:

1) Fresh, frozen and processed meats.
2) Milk, cheese and other dairy products.
3) Poultry and egg products.
4) Manufactured foods that include the above.

Animal agriculture also makes important contributions to nonfood by-products. Significant applications are in the feed, chemical, cosmetic, pharmaceutical and clothing industries.

Food conservation is an integral part of animal agriculture. It includes those activities involved in converting raw food materials into food and food ingredients acceptable to the consumer. The primary functions of food conservation are to:

1) Extend the shelf life (useful storage life) so that the product is available for a longer period of time in a form usable by the consumer.
2) Improve the nutritional value and safety of food products.
3) Alter various characteristics of animal foods to provide a wider variety of products for the consumer.
4) Separate animal food components so that only the desirable, nutritious or readily utilizable parts need to be transported to the consumer for subsequent storage and preparation.
5) Formulate and manufacture new food products from a variety of raw food materials and ingredients.
6) Decrease the food costs to the consumer.

The above functions of food processing are consistent with universal human needs and the goals of a modern, industrialized society. However, the method of meeting these needs and societal goals must be monitored periodically to ensure that new societal, economic and health realities are balanced by new technologies and practices. A reassessment of current practices in food processing may be required in view of stresses such as: a) development of new directions for preserving foods; b) conserve resource utilization; c) costs of processing; and d) data which suggest an association between animal products in the total diet with chronic degenerative diseases.

Technologies used in animal food processing developed in the first part of the 20th century capitalized on cheap energy supplies, relatively inexpensive manufacturing methods and the need to fortify foods with nutrients which were limiting in the U.S. diet. For example, vitamin-D-fortified milk has effectively eliminated rickets in the U.S. population. The concept of modifying an animal food to achieve a specific nutritional goal during the early years of the 20th century can be applied to newer technologies in order to meet the nutritional and health needs of U.S. consumers in the 21st century.

Processed animal foods supply a variety of wholesome and nutritious products which were lacking in the grocery stores in past generations. Several of these processed animal foods include: a) pasteurized, homogenized milk of known nutrient quality with a specified fat level; b) poultry products ranging from fresh, ready-to-cook poultry to turkey salami to dried eggs as part of numerous bakery mixes; and c) numerous sausages and cured meats of uniform composition, appearance and nutritive value. A wide variety of animal foods is now available at all seasons of the year. These products are prepared and packaged in a manner which ensures nutritional quality, uniformity and minimal human risk.

Centralized processing of animal foods into retail portions is widely practiced today and will continue in many different forms in the 21st century in order to meet the needs of changing family modes, lifestyles and resources. Consumers have shown an increasing demand for more completely processed food products, ready-to-eat convenience foods and foods for consumption away from home. Food products processed for this market are portion-controlled to precise sizes of uniform composition and quality standards. This market is continuing to expand and be a significant portion of the animal products market in the 21st century.

Sociodemographic projections indicate that family units of the 21st century will be small. Small family units may alter preparation of foods in the home. In future food delivery systems, centralized processing may be used more extensively, and research is needed to assure nutritional value, sensory qualities, convenience, a uniform distribution of ingredients in all portions and minimum human risk.

A major challenge to food processors in the 21st century will be to deliver products of the desired quality in sufficient quantity to meet consumer needs. Consumers now purchase animal products that are consistent with their lifestyles, disposable income and health values, i.e., products in convenient forms and with high levels of protein, vitamins, minerals and other micronutrients. Consumers are concerned that food costs be low and that resources including water and energy be efficiently utilized. Current losses of potentially edible foods from animals through spoilage or poor processing procedure should be minimized. Conservation of the total supply of animal food products will be necessary to meet demands of the future. It has been demonstrated that the application of present technological knowledge to the food processing industry could reduce the energy requirements in some processing plants by as much as 20-30%.

MEAT

Processing of meat which includes poultry may be as simple as the preparation of retail cuts of meat, or it may involve grinding, flaking, sectioning, seasoning, salting, curing, forming, smoking, heating, fermenting, drying or combinations of these treatments. Regardless of the processing method, the primary goal of processing is to extend shelf life and to convert the product into a form that provides greater variety and convenience to the consumer. In general, the processing of meats does not significantly lower nutritional value of the product. For example, salting does not destroy nutrients in meat, nor does the addition of nitrite to cured meat. However, these chemicals may affect the acceptance of animal products by certain population subgroups. In most instances, processing has been shown to increase the availability of nutrients by preventing spoilage and by transforming products not ordinarily consumed into forms that are more aesthetically acceptable. Retort processing of canned animal foods can cause significant destruction of heat labile vitamins such as thiamine, folate and B_6. The physiological availability of some proteins and amino acids may be reduced by browning reactions involving carbohydrates and secondary lipid oxidation products, or by inter- and intra-protein reactions in the presence or absence of oxygen. Although researchers have reported losses in the gross protein value as a result of heat destruction of amino acids, these losses are not considered an important practical problem unless the amino acids affected are the growth-limiting amino acids in the protein, and provided that this protein is the sole dietary protein source. Normally, changes affecting the availability of amino acids would override any small change in amino acid composition as a result of thermal processing. Such changes would not seriously affect the nutrient value of the protein or protein rich foods.

REFRIGERATION AND FREEZING

Fresh and processed meat is generally moved through market channels under refrigerated and/or frozen conditions. Cold temperatures reduce microbial and enzymatic spoilage, and loss of nutrient content. Maintenance of cold storage conditions throughout the food system requires a highly intensive energy input. There is a need to consider alternatives in product processing, packaging and distribution that would permit the use of less energy-intensive technologies.

CURING

Meat is presently cured with sodium nitrite, sodium nitrate (certain products), salt and curing adjuncts, such as ascorbate and erythorbate, etc. Nitrite: a) inhibits the outgrowth of Clostridium botulinum spores in or on meat; b) helps in the development of characteristic flavor and fixation of the inherent red color of cured meats; c) prevents warmed-over flavor in reheated meat products; and d) prevents rancidity from developing during the storage of the product. The antibotulinal effect of nitrite is by far the most important factor. To date, no other chemical has been identified that can replace nitrite for inhibiting the growth of Clostridium botulinum bacteria that cause botulism.

The risk of botulism is currently being weighed against some evidence that nitrite may be harmful in itself or as it may be involved in nitrosamine formation. This research is being carefully scrutinized and the outcome of the reviews must be considered in future research on the production of cured meat products.

SALTING

Sodium chloride added to processed meat products serves as a preservative and makes it possible to distribute perishable meats through complicated distribution channels. If the amount of salt used in manufacturing processed meat products was reduced, greater reliance would have to be placed on refrigeration and the margin of safety in product handling would be reduced. In emulsion meat products, such as weiners and bologna, salt serves a functional role in the formation of the protein-fat emulsion. Salt also contributes to the sensory attributes of these products. Attempts to partially or totally replace sodium chloride with other salt mixtures have not proved successful.

SMOKING

There is no evidence that meat products that have been smoked influence the incidence of cancer in humans. Liquid smoke is made from natural smoke and may be substituted for wood smoke in the processing of animal products. Recent studies reported at the 64th annual FASEB meeting indicate that benzopyrene, a component of smoke, may be related to the development of atherosclerosis. This potential relationship warrants further investigation.

MECHANICALLY PROCESSED MEAT PRODUCTS

Mechanical deboning techniques are being used to remove meat from irregularly shaped bones. The mechanically deboned product has a finely ground appearance because the meat is forced through small holes or slots to separate it from bone. In 1978, the FSQS of USDA approved the term "mechanically processed (species) product" to describe mechanically deboned meat. Mechanically deboned poultry products have been approved and used in many processed meat items for the past 15 years. Mechanical deboning of fish has been in practice for approximately 30 years.

Mechanical deboning procedures increase the supply of meat for further processing, reduce waste of high quality protein, conserve energy and reduce labor costs. Mechanically processed products contain both bone powder and bone marrow. If used indiscriminantly, these ingredients adversely affect flavor, color and texture of formulated meat products, and may have a nutritional effect.

EGGS

Consumption of eggs and egg products in 1979 was about 283 eggs per capita. Eighty-eight percent (250) were consumed as shell eggs and 12% (33) were in liquid, frozen or dried-product form. Most eggs are packaged into consumer-sized cartons for distribution. Breakage of eggs during cleaning, grading, sizing, packaging and distribution exceeds 5%. The need for improvements in equipment, packaging and handling procedures is evident.

Eggs for processing are washed and broken, and albumen and yolks are separated. Products are homogenized, clarified, blended, pasteurized, packaged and delivered for use as either liquid or frozen products. Albumen to be dried is desugared prior to drying to eliminate browning.

DAIRY PRODUCTS

The dairy industry has evolved to large processing plants that produce a wide variety of products, including an array of fresh dairy products. Modified dairy products such as low lactose, low fat, low sodium, low calorie, reduced carbohydrate, etc. are being manufactured to meet special dietary requirements. Cheese consumption continues to increase, and additional varieties are being developed to meet the appetites of the consumer. Butter consumption has decreased in volume in recent years. The primary role of nonfat dry milk is its use as a constituent in other food products. Whey has become a major disposal problem, and butterfat is becoming a more difficult constituent for which to find a market.

The dairy industry is energy intensive because of the perishability of its products, the need for heating and cooling during storage and transportation, and the concentration and drying of various dairy products. Progress has been made in mechanizing and automating processing technology in the dairy industry. These developments have been estimated to save approximately 10% of the total energy costs. Significant savings can be made if innovative processing technologies can be developed that are vastly different from those commercially available today.

Packaging of dairy products is expensive and energy intensive. Each of the major types of packages, i.e., glass, paper, plastic and laminates, has advantages and disadvantages which indicate that better packaging materials and equipment are needed.

There is increasing recognition for the need of dairy foods to meet special dietary requirements. These nutritional demands can be met by altering the ratio of certain macro-constituents through processing; however, changes in the physical properties may occur.

PACKAGING

The function of packaging is to protect the food from environmental hazards between the point of production and the point of consumption. Over 80% of the food purchased in the U.S. is packaged. More specifically packaging is utilized to:

1) Prevent further exposure of the food to undesirable microorganisms.
2) Control undesirable effects of oxygen such as off-flavor production, formation of potentially harmful oxidation products, or nutrient loss.
3) Maintain appropriate moisture content through exclusion or retention of water vapor.

4) Protect the food from pests such as insects or rodents.
5) Extend the useful (acceptable) life of food products through modification of the atmosphere surrounding the food.
6) Protect the food from mechanical damage.
7) Prevent harmful effects of light on nutrients and quality.
8) Provide convenience and information.

Packaging requirements for animal foods are specific for each product. In meats, for example, permeability of packages to water, oxygen and other gases, such as carbon dioxide, depends on the product. In addition, some packages must be opaque, while others are transparent. These properties are necessary to enhance shelf life and protect the product from the external environment, and provide customer appeal.

NEW CONSERVATION TECHNOLOGIES

The natural evolution of events in food processing includes:

1) Obtaining basic information leading to the development of new technologies.
2) Developing new technologies.
3) Using these technologies to: a) provide greater consumer acceptability; b) provide sensory properties; c) provide nutritional attributes to products; and d) conserve energy and food products through improved handling, transportation and storage techniques.

New conservation technologies developed for animal product processing may include heating, cooling, freezing, fermenting, concentrating or altering by other methods in order to conserve resources, provide greater utilization of animal products or create new product forms. Considerable research including data acquisition and analysis is needed to develop and apply these technologies to a full range of animal products.

DESIRABLE GOALS FOR ANIMAL FOOD CONSERVATION

The goal of research in food conservation in animal agriculture is to develop food processing systems that optimize the use of available raw supplies, energy, water and other resources. Achievement of this goal will make it possible to provide nutritious and acceptable foods at reduced cost to consumers with negligible risk and environmental impact. Achieving this goal includes:

1) Developing technologies which permit production of animal food products that reduce health risks.
2) Developing energy-efficient food preservation methods.
3) Increasing the efficiency of water conservation techniques.

4) Improving processing techniques to more fully utilize raw materials.
5) Improving technology that reduces costs of processed animal products to consumers.
6) Improving packaging, storage and distribution systems while reducing transportation.
7) Developing systems to deliver foods to help reduce waste of food and energy by the consumer.

CONSTRAINTS

Modification and improvement in food conservation systems for the 21st century will meet with resistance for a variety of reasons. The more probable constraints which must be overcome include the following:

1) Resource availability. Efficient animal food conservation in the United States in the 21st century will depend on an adequate supply of animals. Research on improving and controlling meat quality must be conducted for its optimal use. To accomplish this, it is essential that regulating mechanisms modulating protein and fat synthesis during growth of the animal be identified and understood. Provided that an adequate supply of animals is available, food conservation will be further constrained by the ability to efficiently preserve animal foods and recycle water, air and solid materials, and by the availability of energy. Availability of trained, reliable scientists and technologists is also a critical resource. Animal food processing is a high technology industry and demands complementary qualities in the labor supply. As the food industry continues to develop technologically, capital for establishment of the most efficient conservation systems may become limiting.

2) Consumer acceptance. This constraint involves a combination of attitudes and perceptions, product performance characteristics and economic realities. There is a lack of consumer information with respect to health risks, nutrition and wholesomeness. Emphasis on conservation of food resources should be given high priority in food information programs to encourage utilization of wholesome foods available from organ meats, mechanically processed meat products and other new technological innovations in the conservation of food products. Quality, grading and labeling information should emphasize conservation in addition to other traditional factors. Objective constraints on consumer acceptance of foods of animal origin include price, convenience, product stability, home energy conservation, nutrient content and health risk factors.

3) Nutrition. Foods of animal origin are an important source of high quality protein and other nutrients essential to human health. Conversely, concern with increased caloric intake coupled with decreased physical activity raises questions about the role of dietary fat, and the content and composition of fat in products of animal origin. Similarly, the use of some processed foods of animal origin prepared by present techniques may result in excessive sodium intake for some population subgroups.

4) Economic. Human food needs in the 21st century are expected to be substantially increased. Foods of animal origin (which require more resource input per unit of usable food) will be under severe economic restraint. Availability of additional supplies will be limited by the high cost of critical materials and energy, and by the availability of productive land. Improved food conservation systems can increase the amount of usable food of animal origin.

5) Regulatory. New processes and technology that have the potential to significantly increase the utilization and conservation of nutrients from foods of animal origin are in some instances slowed by outdated regulations. Emphasis should be placed on reducing the "lag time" between development of new or improved processes and the development of appropriate regulations to permit use of these processes. Another aspect of regulatory constraint involves the overlapping and sometimes conflicting regulations imposed by multi-regulatory agencies (i.e., federal, state, county and municipal).

6) Human risk. Substances in foods of animal origin that may have public health significance may be inherent (naturally occurring) or may result from contaminants of microbial or chemical origin. Risk results from inability to identify or quantify these substances and from the inability to avoid a substance because it cannot be removed or is deemed critical to production or processing of the food. Use of direct additives poses no significant constraint because their use is regulated and can be controlled.

7) Biological. Animal breeders require 5-20 years to make changes of major impact. As societal or consumer attitudes or desires change, the selection process may have to be adjusted, where possible, to meet the new demands. Both the food processor and the consumer must cope with product variation and with naturally occurring biochemical changes in animal products. For the processor, variations in body size, ratio of lean to fat tissues, etc. decrease processing efficiency. The consumer must accept biological variation which affects product size, quality and cost of the finished product.

8) Technological. This constraint can be classified into three main categories: a) lack of basic information on materials, biological processes and systems for conservation and distribution of foods of animal origin; b) lack of rapid, accurate and quantitative techniques for measuring physical, chemical, microbiological and sensory parameters; and c) lack of up-to-date methodologies for performing the physical operations associated with conservation of foods of animal origin and maximizing the efficient utilization of resources.

9) Ideology. Individual or group concepts, notions and definitions of edible and inedible items can facilitate or hinder acceptance of new or reformulated dietary items. Most individuals and groups have idealized notions of the appropriateness of certain foods, processing techniques and technologies. Food choice selection criteria may have little to do with science and include factors such as: a) familiarity; b) status appeal; c) cultivated tastes; d) convenience; e) group identity; f) keeping quality; g) availability; h) environmental concerns; and i) safety, nutritive value, wholesomeness, cost and energy conservation are of varying degrees of importance to consumers.

These concepts are derived from people's notions of reality and will limit innovative food conservation measures, as well as the application of new or different technologies. The "food conservation" theme can be used to overcome many traditional constraints incorporating conventional "nonfood" items into selected areas of the food system.

RISK STATEMENT

It is a basic premise in food processing that the resulting product be "safe" (neglible human risk) for the consumer. It should be clear that a product that does not present a health risk can be defined only in terms of current research methodology and available data. This means that a product or process considered "safe" today may be identified as representing a hazard to human health in the future as new information and research methods become available and exposure modalities change. The total concept of "safety" or risk associated with animal food products is examined in much greater detail elsewhere in this document. It is mentioned here only to inform the reader that the committee developing research priorities concerning "food processing and acceptability" had this concept foremost in their minds.

RESEARCH PRIORITIES

FOOD CONSERVATION TECHNIQUES

Development of improved food conservation systems for the 21st century assumes the availability of information from the following major areas of research:

1) Conservation of animal products from harvest to consumption.
   - There is a need for basic research on the mechanisms of deterioration of foods of animal origin in order that sensory and nutritional characteristics can be conserved by new improved or combined processes.
   - Research is needed to understand pathogenic and spoilage microorganism ecosystems in order to improve conservation of foods from animal origin.
   - A need continues to exist for basic research on functional properties of animal and plant proteins to enhance the incorporation of animal products in human foods.
   - Basic research is needed on the physical and chemical properties of heterogeneous food systems containing water, salts, proteins and fats as influenced by composition and processing conditions to permit formulation and processing. Research results could be applied to foods with modified sodium, cholesterol, fat and phosphorous levels.
   - Basic and applied research on procedures for the total use of animals with emphasis on formulating foods for humans and conversion of animal waste to energy is needed.

- Basic research is needed in packaging materials, and in systems to protect the sensory, nutritional and safety characteristics of animal products.

2) Conservation and recycling of resources.
   - An imperative need exists for new processes to separate water from other animal components with minimal energy input.
   - Research is needed to identify constraints to the development of new methods to reduce or eliminate energy and water use during slaughter, further processing and clean-up.
   - Development of transducers for on-line monitoring and computer-based control of processing of foods is needed.
   - Basic research is needed to develop procedures to assess quality, yield and safety with greater efficiency and effectiveness.
   - Basic research on sensory, nutritive and functional properties is required before fuller utilization of animal products for human foods can be accomplished.

FOOD ACCEPTABILITY AND QUALITY ASSESSMENT

Competition for scarce resources will place increased, possibly severe, restrictions on society's ability to provide the consumer with adequate, wholesome, high quality food. Innovative and ingenious techniques will have to be used to preserve high food quality during processing and distribution. Research needs include:

1) Development of accurate, rapid and economical assays that minimize loss and ensure wholesomeness and safety in animal foods.
   - Development of rapid, accurate techniques to measure gross composition, macro- and micro-nutrients and potential yield of products and by-products.
   - Development of rapid, accurate assays for total pathogenic and toxigenic microorganisms, including viruses.
   - Development of rapid, accurate assay procedures for determination of contamination by chemical and other foreign substances (e.g., pesticides, antibiotics, growth regulators, microbial toxins, industrial and agricultural chemicals, heavy metals, etc.).
   - Research is needed to develop new detection methods which are applicable to on-line detection and control of physical, chemical and microbial attributes of foods.
   - Development of rapid, accurate methods for objective assessment of sensory and physical qualities (i.e., tenderness, consistency, color, flavor and odor).
   - Development of appropriate sensors to incorporate into systems for control and identification of food quality factors in food processing operations.

2) Improving the scientific basis of regulatory activities that affect the conservation of animal foods by:

- Studying objective methods of postmortem inspection of animals to determine if certain manual subjective procedures might be replaced. Data are also needed to validate or update inspection procedures. Periodic studies are needed to reevaluate procedures on a risk assessment and cost/benefit basis.
- Reevaluating the human health significance of inspection criteria currently being used to determine if animal products are fit for human use. This reevaluation should include the need to investigate procedures for utilizing animal products for human use that exceed existing residue tolerance.
- Determining the cost of multi-agency involvement in regulation and inspection. The cost/benefit ratio of implementing changes in inspection procedures as well as efficient systems should be investigated.
- Developing improved animal product grading systems to accurately reflect palatability and consumer acceptance characteristics.

3) Determining the effects of processing, storage, distribution and preparation on direct, indirect and unintentional chemicals added to foods.
   - Basic research is needed to characterize the changes in animal foods during processing, storage and distribution, including the fate of intentional and unintentional food additives.
   - Development of methodology and instrumentation for analysis of food additives and their reaction products in animal food products is needed.

4) Development of methods for accurate, cost effective and representative evaluation of consumer food interests, concerns and expectations.
   - Development of methodologies and measuring instruments for identifying major food concepts and ideologies, and for detecting changing food concepts, food preferences and lifestyles.
   - Determining the food ideologies, concerns and behaviors of various subgroups in the heterogeneous U.S. society.
   - Determining shifts, both geographically and at selected time intervals, in consumer food ideology, food concerns and food behaviors.

REFERENCES

Baker, R.C. and W.B. Robinson. 1975. Food Science Research Needs for Improving the Utilization, Processing, and Nutritive Value of Food Products. A report to the Science and Technology Policy Office, NSF. Nov. 1975.

Bender, A.E. 1978. Food Processing and Nutrition. Academic Press, New York. 243 p.

CAST. 1980. Foods from Animals: Quantity, Quality, and Safety. CAST Rep. No. 82.

Chandler, C.A. and R.M. Martson. 1976. Fat in the U.S. Diet. Nutrition Program News, May-Aug., 1976, 1. USDA.

Cross, H.R., J. Stroud, Z.L. Carpenter, A.W. Kotula, T.W. Nolan and G.C. Smith. 1977. Use of mechanically deboned meat in ground beef patties. J. Food Sci. 42:1496.

Dill, C.W. 1976. Use of plasma in edible meat products. Proc. Recip. Meat Conf. 29:162.

Drake, T.G.H., S.H. Jackson, F.F. Tisdall, W.M. Johnstone and L.M. Hurst. 1949. The biological availability of the calcium in bone. J. Nutr. 37:369.

Dvorak, Z. and I. Vognarova. 1965. Available lysine in meat and meat products. J. Sci. Food Agr. 16:305.

Enig, M.G., R.J. Munn and M. Keeney. 1978. Dietary fat and cancer trends -- A critique. Fed. Proc. 37:2215.

Expert Panel on Food Safety and Nutrition. 1979. Mechanically Deboned Red Meat, Poultry, and Fish. Institute of Food Technologists, 221 N. LaSalle Street, Chicago, IL.

Expert Panel on Food Safety and Nutrition and the Committee on Public Information, Institute of Food Technologists. 1974. The effects of food processing on nutritional values. Food Technol. 28 (No. 10):77.

Expert Panel on Food Safety and Nutrition and Committee on Public Information. 1975. Naturally Occurring Toxicants in Foods. Institute of Food Technologists, 221 N. LaSalle Street, Chicago, IL.

Field, R.A. 1976. Mechanically-deboned red meat. Food Technol. 30 (No. 9):38.

Food Safety and Quality Service Meat and Poultry Inspection Program, USDA. 1979. Federal Meat Inspection Regulations and Humane Slaughter Regulations. Federal Register 44:37954.

Greenberg, R.A. 1972. Nitrite in the Control of Clostridium botulinum, 25. Proc. Meat Ind. Res. Conf., American Meat Institute, Washington, DC.

Hirst, E. 1974. Food-related energy requirements. Science 184:134.

Hoagland, R., O.G. Hankins, N.R. Ellis, R.L. Hiner and G.G. Snider. 1947. Composition and nutritive value of hams as affected by method of curing. Food Technol. 1:540.

Høyem, T. and O. Kvale. 1977. Physical, Chemical and Biological Changes in Food Caused by Thermal Processing. Applied Science Publishers, Ltd., London. 398 p.

Intersociety Council. 1979. Priorities for Animal Research. Communications from Intersociety Research Committee for Animal Agriculture.

Intersociety Research Committee. 1978. Animal Agricultural Research Needs. Intersociety Research Committee, 309 W. Clark Street, Champaign, IL

Intersociety Research Committee. 1979. Food from Animals: Research to Feed the 21st Century. Intersociety Research Committee, 309 W. Clark St., Champaign, IL.

Kirk, J.R. 1977. Research priorities in food science. Food Technol. 8:66.

Knight, S. and E.A. Winterfeldt. 1977. Nutrient quality and acceptability of mechanically deboned meat. J. Amer. Diet. Assoc. 71:501.

Kolbye, A. and M.A. Nelson. 1977. Health and Safety Aspect of the Use of Mechanically Deboned Meat. Final Report and Recommendations. Select Panel. Meat and Poultry Inspection Program, Food Safety and Quality Service, USDA, Washington, DC.

Kramlich, W.E., A.M. Pearson and F.W. Tauber. 1973. Processed Meats. AVI Publishing Co., Westport, CT.

Layrisse, M., C. Martinez-Torres, J.D. Cook, R. Walker and C.A. Finch. 1973. Iron fortification of food: Its measurement by the extrinsic tag method. Blood 41:333.

Lechowich, R.V., R.G. Cassens, P. Issenberg, D.I. Padberg, J.G. Sebranek, J.V. Spencer and R.N. Terrell. 1978. Nitrite in Meat Curing: Risks and Benefits. CAST Rep. 74.

Levin, E. 1970. Conversion of Meat By-products into Edible Meat Protein Concentrate Powder, 29 p. Proc. Meat Indust. Res. Conf., American Meat Institute Foundation, Washington, DC.

Paulsen, A. 1978. Feeding Animals Versus Feeding People. In: G.H. Koerselman and K.E. Dull (Ed.) Food and Social Policy, I. Proceedings of the 1976 Midwestern Food and Social Policy Conference. The Iowa State Univ. Press, Ames, 200 p.

Priestley, R.J. 1979. Effects of Heating on Foodstuffs. Applied Science Publishers, Ltd., London. 417 p.

Tannenbaum, S.R. 1979. Nutritional and Safety Aspects of Food Processing. Marcel Dekker, Inc., New York. 448 p.

USDA.1979. Health and Safety Aspects of the Use of Mechanically Deboned Poultry. Supt. Documents Stock. No. 1001-000-03985-7.

"Consumers are much more concerned than they have ever been about the relationship between the food they eat and their long-term well-being."

Anson Bertrand.

4

T. M. Farber (Co-Chairman), S. Friess (Co-Chairman), A. M. Pearson (Rapporteur)

S. Aust, J. F. Bauerman, J. Bond, C. R. Burbee, R. G. Cassens, M. K. Cordle, E. M. Foster, J. C. Street, D. G. Topel, W. J. Visek, A. Wasserman, V. O. Wodicka, C. Woteki, and J. Wyatt

FOOD SAFETY

SUMMARY

STATE OF THE ART AND SCIENCE

In the 21st century the consumer will expect animal food products to be even more safe and nutritious than at present. The research to achieve this goal must be broadly based to prepare for present and emerging safety problems associated with eating foods containing hazardous chemicals, either man-made or of natural origin. The risk to man from such chemicals and biological agents in food may cause an early onset of toxic symptoms in the consumer, or lead to long-term development of adverse effects. The possibility of irreversible effects, such as birth defects, organ damage, or cancer, adds to the urgency for solving food safety problems.

The food safety problems can be formulated as a series of constraints which will limit the growth of animal agriculture in the future, and of a series of research efforts to eliminate these constraints.

PRIORITIES OF RESEARCH IMPERATIVES

The research imperatives are as follows:

1. To develop effective and reliable methods to predict toxicity of animal products to humans and to evaluate the safety of animal products.
2. To develop effective methods for safe disposal or salvage of animals and animal products with persistent hazardous constituents.
3. To develop early identification of and containment procedures for hazards in animal food products.
4. To develop efficient methods for producing animal products free of infectious and toxic agents.

STATE OF THE ART AND SCIENCE

Although animal products in the American diet are as safe as the current state of knowledge permits, animal agriculture in the 21st century must be planned and administered to assure consumers that animal products will be both safe and nutritious. This assurance must be derived from research on the full spectrum of human risks that might result from ingestion of foods containing new chemicals, or foods prepared by new processing techniques that might be contaminated with pathogenic and toxigenic microorganisms, hazardous chemicals or toxicants of natural origin. The risk to man from these chemicals and biological agents in food may either be acute, leading to early onset of toxic symptoms in the consumer, or chronic in the sense of very long-term development of adverse effects. These effects are well documented and have been extensively reviewed (NAS/NRC, 1973; Campbell and Stoloff, 1974; Coon, 1974; Liener, 1974; Agricultural Research Policy Advisory Committee, 1975; IFT Expert Panel on Food Safety, 1975; Irving, 1978; Scientific Committee of Food Safety Council, 1978; Hui, 1979; Study Group on FDA Residue Programs, 1979). Starting with analysis of the potential problems, it is vital that we move to develop well-planned and coordinated research programs for quantitation of the risks involved, and for the design of measures to protect the health and well-being of consumers of animal food products. The Food and Drug Administration has categorized food associated hazards into six groups (Schmidt, 1975): 1) microbiological contamination; 2) malnutrition; 3) environmental contamination; 4) naturally occurring toxicants; 5) pesticide residues; and 6) food additives. Foods of animal origin (meat, poultry, eggs and dairy products) are subject to all of these hazards. Potentially harmful chemicals may enter animal products at numerous points during production, processing, transportation, storage and food preparation, through contaminated water and feed, as a result of unsanitary conditions at the point of collection or slaughter, during food processing, or through improper handling at home, or in public eating establishments. In addition, contamination may arise from unexpected sources such as livestock back rubs and barn siding.

MICROBIOLOGICAL CONTAMINATION

Botulism and staphylococcal food poisoning are of constant concern to manufacturers and consumers of cured meats, fermented sausages and certain types of cheese. Poisoning by Clostridium perfringens and Bacillus cereus is often associated with cooked meat and poultry foods. These diseases are distinguished by one common feature - they are easy to prevent simply by following good food manufacturing and handling practices.

Not so easily controlled, however, is salmonellae infection from foods of animal origin. Domestic animals serve as an enormous reservoir of salmonellae, and a few serotypes are pathogenic to animals. By far the greatest number of them simply live in the animal, contaminating meat and poultry during slaughter and constitute a hazard to humans. Proper cooking will kill salmonellae, but the cooked product frequently can also be recontaminated during subsequent handling and preparation.

Although the acute problem with salmonellae contamination in poultry and pork has been widely recognized (NAS/NRC, 1969), no progress has been made in reducing the incidence of contamination. A recent survey by the USDA showed that 37% of retail packaged chickens carry salmonellae (Anomymous, 1980). It is unreasonable to expect a substantial reduction in human infections as long as meat and poultry products are so heavily contaminated. The American people will continue to suffer from food-borne salmonellosis until it is learned how to control infection in breeding stock, how to rear animals with minimal contamination and how to slaughter them without spreading contamination from one carcass to another. This research is badly needed and the need will grow as we move into the 21st century.

Less well understood, however, are the viral animal diseases and their impact on man (Kostenbader and Cliver, 1977). Swine vesicular disease, Newcastle disease, ornithosis and swine influenza are four important zoonoses that may be transmitted to humans by animal products. Research is needed to evaluate the hazard to man of these and other causes of animal disease.

The problems relating to assessment of hazard from microbiological-viral contaminants of animal products can also be expected to be quite complex in the future. In part, the complexity is derived from the diversity of potential disease-causing organisms which include bacteria, viral agents, fungi, etc. Viruses have been found in raw comminuted meat, but are not found in heat-processed meats such as bologna, frankfurters and luncheon meats (Kostenbader and Cliver, 1977). With the evolution of new types of animal products in the future, it will become increasingly important to study the mechanism by which virulent microorganisms invade, propagate and generate hazards to human health by their presence in these products.

MALNUTRITION

Malnutrition results when body tissues receive an inadequate or an excessive supply of essential factors, including energy. Studies in animals show that diets varying in the quantities of certain essential vitamins or minerals cause variation in the number of cancers in animals given cancer-producing chemicals (Visek et al., 1978). Restriction of caloric intake in animals decreases the number of spontaneous and chemically-induced cancers and also decreases the changes associated with aging (McCay et al., 1939). Epidemiologic evidence for humans shows that persons who are overweight due to intake of excess calories have a higher incidence of certain cancers and other diseases. Although the role of specific nutrients in these relationships is not understood, it is clear that dietary constituents, including essential nutrients, modify the incidence of infections, the diseases enumerated above and the response to toxic substances. Thus, any assessment of hazards associated with the food supply must take cognizance of the nutritional status of target populations.

ENVIRONMENTAL CONTAMINANTS

Environmental contaminants include: 1) industrial chemicals accidentally contaminating animal feeds, such as the PBBs (polybrominated biphenyls) and PCBs (polychlorinated biphenyls), which have very long biological half-lives in living systems; 2) industrial chemical pollutants, toxic trace metals, etc. which may enter ground water supplies and ultimately become incorporated into animal-derived products; 3) trace chemical contaminants in animal products resulting from ingestion of feeds containing residues of agricultural chemicals, such as pesticides, fungicides and herbicides, as well as degradation products of these chemicals caused by light; and 4) classes of chemicals such as plasticizers and residual monomers which may enter food products by diffusion transfer from polymeric package materials used for wrapping, storage and display of these products. The hazards to humans from consumption of a wide variety of contaminants in animal products over large fractions of their lifetime will require extensive study, especially from the standpoint of additive or possible synergistic effects on health due to multiple chemical insult.

NATURALLY OCCURRING TOXICANTS

A closely related set of chemical risks is posed for the human by animal products which may have become accidentally contaminated by natural toxicants in the animal feed. These toxicants may rise to significant concentrations in animal-derived products. Included in this group are toxicants of plant, animal and fungal origin. Notable examples are the potent toxicants such as crotalaria, aflatoxin, saxitoxin and ciguatera toxin, which may find their way into meat, milk and eggs through animal feeds. The risks from such substances may be either acute or chronic and will require careful quantitation, followed by studies of techniques for neutralization or reversal of their toxic actions.

PESTICIDE RESIDUES

Livestock and poultry can be contaminated when application or manufacturing of pesticides occurs in their vicinity or when residues are transported through the food chain. Railroad cars, trucks and ships used for transport or buildings used for storage of pesticides and later used for animal feed or animal products are also potential sources of contamination (Office of Technology Assessment, 1979).

FOOD AND FEED ADDITIVES

Another category of potentially hazardous substances involves a large number of additives intentionally used for food preservation, flavoring and other desired effects. The trend toward purchase of ready-prepared or convenience foods has expanded the use of additives in association with finished animal-derived food products. Since 1958, federal law has required premarket testing of additives for safety.

However, the concepts and methods used for safety evaluation have changed dramatically since 1958. There is a need to reexamine the supporting safety base in view of these changes and changes in human exposure levels.

Risks also arise from chemicals deliberately added to animal feed for purposes of growth enhancement, for control of disease and for the general health of livestock. These chemicals include both natural and synthetic homologs of hormones, vitamins, antibiotics, feedstuffs and medicinal agents. The levels of these chemicals and of their metabolic degradation products in meat, milk, cheese, eggs, etc. must continually be assessed; bioaccumulation pathways in animals and humans studied; and the mechanisms of combined chemical effects on the health of people measured during both acute and chronic consumption conditions.

Advances in analytical chemistry and toxicological research have led to vexing new problems in hazard assessment of intentionally used chemicals (e.g., estradiol and nitrite). Such additives can cause low level residues of carcinogenic substances in food. However, humans also sustain exposure endogenously to such chemicals, often at substantially higher levels, raising serious problems in the total hazard assessment process. Methods to assess the hazards associated with incremental small increases in the normal background burden of potent biologically active substances are currently lacking.

TRANSFORMATIONS DURING PROCESSING AND COOKING

Potentially toxic chemicals may be produced during preservation, curing or storage of animal products. Breakdown of endogenous components of these products and reactions among these components or with compounds added for technological purposes may lead to formation of hazardous substances. Of particular significance are the compounds formed during high temperature thermal reactions occurring in processing and cooking. The development of new processing techniques and the introduction of nonanimal components into animal products may lead to the formation of unforeseen toxicants. Recognition also must be given to the potential for destruction of toxic material as a result of heat processing or for even greater toxicity on heating.

RISK ASSESSMENT

Risk estimation for chemicals in food products must be viewed at present as a very rudimentary branch of toxicological science. There is an urgent requirement for development and international acceptance of a body of principles for risk assessment for food chemicals. A major complication stems from the possible presence of multiple toxic agents in a given food source. The ability to estimate risk from a single chemical contaminant by use of toxicological and epidemiological data is evolving, yet little has been done to generate risk estimation theory for toxic effects from multiple chemical stresses from foods.

The two principal components with respect to effects of a toxic agent on an individual are exposure and potency. What is lacking is an accepted basis for combining these factors into an estimate of hazard. Another major issue deals with the validity of methods currently being used in animal-model testing to calculate the effects to be expected from low-dose/long-term exposure from the effects observed at high dosages. A third issue concerns the basis for translation to safe levels in the human from toxicological test results in laboratory animals.

Economics is an important part of the food safety issue. In the case of pesticides as regulated under the Federal Insecticide, Fungicide, and Rodenticide Act and other chemicals under the Toxic Substances Control Act, Congress requires that assessed risks be judged against the benefits resulting from the use of such chemicals. It is important to recognize that determining estimates of the benefits and costs for making such a judgement is imperfect at best. No generally accepted basis has been developed for comparing benefits, as perceived in dollar values, to risks perceived in terms of human morbidity and/or mortality values. The judgement is less difficult when the benefits are perceived in terms of direct effects on health, such as those provided by beneficial drugs. The chief concern is that no consistent or coherent policy to guide society's food safety decision-making process exists so as to provide the highest possible level of well-being to the general public.

Public reaction to involuntary consumption of contaminants in foods may not be directly proportional to the hazards. Some of the public reaction to the disclosure of incidences of food contamination arises from the assumption that some government agencies will protect people from involuntary consumption of toxicants. However, protection cannot be assumed when analyses are not conducted, or when toxicity is unknown. Monitoring and toxicity testing programs must, therefore, be developed simultaneously with the development of toxicology education programs designed to alleviate perceived, but unjustified concerns.

Other facets of the hazard assessment issue concern the reactions of society-at-large to the concept of risk, and to acceptance of degrees of risk in return for the obvious benefits to be derived from animal food products. These societal considerations include: 1) the concept of absolute safety vs the competing concept of usage patterns based on trade-offs between benefits and risks, with ancillary consideration of controls and costs; 2) societal attitudes for toxic effects (e.g., cancer), which may be irreversible, vs the benefits of usage of valuable resources and food products; 3) identification and protection of sensitive population sectors (e.g., fetus, newborn, children and the aged); and 4) possible human adaptation to existence under increasing chemical and microbiological stresses from contaminants in foods.

CONSTRAINTS THAT MAY INFLUENCE FUTURE DEVELOPMENTS IN ANIMAL AGRICULTURE

In the discussion of the state of the art and science with respect to food safety, knowledge gaps have been identified in this document which act as constraints on the future development of animal agriculture. Some of these constraints are in the area of technology, and some in societal perception of hazard assessment and safety. It is important that these constraints be removed by appropriate efforts in research and development that will generate the new knowledge and technologies required to assure safety and nutritional quality of animal foods.

It is also important to note that certain regulatory reforms may be achieved prior to the beginning of the 21st century, which will have the effect of eliminating some of the constraints on growth of animal agriculture. Specifically:

1. Regulatory reform is expected to resolve the existing problems of gaps and overlaps in jurisdictional authority between and among regulatory agencies with elimination of outdated regulations and provision for uniform food safety standards and enforcement procedures. Examples of reform are: a) Requiring inspectors to focus attention on more relevant hazards; b) Allowing inspection of meat animals and of health and feeding records at the point of production to improve inspection efficiency and effectiveness.

2. It is expected that USDA and FDA regulations, which do not permit the blending of contaminated feeds or foods to achieve safety, and which may unnecessarily waste feed or food, will be revised.

3. Modification of the Delaney clause relating to the presence of potential carcinogens in food-producing animals will be achieved.

In the preceding discussion, a set of gaps in present knowledge has been identified, which may constrain future developments in animal agriculture if not offset by research and new knowledge. These gaps are identified below with no attempt at ranking or prioritization and with realization that the list is not totally comprehensive:

1. The state of development of risk/hazard assessment and safety evaluation for the consumer of animal products is inadequate. This situation results in a considerable time lag before the scientific community comes to a consensus concerning the safety of a drug, ingredient or process, and in the reluctance of the governmental agencies to make a decision. Specifically:

a) There is no generally agreed way to estimate risk at low doses from toxicity data at high doses.

b) There is no generally agreed way to translate findings on test animals to man in quantitative terms.

c) There is insufficient information on details of consumption of food products, particularly by vulnerable segments of the population.

d) There is no agreement on definition of a socially acceptable risk associated with specific foods or food ingredients, and there is no mechanism established to determine such a risk. This is a public policy issue and involves societal values.

e) There is insufficient knowledge of the metabolism and pharmacokinetics of animal drugs and active feed ingredients and there are no simplified, inexpensive standard methods for developing this information. It is difficult to establish safety of animal foods for man without this information.

f) There are no ways to determine at reasonable cost the interactions of foods or food ingredients consumed together, the influence of diet, of age, of disease state and of other factors that might influence response to substances being evaluated for safety in the animal and human.

g) Controversy exists whether safety may be impaired by excessive intakes of normal constituents, including essential nutrients, from the diet. The potential impairment of safety is not restricted to animal products. The existence and characterization of these hazards and their potential effects are poorly established and require further clarification.

2. The public expects regulatory agencies to protect people from food-borne hazards that they cannot detect for themselves. There are constraints which limit this protection, with the result that there are frequent problems leading to public disappointment and indignation. Examples are:

a) A number of toxic manifestations, such as behavioral effects, altered immunocompetence, allergic effects and other effects that are not usually studied in present procedures.

b) There is a lack of efficient analytical procedures to support monitoring of foods for a wide spectrum of toxic residues.

c) In that many episodes will arise through accidental contamination, the lack of efficient methods for identifying toxicants of unknown nature imposes delays in characterization and control of the problems.

d) Lack of adequate quality control system methodology leaves a gap between the number of monitoring samples examined and the level of control expected by the public. A part of this problem stems from the loss of identity or origin of animal foods moving through the procurement, processing and distribution chain.

3. The effects on food safety of using nonconventional sources of feed for animal production are not known. The magnitude of this problem will increase as competition for resource use intensifies. Diversity and geographic specialization make it difficult to identify and control potentially unsafe feeding practices.

4. There is a lack of information on new processing techniques that may have beneficial or deleterious effects. The impact of new technologies needs to be evaluated prior to implementation. Examples include use of new enzymes in cheesemaking, new packaging materials, altered heat treatments and replacement of currently used ingredients.

5. Requirements to demonstrate food safety are of sufficient economic disincentive, especially for small volume products (some of which are essential in agriculture), to keep them off the market. This is a public policy issue and may require governmental underwriting with full or partial support.

6. Safety evaluation techniques are not only costly, but also slow. If a chronic effect is at issue, roughly 3 years is the minimum time for receipt of test results. If the testing process does not give a definitive answer, an additional 3 years are required to resolve the question. A gamut of in vitro tests, arising from better understanding of the mechanisms of toxicity, would alleviate this problem.

7. The state of animal health and the level of production and/or behavior can be indicators of the presence of hazards in the feed or environment impacting on food safety. There is no system for reporting and effectively using these diagnostic indicators. Development of a reporting system would require public policy action.

8. The relative hazard of foods and food ingredients may differ widely between the factual situation and public perception of it. This situation can lead to distorted emphasis in application of public resources in both research and regulatory enforcement. The educational facet of this problem makes it a public policy decision.

9. The technology employed in mixing and distributing animal feeds and feed components lacks the refinement which may be necessary to prevent cross contamination of animal feeds.

10. The number of safety evaluations will be limited by the number of qualified personnel and qualified laboratories available. A correction of this situation would require additional emphasis on training and would need public policy action.

11. A major problem in maintaining food safety is at the food preparation, serving and storage levels. Improper handling through ignorance and/or indifference can lead to problems with pathogenic bacteria and their toxins. Improvement of handling procedures would require an educational effort and, therefore, public policy action.

12. Adequate national reporting surveillance systems for outbreaks of food-borne diseases, chemical contamination and animal diseases do not exist. The complex structure of the agricultural system is such that an unfortunate incident in an isolated area or industry may introduce serious hazards in the food production and distribution system. Development of such a surveillance system would require public policy action.

13. Adequate technology and facilities for disposal of contaminants and contaminated materials do not exist. Establishment of such facilities would require public policy action and funding.

14. Genetic selection for a desired characteristic in domestic animals may result in unanticipated changes in the animal. Specifically, physiological and biochemical functions of the animal and the characteristics of the animal products derived from it may be greatly altered. Therefore, awareness of this situation is important for safety assessment.

15. The animal production and food processing industries are constrained by a lack of education and management advisory support directed toward protection of food at the major entry points of microbial and chemical toxicants. Such support should help promote both the means of anticipating food safety problems and rapid solution of problem episodes. The provision of education and support is a public policy issue.

16. The existing regulatory systems to evaluate and control food safety stop at identification of potential toxicants in the raw animal products (i.e., USDA monitors meat and poultry at the point of slaughter, while tolerances are established in the raw product by FDA and/or EPA). However, consumers do not normally eat animal products without further processing. Information is lacking on the effect of such treatment and storage on the alteration of toxicants to more or less hazardous constituents. Practical standardized techniques for such evaluations do not exist.

17. Improving crop production technology may result in animal feedstuffs that present new problems in food safety. For example, plant breeding and selection programs may introduce varieties having altered quantities or types of natural toxicants, while changing practices in fertilization, pest control, harvesting, processing and storage may cause altered quantities or types of chemical contaminants in feeds.

18. Information to understand how chemicals move in the environment and the impact of environmental recycling is severely limited.

RESEARCH PRIORITIES

The Food Safety Working Group has concluded that food safety in regards to animal products in the 21st century is dependent upon four major research imperatives in the following priority sequence:

1. To develop effective and reliable methods to predict toxicity of animal products to humans and to evaluate the safety of animal products (Risk Assessment and Safety Evaluation).

It will be necessary to develop, improve and validate reliable biological tests of animal drugs and feed additives, agricultural chemicals and their residues in feedstuffs, and other contaminants occurring in animal products. A basis of understanding is needed to improve the interpretation of data derived from toxicity testing in evaluating consumer safety.

Research thrusts related to this objective are to:

a) Develop sound procedures for the estimation of risk at low levels of exposure, based upon toxicity data obtained from high level exposure in the same species.

b) Develop epidemiological methods to be applied to populations with different diets and habits to establish the quantitative role of nutrients and other natural constituents in the diet in relation to health and disease.

c) Develop a basis for predicting dose response relationships for humans from observed dose response relationships for food constituents in animals.

d) Develop research methodologies for the prediction of the impact of disease, dietary state, strain and the interaction between two or more agents on the dose response relationship in a test animal, the food producing animal and humans.

e) Develop rapid, reliable methods for collecting, analyzing and reporting food consumption data on a demographic basis.

f) Develop simpler and more efficient techniques for studying the metabolism and pharmacokinetics of food contaminants, both as individual components and in different combinations.

g) Develop sensitive, standardized methods to study neurobehavioral effects, altered immunocompetence and allergic reactions which can be used in the safety evaluation of animal-derived products.

h) Establish a normal baseline of natural hormones in animal products for use in judging the safety of hormone administration in animal production.

i) Develop simple, rapid and inexpensive in vitro test procedures which are predictive of well defined classes of toxicity in animals and humans. The definitions of these classes remain to be established. These simple tests should have potential for predicting chronic in vivo effects.

j) Determine the potential toxicity of any new or unconventional animal food source prior to its introduction into the animal production system and establish safe ingestion levels for the toxicant(s) in these feeds in terms of human safety.

2. To develop efficient methods for safe disposal or salvage of animals and animal products with persistent hazardous constituents (Salvage or Disposal of Unsafe Products).

Adequate technology and facilities for the disposal of contaminated materials do not exist. The problem lies in the fact that many of the contaminants are very long lasting, and most present disposal methods would result in further contamination of the environment. Ideally, disposal would include removal of the contamination in ways that would allow salvage of the the purified food, and thereby avoid the economic loss from destructive disposal of the food.

Research thrusts related to this objective are to:

a) Develop cost effective methods to destroy or contain contaminants or contaminated wastes.

b) Develop cost effective techniques to reduce or remove contaminants from feeds, animals and animal products.

c) Study physical and biological transformations and movements of chemicals in the environment.

d) Conduct modeling studies and other cost-effective methods for establishing the probable types of contaminating substances and the amounts recycled into animal food production.

3. Develop early identification and containment procedures for hazardous substances in animal food products (Early Hazard Identification and Containment).

The complexity and size of food production and marketing systems present monitoring problems. Because the number of potential contaminants is so large, new analytical techniques are needed that will detect a wide spectrum of toxic residues. The ultimate goals of monitoring feeds and animal derived foods are the early identification of contaminants and the prevention of contaminated food from reaching the public.

Research thrusts related to this objective are to:

a) Determine the identity of any potential toxicant in any new or unconventional feed source prior to its introduction into the animal production system.

b) Isolate, identify and quantitate any potential toxic components.

c) Investigate procedures for reducing or eliminating the toxicants.

d) Study the metabolism of feed ingredients in animals with respect to human safety.

e) Identify appropriate sentinel animals and plants for use in environmental monitoring.

f) Develop techniques to determine the bioavailability of hazardous chemicals in their bound or conjugated forms that might occur in feed or food products.

g) Develop systems-modeling methodology to define the uptake and distribution and the elimination kinetics of chemicals and their metabolites in food-producing animals.

h) Isolate, identify and quantitate components of new packaging materials transferred into animal products, and establish the safety limits for these compounds.

i) Develop sensitive, reliable, rapid analytical procedures for multi-residue detection of a wide spectrum of toxic compounds in feeds, animals and animal products.

j) Develop procedures for establishing safety of products by studying the chemical similarity of their composition to those of chemical counterparts with known safety profiles.

k) Develop new analytical techniques applicable to the study of metabolism of potentially toxic materials in the animal.

l) Evaluate the effects of storage and of processing in the factory or in the home on toxicants in animal products.

m) Investigate means for improved reporting of outbreaks of food-borne diseases, animal diseases and chemical contamination.

n) Develop a regional reporting system that uses responses of biological species as indicators of food contamination.

o) Identify the critical points for hazard control of animal food products in the total distribution system from producer to consumer in order to serve as a basis for quality control.

4. To develop efficient methods for producing animal products free of infectious and toxic agents (Production of Safe Animal Products).

Although animal products properly handled and prepared are generally safe, they often contain harmful bacteria and may contain other harmful agents. Reducing this threat to public health can best be done by development of methods to protect domestic animals from contamination before slaughter. Effective ways are also needed to make meat, milk and poultry products safe when preventive measures have failed.

Research thrusts related to objective 4 are to:

a) Develop information on the mechanism of genetic control of microbial toxin formation and other pathogenic manifestations to provide a basis for reduction of microbial hazards.

b) Develop diagnostic screening tests to evaluate the deleterious effects of microbial contamination occurring during food processing.

c) Devise procedures for destroying harmful bacteria in raw meat and poultry.

d) Develop a better understanding of the salmonellae carrier state among domestic animals and find ways to prevent it.

e) Develop more effective means of educating the general public on how to handle raw animal products without spreading bacterial contamination.

f) Investigate the products of genetic changes in plants and food-producing animals that enable them to introduce or initiate development of hazardous chemical or biological compounds in animal food products.

REFERENCES

Agricultural Research Policy Advisory Committee. 1975. Food safety. In: Research to Meet U.S. and World Food Needs. Vol. 1, p. 89.

Anonymous. 1980. Salmonella incidence in chickens at 37% in FSQS survey. Food Chem. News, Feb. 25, 1980, p. 12.

Campbell, T.C. and L. Stoloff. 1974. Implications of mycotoxins for human health. J. Agr. Food Chem. 22:1006.

Coon, J.M. 1974. Natural food toxicants - A perspective. Nutr. Rev. 32:321.

FAO. 1967. Toxicological Evaluation of Some Antimicrobials, Antioxidants,Emulsifiers, Stabilizers, Flour Treatment Agents, Acids and Bases. Joint FAO/WHO Expert Committee on Food Additivies, WHO/Food Add. 67.29. United Nations.

Hui, Y.H. 1979. United States Food Laws, Regulations and Standards. John Wiley and Sons, New York.

IFT Expert Panel on Food Safety. 1975. Naturally occurring toxicants in foods: A scientific status summary. J. Food Sci. 40:215.

Intersociety Research Committee. 1979-1980. Safety and Nutritional Value of Animal Products. Animal Agriculture Research Priority Recommendations to American Society of Animal Science, American Dairy Science Association, Poultry Science Association and American Meat Science Association.

Irving, G.W. 1978. Safety evaluation of the food ingredients called GRAS. Nutr. Rev. 36:351.

Kostenbader, K.D., Jr. and D.O. Cliver. 1977. Quest for viruses associated with our food supply. J. Food Sci. 42:1253.

Liener, I.E. 1974. Toxic Constituents of Animal Foodstuffs. Academic Press, New York.

Lillis, R., H.A. Anderson, J.A. Valcinkas, S. Freedman and I.J. Selikoff. 1978. Comparison of findings among residents of Michigan Dairy farms and consumers of produce purchased from these farms. Environ. Health Perspect. 23:105.

McCay, C.M., L.A. Maynard, G. Sperling and L.L. Barnes. 1939. Retarded growth, life span, ultimate body size and age changes in the albino rat after feeding diets restricted in calories. J. Nutr. 18:1.

Meester, W.D. and D.J. McCoy, Sr. 1977. Human toxicology of polybrominated biphenyls. Clin. Toxicol. 10:474.

NAS/NRC. 1969. An evaluation of the Salmonella problem. Committee on Salmonella. Pub. No. 1683, National Academy of Science-National Research Council, Washington, D.C.

NAS/NRC. 1973. Toxicants Occurring Naturally in Foods. Committee on Food Protection. National Academy of Sciences-National Research Council, Washington, D.C.

NAS/NRC. 1978. An Assessment of Mercury in the Environment. National Academy of Science-National Research Council, Washington, D.C.

Office of Technology Assessment. 1979. Environmental Contaminants in Foods. OTA Rep., Congress of United States, Washington, D.C.

Schmidt, A.M. 1975. Food Safety. A Century of Progress. Talk given at Conference celebrating the 100th anniversary of the Food and Drug Act. London, Oct. 20-21, 1975.

Scientific Committee of Food Safety Council. 1978. Proposed system for food safety assessment. Food Cosmet. Toxic. 16 (Suppl. 2):13.

Social and Economic Committee of Food Safety Council. 1980. Principles and processes for making food safety decisions. Food Technol. 34 (2):79.

Study Group on FDA Residue Programs. 1979. FDA Monitoring Programs for Pesticide and Industrial Chemical Residues in Food. Rep. of Study Group on FDA Residue Programs, Food and Drug Administration. Rockville, Md.

Visek, W.J., S.K. Clinton and C.R. Truex. 1978. Diet and experimental carcinogenesis. Cornell Vet. 68:1.

"...producers and consumers need each other. They are natural allies, not born antagonists."

Carrol Tucker Foreman

"...broad based perspectives brought to bear on the urgent issues... lead to clearer definiton of the needs and researchable areas that can benefit the public..."

Maryln K. Cordle

5

R. R. Oltjen (Co-Chairman), D. E. Ullrey (Co-Chairman),
C. B. Ammerman (Rapporteur)

D. R. Ames, C. A. Baile, T. H. Blosser, G. L. Cromwell,
H. A. Fitzhugh, D. E. Goll, Z. Helsel, R. E. Hungate,
L. S. Jensen, N. A. Jorgensen, L. J. Koong,
D. Meisinger, F. N. Owens, C. F. Parker, W. G. Pond,
R. L. Preston, and H. S. Teague

ANIMAL NUTRITION AND DIGESTIVE PHYSIOLOGY

SUMMARY

The period of the industrial revolution has been paralleled by a biological revolution of comparable significance. In the last century, 50 nutrients have been identified as essential for animals, and the integration of nutritional knowledge with genetic advances and modern husbandry systems has produced remarkable achievements. A notable example is the production of 25,000 kg (55,000 lb) of milk in a single lactation by a Holstein cow. Unfortunately, this and comparable record achievements by other "food-producing animals of the future" are not typical. Limits to maximum production may be partially identified, but the means by which average production can be pushed near those limits still must be defined. Peak feed consumption by the record-setting cow required 64 kg (140 lb) of dry matter intake per day. The proportion of her nutrient intake used for milk compared to that used for maintenance was markedly greater than that of the average cow. We do not know how these remarkably efficient milk synthesizing processes are regulated, what governs the intake of those prodigious amounts of feed, and how we may better meet the nutrient needs with different and perhaps limited future resources to achieve the same end result. If these discoveries could be made for all classes of food-producing animals, the efficiency of meeting human food needs with animals would be enhanced enormously.

We are presently constrained by our limited understanding of factors governing protein and fat synthesis, interactions between alimentary microorganisms and their host, and the means by which nutrient needs for sustained and efficient production may be met. Research imperatives have been singled out and successful research on several critical areas will enable a more efficient utilization of feedstuff nutrients for animal production in the 21st century.

1. Cellular processes associated with protein synthesis and animal growth. Basic factors that limit the rate and extent of muscle growth and protein synthesis into animal products, as

well as those concerned with the development of adipocytes and fat deposition have not been defined. An understanding of these mechanisms will permit manipulation by nutritional, endocrinological, selective breeding and other means to maximize product output.

2. Interrelationships between alimentary microbial ecosystems, digestive processes and the host animal. There is an urgent need for knowledge on the nutritional requirements of the microbes in all sections of the alimentary tract and on how they may be manipulated to improve animal performance. Successful research will increase the utilization of competitive and noncompetitive feedstuffs, provide new information on antimicrobial agents and offer new methods to utilize the positive features of the microbes in nonruminant as well as ruminant animals.

3. Nutrient conservation in meeting animal needs. Nutrient requirements have not been established adequately for various states and levels of performance, sex (intact males), age, environment, stress and disease conditions. Techniques for measuring metabolic profiles which quickly detect and accurately evaluate the nutritional status of all farm animals need to be perfected.

4. Controlling feed intake to maximize animal efficiency. Unrecognized physiological factors control nutrient intake, thus limiting the utilization of feedstuffs for maximum conversion into animal products. Research is needed to provide the most economic nutrient supply during various physiological states of the animal and to precisely limit feed intake in keeping with the desired productive output. A basic understanding of the mechanisms involved is paramount to feed intake control.

5. Integrated nutrient management to enhance production efficiency. Availability and costs of conventional and nonconventional feedstuffs and other nutrient resources must be integrated for cost-effective animal production. Research is needed to determine replacement values for nutrient resources and the economics of animal tissue nutrient storage for later use during periods of low productivity. Appropriate models to optimize this concept are needed.

6. Nutritional characterization of feedstuffs. There is an urgent need for rapid and economical techniques which accurately characterize all nutrients in feeds as well as their bioavailability to animals. Especially important will be a high relationship between compositional attributes and animal performance. As data become available, improved methods of data capture, cataloging and retrieval must be developed.

PRESENT STATE OF THE ART AND SCIENCE

QUALITATIVE NUTRIENT NEEDS

During the past 4 to 5 decades, excellent progress has been made in determining which nutrients are essential for food-producing animals. This list now includes 50 items (table 5.1). Some, like nitrogen and sulfur, are used by microorganisms in the ruminoreticulum in the synthesis of protein which is degraded to amino acids in the abomasum and small intestine. These organisms also synthesize vitamin K and the vitamin B-complex so that functional ruminants are less dependent than nonruminants on dietary supplies of these nutrients. On the other hand, nonruminants require a long list of preformed organic nutrients including essential amino acids and the vitamins. Both ruminant and nonruminant tissues probably require the same mineral elements, although for some minerals the evidence is quite indirect. Fiber also is important for proper function of the digestive system of many species.

Although excellent progress has been made on the qualitative needs for nutrients, areas of uncertainty still exist. Typical examples needing further research include the following inquiries: Are glycine and serine dietary essentials for the fast-growing pig as they seem to be for the fast-growing chicken? Is the essential amino acid composition of dietary protein critical for fast-growing and/or high-producing ruminants? Do nonruminant animals have a dietary requirement for cobalt independent of its presence in vitamin B_{12}? Are chromium, fluorine, nickel, silicon, vanadium, tin, arsenic or molybdenum deficiencies a potential practical problem for ruminants or nonruminants? Do fast growing or high-producing ruminants have dietary requirements for B-vitamins? Is inositol required for lactation in swine as suggested in rats? Do gastrointestinal microorganisms have unique nutrient requirements which, when supplied, would also benefit the host? Do certain stresses exhaust the ability of farm animal tissues to synthesize adequate vitamin C and, thus, make it a dietary essential? Are there unidentified nutrients that are required for optimum health and performance of animals?

QUANTITATIVE NUTRIENT NEEDS

Present knowledge of the general quantitative nutrient requirements of food-producing animals is documented in the nutrient requirement series (NRC, 1975, 1976, 1977, 1978, 1979). Since their initiation in the 1940s, the individual publications in this series have been regularly updated to incorporate current research findings. A review of this series indicates that there are information gaps for all species but, in general, nutrient requirements are most thoroughly quantitated for chickens and turkeys, fairly well-defined for swine, and least extensively defined for ruminants. Two major reasons for this are 1) the shorter life cycle, smaller size and lower cost of conducting research trials with poultry and swine as compared to ruminants, and 2) the greater ease of controlling environmental variables which influence nutrient needs in poultry and swine. The latter are commonly reared in confinement, while ruminants are usually reared outside and have anaerobic bacteria and protozoa interposed between their diet and their tissues.

Quantitative nutrient needs are less well-defined for lactation in mammals and for reproduction in both poultry and mammals than for growth and finishing. Nutrient intakes appropriate for the mother of three or more lambs have not been adequately studied. Genetic selection for very high milk production in dairy cattle and the introduction of exotic germ plasm into beef cattle have not been consistently accompanied by research to identify effects on nutrient requirements. Certain metabolic disorders related to this lack of information include acidosis, milk fever, displaced abomasum, ketosis and fatty liver syndrome. The mastitis-metritis-agalactia syndrome in swine also has a dietary component. Accurate quantitation of nutrient needs for reproduction and lactation is lacking because of the large numbers of animals required and the long-term commitment of facilities and funds. Research in this area must involve life cycle studies. Ultimate answers can be derived only from the study of several generations. Complicating such studies is the tendency for very productive animals to go "off feed" and for production to decline when they are placed in research environments.

Information on nutrient requirements from birth to weaning is very limited in ruminants, effectively restricting successful development of new and potentially more efficient management systems. While this information base is larger in swine, several poorly understood factors limit early weaning. A means of providing nutrients without inducing diarrhea and growth setback has not been satisfactorily established. Dietary needs for a specific nutrient are frequently complicated by levels of other dietary nutrients which may alter absorption or utilization of the nutrient in question.

Intact males have great potential for meat production because of their faster and more efficient growth and decreased carcass fat as compared to castrates. For example, beef bulls will yield 12% more lean at the same weight than steers (Warwick et al., 1970). Similar advantages accrue to ram lambs and young boars. The information available suggests that quantitative requirements for these males are higher than for castrates or females (Liptrap et al., 1970). However, such information is limited and inadequate to maximize their physiological advantage. Behavioral factors may intermittently reduce growth rate of intact males and will require control before commercial production becomes feasible.

GENETICS AND NUTRIENT REQUIREMENTS

Remarkable progress has been made in improving animal production through genetic selection. Little concomitant effort has been expended to establish whether these improvements alter nutrient requirements. Much of the current knowledge on this subject was summarized in The Effect of Genetic Variance on Nutritional Requirements of Animals (NRC, 1975). While data on poultry are more extensive than those for other farm animals, there is relatively little information for any species. The issue is particularly important when nutrient resources are limited and expensive. There are two important questions: 1) Do the nutrient requirements change for animals bred for increased productivity? 2) Can one effectively select for strains of animals with lower than average nutrient requirements while maintaining maximum production?

ENVIRONMENT AND NUTRIENT REQUIREMENTS

The environment has profound effects on farm animals, including rate of productivity and nutrient requirements. Although many aspects of the total environment alter animal response, the climatic environment is of major importance, particularly when described in terms of effective temperature. Studies to characterize animal responses to thermal zones have been reported, although much remains before animal-environment interactions are adequately understood. Existing nutrient requirements have generally been developed in the absence of environmental stresses and may not be applicable to animals exposed to climatic extremes or other stresses. Increased feed intake during cold weather and depressed feed intake during hot weather have not been described quantitatively. Nutrient digestibility is inversely related with ambient temperature but needs to be quantified for specific nutrients. The information on increased rates of heat production during cold weather requires refinement, and little is known regarding energy required to rid the animal of heat during heat stress. Although animals do acclimate to environmental stress, little is known regarding this mechanism or its role in minimizing effects of environment on nutrition. The impact of thermal stress does not uniformly alter requirements for all nutrients, and some nutrients may be utilized more efficiently, as is the case of fat in poultry and swine diets during heat stress. Consequently, percentage approaches to diet formulation must be altered when specific diets are designed for heat or cold. In general, more is known regarding effects of cold than is known relative to effects of heat. Little information is available suggesting that qualitative and quantitative nutrient adjustments should be made for animals exposed to climatic stress, although more is known for poultry compared with cattle, sheep and swine.

WATER AND NUTRIENT REQUIREMENTS

Extensive analyses of surface water and groundwater have demonstrated the presence of all known essential dietary mineral elements for livestock and poultry (NRC, 1974). There are few published data, however, indicating whether these nutrient concentrations are sufficient to be useful as supplements to the dry diet. Seldom is the presence of mineral nutrients in water taken into account when diets are formulated. It is possible that analyses of local water supplies may support adjustment of diet mineral fortification.

A second relevant issue relates to excessive concentrations of water-borne substances that may impede performance. Because high quality water may not be available in some areas, it may be necessary to devise means of dealing with this problem. Several antagonistic relationships between mineral elements are known (Underwood, 1977). It is conceivable that the adverse effects of excesses of certain minerals may be controlled by adjusting levels of these and other elements in the diet. Finally, a new research thrust concerning the influence of restricted or contaminated water intake on animal production is needed.

DISEASE AND NUTRIENT REQUIREMENTS

The interrelationship between nutrition and infectious disease is of great importance but has been little researched. Nutritional deficiencies, by themselves, result in disease, but they may also influence the course of an infection in one of the following ways: 1) by exerting a direct effect on the host once the infective organism has gained entry to the tissues; 2) by facilitating invasion by secondary infective agents; 3) by delaying recovery from the disease; or 4) by specifically inhibiting the defense mechanisms of the host. Researchers have noted that animals with certain nutrient deficiencies seem more susceptible to bacterial infections but less so to viral infections. The proposed explanation is that bacteria have simple nutrient requirements for their own replication, but the missing nutrient may be necessary for proper function of the immune system and protection of the host. Virus replication, on the other hand, is dependent on functional metabolic machinery in infected host cells. If a nutrient deficiency impedes the metabolic events programmed by the virus, then virus replication is impaired. This theory is interesting but has not been widely tested. In addition, some experimental data are not consistent with this view. In any case, viral infections are accompanied frequently by secondary bacterial invasions. Whether nutrient concentrations can be altered to prevent or control infectious diseases is largely unknown.

The same shortage of information prevails in relation to dietary systems that may be helpful in reducing the adverse effects of parasitism, trauma or surgical procedures. For example, would animals afflicted by any of these conditions be restored to health more quickly by elevated levels of certain nutrients? If so, what level and which nutrients? What about correcting simple nutrient deficiencies? Should nutrients be simply restored to minimum requirement concentrations or would elevated levels be more effective?

NUTRITIONAL CHARACTERIZATION OF CONVENTIONAL FEED INGREDIENTS

Scientific formulation of animal diets can be no more accurate than information on nutrient concentrations in the feedstuffs. This is a major weakness in efforts designed to improve the efficiency of animal production. The published data, while massive, are still so general that many regional, climatic and soil factors which affect plant composition may be overlooked. With present technology, it is impractical to analyze all feed ingredients before diets are formulated, and yet recorded values seldom represent actual nutrient concentrations.

Crude protein values are often considered among the most reliable and useful of analytical data, but the determination does not distinguish between protein and nonprotein nitrogen and assumes that all protein contains 16% nitrogen and is of equal biological availability. Amino acid analyses are limited, expensive and frequently suspect in regard to cystine and tryptophan values.

Mineral analyses have been greatly improved by the advent of practical atomic absorption/emission spectrophotometry, and inductively coupled plasma techniques should provide further help. Unfortunately, few recent values have been incorporated into standard analytical tables, and some elements such as iodine and chromium are still very difficult to assay. In addition, it is important to know mineral levels of feedstuffs, since plant concentrations of certain elements may differ 10- to 100-fold.

Vitamin analyses are very limited because the technology is expensive, slow and tedious. As a consequence, animal diets may be supplemented with certain vitamins without regard to natural vitamin concentrations in the feedstuffs. While wasteful, this approach may be satisfactory as long as supplemental vitamins are inexpensive. For those that are not, this deficiency in technology introduces an unnecessary cost.

Improved systems for fiber analysis (Goering and Van Soest, 1970) have been widely adopted in research laboratories as replacements for the older crude fiber technique. They offer the advantage of generally greater accuracy in defining cellulose, hemicellulose and lignin fractions in forages. However, they do not work consistently well for concentrates, and some plant fractions, such as tannin, may produce erroneous results. None of the state-control laboratories has adopted these methods for regulatory purposes, and they require further refinement and cooperative testing.

A variety of systems to define the useful energy concentration of feeds has evolved. Expressions of energy value in terms of total digestible nutrients (TDN), digestible energy (DE), metabolizable energy (ME) and net energy (NE) are currently in use. While it may not be necessary or appropriate to have a single system for all species, data on any one system tend to be fragmentary and unconfirmed. This area would benefit from a directed, concerted effort by specialists in energy metabolism.

Despite the importance of accurate and precise nutrient analyses of feedstuffs, nutrients are not useful unless they are biologically available. Information in this area is extremely limited and is restricted largely to considerations of dietary gross energy availability for several species and to limited information on dietary phosphorus availability for poultry and swine.

Antiquality factors such as tannins, alkaloids, silica and toxic fatty acids limit intake, digestibility and other nutritionally-related functions in animals. Determination of their concentration and characterization of their significance is needed in all feedstuffs.

NUTRITIONAL CHARACTERIZATION OF NONCONVENTIONAL FEED INGREDIENTS

All issues of concern relative to conventional feed ingredients are more extreme relative to nonconventional feedstuffs. While nonconventional feedstuffs will likely become more important for animal production in the future, relatively little is known of the concentration or bioavailability of nutrients which they contain. Possibilities include crop residues, by-products of the food processing industry, animal waste, algae, aquatic weeds, fast-growing trees, scrap paper, industrial and alcohol distillation by-products and other products of microbial fermentation. Some of these feedstuffs may require further processing to improve nutrient utilization. Others may require special attention because of toxic contaminants. The associative influence of these materials with natural feedstuffs is largely unknown.

Nationally, crop residues comprise less than 5% of the dairy cow diet and about 15% of the diet for beef cattle and sheep. Vast tonnages are available for animals (Klopfenstein, 1973). Long-term use of these materials in animal production systems has not been studied nor has their compositional variability. With most of these materials, energy is the most useful nutrient, yet little is known as to content or availability. The following information is needed on nonconventional feedstuffs:

- The extent to which nonconventional feeds can substitute in the animal diet for grain or protein supplements that may best be used for direct human consumption.
- The extent to which nonconventional feedstuffs can substitute for conventional forages.
- The optimum dietary level at which nonconventional feedstuffs can be fed to animals at different stages in the life cycle (growth, gestation and lactation).
- The influence of nonconventional feedstuffs on animal productivity and animal product quality.
- The status of nonconventional feedstuffs relative to residues or toxins that would adversely affect animal productivity and animal products.

FEED INTAKE

Maximal animal performance is dependent on high levels of feed intake. The amount of feed consumed each day is the result of a balance of hunger drives in concert with other drives under physiological regulation. Hunger and satiety behaviors are controlled by various central nervous system structures, including those in the hypothalamus (Sclafani, 1976). In nonruminants, certain metabolites and hormones influence neural activity of these centers (Bray and Campfield, 1975). In ruminants, volatile fatty acids from microbial fermentation in the ruminoreticulum are major energy sources, and these metabolites may influence feeding behavior by acting on peripheral or central nervous system tissues (Baile, 1979). The role of gastrointestinal peptides, e.g., cholecystokinin (Gibbs and Smith, 1979; Stacher et al., 1979), and central nervous system peptides, e.g., endorphins and cholecystokinins (Della-Fera and Baile, 1979) in the control of feeding behavior and energy balance regulation is now being studied in both ruminants and nonruminants.

Physical limitations of the digestive tract in ruminants may restrict intake of coarse feeds (Baile and Forbes, 1974). Difficult to digest feeds are usually retained longer in the ruminoreticulum than are easily digested feeds, limiting the amount of dry matter consumed per day. Thus, indigestible material, such as plant cell wall, may be a major constraint on feed intake when forage of low digestibility is fed (Mertens, 1980). When high concentrate diets are fed, systems which control feeding behavior and energy balance regulate energy intake to energy output, and dry matter intake declines with increasing energy density. Fermented feeds, such as silage, introduce another factor into the intake equation, since dry matter consumption is generally less than would be expected for unfermented feeds of similar digestible energy densities. Because anaerobic fermentation is an effective and economical way to preserve livestock feeds, it is important that the factors limiting intake of fermented feeds be identified. This research must be conducted with target species, since laboratory animals are not adequate models.

Many disease conditions decrease normal feeding behavior in spite of increasing nutritional demand. The reasons for this inhibition are not known, although the resulting anorexia may contribute greatly to economic loss.

ALIMENTARY MICROBIOLOGY AND DIGESTIVE PHYSIOLOGY OF THE HOST ANIMAL

Microbes (bacteria, protozoa and a few fungi) in the rumen ferment and grow on feedstuffs consumed by the ruminant. Their cell mass and the end products from their fermentation are digested and absorbed to nourish the host. Thus, ruminant growth depends greatly upon microbial growth, and feedstuffs must be designed to nourish both the alimentary microorganisms and the host animal. Because microbes can synthesize protein from nonprotein nitrogen and cellulose, net synthesis of protein may result from use of feeds unsuitable for human consumption.

Microbes also thrive in the small and large intestine. The numbers in the small intestine are usually not large, but parts of the epithelial surface may be coated with bacteria in sufficient numbers to be important to host nutrition. Microbes (chiefly bacteria) growing in the cecum and large intestine ferment significant quantities of cellulose and other cell wall materials. Their activities increase the thickness and strength of the gut wall, which is an aid in diminishing some malfunctions in the lower alimentary tract.

Many species of microbes in the alimentary tract of lambs have been characterized, but when inoculated into germ-free lambs the artificial population of microbes does not support growth. Additional unidentified microbes in the normal population, but not included in the inoculum, must serve an unidentified function. In mice, a very large number of bacterial strains must be reinoculated into the germ-free animal in order to achieve a normal population of E. coli; presumably the microflora of other nonruminants behave similarly. Modification of the gut microbial population by feeding anti-microbial agents usually increases the efficiency of feed use by food-producing animals. Many important interactions between microbial species and the host remain unexplained.

The stomach and small intestine of ruminants and nonruminants sense the substances presented in the alimentary tract, integrate appropriate signals and release gastric juices, enzymes, hormones and other secretions necessary for digestion and absorption of nutrients. The level of feed intake may alter these processes. Concurrent changes in gut motility and blood flow, as well as endocrine integration of these events, requires study.

To understand digestive physiology, it will be necessary to identify the hormones, enzymes and other factors (passage rate, electrolytes) concerned. Availability of pure hormonal peptides will allow development of highly sensitive techniques -- radioimmunoassay and immunocytochemistry --for their assay. In humans there are several peptide hormones involved in regulation of nutrient intake, digestion and absorption. Not all have been purified, nor are sites of release and action known. The hormones originate in the alimentary tract, primarily in the stomach and small intestine. There may be several undiscovered gut hormones. Knowledge in this area is incomplete but is needed to modify host digestive processes.

Current chemical methods for determining the composition of plant cell walls do not adequately predict their digestibility by either animals or microbes. Both the hemicellulose and cellulose fractions are incompletely digested. Although lignin encrustation is involved, the chemical bonds concerned are imperfectly understood. A rapid and economical method of separating cellulose from lignin would greatly increase the quantities of carbohydrates available for animal feeding.

CELLULAR AND SUBCELLULAR PROCESSES ASSOCIATED WITH NUTRIENT UTILIZATION, PRODUCT SYNTHESIS AND ANIMAL GROWTH

Much of the available basic information on cellular structure and metabolism has already been exploited and there is little new information on the molecular biochemical and physiological reactions of cells. Detailed information on cell structure and regulation of metabolic reactions will lead to dramatic increases in the rate and efficiency with which animals convert ingested nutrients into useful products. Such information might also support regulation of the differential synthesis of tissues so that composition and nutrient value of animal products might be modified at will. Knowledge of the factors regulating protein and lipid synthesis in muscle tissue may reveal ways to divert nutrients from synthesis of fat into synthesis of muscle protein. Similarly, knowledge of the regulation of the cellular and subcellular reactions involved could lead to modification of the fat, carbohydrate and protein composition of milk and eggs. The above reactions are depicted in figure 5.1.

Basic research is needed in all areas of animal biology. The present state of knowledge in this area may be summarized conveniently in the following three categories:

Protein Synthesis and Muscle Growth

The mechanism by which various hormones act on cells and increase muscle, milk or egg protein synthesis in food-producing animals is largely unknown (figure 5.1, Steps 3 and 5). It has been shown recently that certain peptides, possibly elaborated by the central nervous system, dramatically increase the rate of muscle protein synthesis while simultaneously decreasing fat synthesis (figure 5.1, Step 6) (Foss and Trygstad, 1975). The nature of these peptides, the stimuli for their synthesis and release and the mechanisms by which they exert their effects are unknown.

At a subcellular level, the general mechanism of muscle protein synthesis is known, but the factors limiting the rate of synthesis are unidentified (figure 5.1, Step 4). Rates of protein synthesis in cultures of muscle cells are up to 10-fold higher than those in muscle cells of food-producing animals. The mechanism of ovalbumin synthesis in eggs has been studied thoroughly and the structure of the ovalbumin gene is partially understood. Control of synthesis of other egg proteins and the coordinated assembly of these proteins into the finished egg is not clear. Regulation of rate of milk protein synthesis is also poorly understood.

Muscle protein is in continual flux, being constantly degraded and resynthesized (Swick and Soong, 1974), but the agents responsible are not known (figure 5.1, Step 4). Turnover of muscle protein may be a critical factor in the efficiency of edible protein deposition because up to 90% of the protein synthesized in muscle cells may be required simply to replace degraded protein, with only 9 to 10% used for net growth (Laurent and Milward, 1980). Clearly, decreasing the rate of protein turnover would increase greatly the efficiency with which nutrients are converted into edible muscle. Similar increases in efficiency of milk and egg protein production would likely result if the mechanism and extent of turnover of these proteins could be limited.

Differences in muscle mass can be attributed partly to differences in muscle cell number (Aberle and Doolittle, 1976; Luff and Goldspink, 1967). The neonate has all the muscle cells it will ever have (Goldspink, 1972), so the factors regulating muscle cell number must operate during embryonic differentiation. The nature of these factors is completely unknown. It has been shown (Konigsburg, 1961) that embryonic muscle cells will grow and differentiate in cell culture. Consequently, it is now possible to study muscle cell differentiation and the factors that regulate it. The ability to hasten the rate with which meat, milk and egg proteins are synthesized will also alter nutrient requirements.

Adipocyte Number and Fat Deposition

Fat deposition in food-producing animals is a natural part of growth (figure 5.1, Steps 6 and 7) and is regulated by nutritional, hormonal, genetic and metabolic factors. Although cell membrane phospholipids and cholesterol are essential for survival and well-being of the animal, excessive accretion of neutral fat in animal products is currently regarded as undesirable and as contributing to excessive calorie intake by humans. In 1976, it was estimated that approximately one-third of the 5.5 billion kg (12 billion lb) of fat obtained from food-producing animals was not essential for their well-being and should have been conserved (Allen et al., 1976).

Fat accretion in animal products depends on both fat cell number and net fat deposition in these cells. The origin of fat cells (adipocytes) in food-producing animals is still unclear. It originally was thought that adipocytes differentiated from fibroblasts, but more recent evidence indicates that adipocytes may originate from endothelial cells. The principle difficulty in studying adipocyte number has been the inability to identify adipocytes until fat has been deposited in them.

By this stage, mitotic division has ceased and adipocyte number has been determined. It recently has become possible to grow preadipocytes in tissue culture so the factors affecting hyperplasia are now amenable to experimental manipulation.

The hormonal and metabolic factors regulating fat deposition in adipocytes are unclear. Although it is widely thought that insulin is a key hormone in fat accretion, the lack of response to insulin in some adipose tissue is not well understood. It is known that enzymes, such as acetyl CoA carboxylase, lipoprotein lipase and fatty acid binding protein are involved in fat synthesis and metabolism. Recent studies have indicated that the Na-K ATPase membrane enzyme in adipocytes may have an important role in regulating fat deposition in these cells.

Cellular Processes and How They Are Controlled

Progress in applied research in this area depends on an underlying base of fundamental information. Consequently, it is important that effort be devoted to research on the general principles of cellular metabolism. For example, fundamental information on cell membranes and how they permit cells to communicate and to respond to their environment, and to take in and excrete substances, would be important to a wide range of topics in animal nutrition and production. Recent studies have indicated that all cells have a cytoskeleton that controls cell shape and that may regulate cell membrane movements and translocation of intracellular materials. New information on properties of this cytoskeleton could influence the efficiency of nutrient utilizaton.

NUTRITION AND PRODUCT QUALITY

The extent to which the composition of foods of animal origin may be changed by altering the diet of the animal is largely determined by the nature of the animal's digestive tract. Thus, products obtained from nonruminant animals (swine and poultry) largely reflect the composition of the diet, whereas products obtained from ruminants (beef and dairy cattle and sheep) are not easily altered by diet because of the moderating influence of the rumen. With both types, however, nutrition offers the most promise for alteration (Oltjen and Dinius, 1975). Certainly, through ruminal bypass and dietary manipulation, it is possible to alter markedly the amount of fat and the fatty acid composition of animal products. Whether these changes are desirable is not clear. However, it seems appropriate to investigate those dietary alterations that will favorably influence the acceptability of animal products in the human diet.

ANIMAL WELL-BEING

Farm animals are raised in environments quite foreign to those experienced by their ancestors. Poultry and most swine are housed in confinement facilities with little or no exposure to soil and green forage. Dairy cattle, too, may be kept in dry lot, and feed, including forages, brought to them. Feedlots are commonly used for finishing lambs and beef cattle. These facilities are designed for maximum economic efficiency, and it has been assumed that if animals are optimally productive, they are comfortable. It is desirable to develop research methods that define animal behavioral needs and provide an objective base for the design and operation of modern systems of livestock production.

CONSTRAINTS THAT WILL INFLUENCE FUTURE DEVELOPMENT IN ANIMAL NUTRITION AND DIGESTIVE PHYSIOLOGY

LACK OF BASIC INFORMATION ON THE CELLULAR AND MOLECULAR PROCESSES REGULATING GROWTH AND METABOLISM OF ANIMAL TISSUES

Factors limiting rate and extent of muscle growth and protein synthesis in food-producing animals are not well known. Also, the cellular processes that divert ingested nutrients into adipocytes and fat deposition have not been defined. Additional understanding of the basic cellular and metabolic processes controlling subcellular reactions and cellular interactions would permit manipulation of protein and/or fat formation using nutritional, endocrinological or other mechanisms. It would also permit their use in selective breeding programs to increase efficiency of converting feed resources into milk, meat, eggs, wool and other animal products.

LACK OF KNOWLEDGE OF THE INTERRELATIONSHIPS BETWEEN THE ANIMAL AND ITS ALIMENTARY MICROBIAL ECOSYSTEMS

There is an urgent need for knowledge on the nutritional requirements and activities of the microbes in all sections of the alimentary tract. Manipulation of microbial action can markedly improve animal performance. Such manipulation requires increased knowledge not only of microbial metabolism but also of digestive physiological processes of the host, since feed intake, passage rate, hormone and enzyme secretions and nutrient absorption alter the microbial ecosystem.

It will be especially important to maximize the utilization of non-competitive feedstuffs via modifications of microbial ecosystems and/or digestive processes. Systems or treatments which effectively by-pass or alter rumen fermentation will be important for preserving the nutrient quality of certain feedstuffs.

Very little information is available concerning the relationship between alimentary microorganisms and nonruminant animals. Dietary constituents, their effect on alimentary microorganisms and subsequent impact on the host, must be characterized. Definition of the mechanisms by which low concentrations of dietary anti-microbial agents improve animal health and performance is needed. Lack of such information limits the development of alternative means for producing these needed improvements.

LACK OF DEFINITION OF REQUIREMENTS FOR ESSENTIAL NUTRIENTS, CONSIDERING INTERRELATIONSHIPS AMONG PHYSIOLOGICAL FUNCTION, GENETIC CHANGE, ENVIRONMENT AND HEALTH STATUS.

The quantitative requirements of food-producing animals for essential nutrients are not constants. As genetic constitution and systems of management of animals change, information on nutrient requirements must be updated. The market age and growth rates of lambs, swine and broilers, and the milk production of dairy cows have changed dramatically in recent years. Many previously established quantitative nutrient requirements are no longer valid.

For certain ages or physiological functions, quantitative nutrient needs have never been established. This is particularly true for the intact males of most species. More information is also needed on requirements of early-weaned animals and for reproduction and lactation.

Quantitative nutrient requirements also are influenced by environmental temperature and space, and by use of anabolic agents or drugs to prevent or control disease. Unfortunately, the nature and magnitude of these influences have not been defined.

INADEQUATE INFORMATION TO CONTROL FEED INTAKE FOR OPTIMUM PERFORMANCE

Further knowledge of factors affecting feed intake is required to adjust nutrient supply appropriately during various physiological stages of the animal life cycle. Physiological factors often establish ceilings on nutrient intake which limit the utilization of feedstuffs for maximum efficiency of production. It may be important to restrict feed intake during periods when weight gain may be undesirable or unnecessary. However, information on mechanisms to increase or decrease appetite of food-producing animals in various productive states is inadequate.

INADEQUATE TECHNOLOGY FOR RAPID AND ECONOMICAL DETERMINATION OF THE CONCENTRATION AND BIOAVAILABILITY OF NUTRIENTS IN FEEDSTUFFS AND OF THE PRESENCE OF ANTI-QUALITY FACTORS

Considerable variation exists in the composition and bioavailability of nutrients in forages, cereal grains, protein supplements and nonconventional feedstuffs which are increasingly important in animal diets. In many instances, compositional data on the latter feedstuffs are not available. Little is known about bioavailability of nutrients, and such information is critically needed to accurately and economically formulate diets for animals.

Analytical techniques must be rapid, accurate and inexpensive to permit optimum diet formulation on the basis of available nutrient supply. In addition, improved techniques are needed to screen feedstuffs for toxins, anti-growth substances or other anti-quality factors.

LACK OF KNOWLEDGE OF NUTRIENT COMPOSITION OF GRAIN, FORAGE RESIDUES AND INDUSTRIAL BY-PRODUCTS AND NONCONVENTIONAL FEEDSTUFFS

With increasing energy costs, more of the nonconventional feedstuffs, such as plant residues and industrial by-products will become economically competitive with conventional feeds for animal production. Most potential unused feedstuffs contain high amounts of cellulose and lignin, and thus are slowly and incompletely digested by animals. Quantitative information is needed concerning the chemical composition of cell wall components, permitting the development of techniques to increase availability and utilization of nutrients.

Increased availability of by-products from energy production (e.g., ethanol fermentation and distillation, methane generation, biomass conversion) is expected. Little is known of their nutritive value and of their nutrient bioavailability. Simulation modeling to optimize usefulness and evaluate the economics of treating and supplementing these nonconventional feedstuffs for food-producing animals should prove particularly beneficial.

INADEQUATE DEVELOPMENT OF NEW FEEDSTUFFS AND SYNTHESIS OF ESSENTIAL NUTRIENTS OR THEIR PRECURSORS

Competition for feed grains traditionally used for animal production is being imposed by increasing exports and by domestic fuel production demand. New alternative feedstuffs, not commonly used for animals, must be found. Single cell protein can be produced from a wide variety of waste carbohydrate sources (including cellulose) and offers the possibility of a high quality protein for animal use without further taxing land resources. However, more economical production methods must be developed if this protein source is to be used widely.

The abundance of cellulosic materials (such as wood, weeds, straw, corn stalks, aquatic plants) in all areas of the United States makes possible their use for synthesis of feeds or individual nutrients. Economical means of synthesizing essential amino acids or suitable precursors would make it possible to utilize more efficiently certain feedstuffs which are relatively abundant but nutritionally incomplete.

EFFECT OF REDUCED AVAILABILITY AND INCREASED COST OF FOSSIL FUEL ON FEED SUPPLY

Limited supplies and excessive costs of fossil energy sources will alter animal production systems. Energy-dependent inputs (irrigation, fertilizer, fuel) will be less available for feed production. Therefore, a greater emphasis will be placed on forages and direct harvest of pastures and range by animals. Nonconventional feedstuffs, including byproducts of fermentation of renewable resources for fuel production, will become more prevalent. Suitable animal production systems have not been developed to accommodate these feeds, and it will be difficult to provide food-producing animals with nutrient sources of sufficient quality to support their genetic capacity for product synthesis.

INADEQUATE TECHNIQUES FOR MEASURING ANIMAL WELL-BEING FOR EFFICIENT FOOD PRODUCTION

Accurate, objective techniques are needed to assess the degree of stress or comfort and the behavioral preferences of animals maintained for food production purposes. At present, judgments in this area are made subjectively on an anthropocentric basis. If decisions on animal environments and husbandry practices are made without adequate information, costs of animal products to the consumer will be increased unnecessarily.

RESEARCH IMPERATIVES IN ANIMAL NUTRITION AND DIGESTIVE PHYSIOLOGY

CELLULAR PROCESSES ASSOCIATED WITH PROTEIN SYNTHESIS AND ANIMAL GROWTH

Research is needed in three main areas: 1) protein synthesis and muscle growth; 2) regulation of adipocyte number and fat deposition; and 3) properties and structure of animal cells.

Protein Synthesis and Muscle Growth

Knowledge of the biological control systems that limit rate and efficiency of animal protein production will lead to methods of modifying these systems to increase significantly the efficiency with which ingested nutrients are converted into protein. Such increases in rate and efficiency of protein accretion would also change nutrient requirements for growth and maintenance.

The effects of modifying the supply of nutrients to muscle tissue should be studied to determine their influence on protein synthesis. Certain hormonal influences on protein synthesis have been identified, but the nature of biologically active peptides that are secreted into the circulation of food-producing animals should be studied to determine how these peptides affect the cell to increase protein accretion. At a subcellular level, the factors that regulate rate of muscle protein synthesis and assembly of synthesized proteins into functional structures need to be determined. The agents responsible for disassembly and degradation (turnover) of muscle proteins and the factors that regulate activity of these agents need to be identified. The embryological factors that control muscle cell number and size in food-producing animals should be identified, with the goal of influencing these properties during muscle cell development. The factors that regulate secretion of milk and egg proteins and the cellular mechanisms regulating activity of these factors should be established clearly. The possible role of lactose and phospholipids in controlling production of milk and egg protein should be elucidated.

Regulating Adipocyte Number and Rate and Extent of Deposition

Control of adipocyte number and fat accretion in adipocytes should be mastered so the fat content of animal products can be manipulated. The role of hormones and other factors regulating mitotic division of preadipocytes and their differentiation into adipocytes should be investigated in food animals. Detailed studies are needed on insulin-binding to adipocytes and preadipocytes as influenced by age, cell size, diet and genetic variation. Attempts should be made to discover characteristics of preadipocytes that would permit their identification prior to differentation and fat accretion. The role of various hormones, peptides and enzymes in regulating the rate of fat accretion in adipocytes should be studied. Uncontrolled activity of these enzymes may cause futile cycles of lipid metabolism that consume energy and affect fat deposition.

Basic Understanding of Cellular Processes and How They Are Controlled

Fundamental studies should be conducted to increase the understanding of the processes and structures involved in normal cell metabolism. The nature of the plasma membrane and how it regulates entry and exit of nutrients and the interactions of the cell with its environment should be elucidated. The composition of the cytoskeleton and dynamic interactions with other cellular proteins and with cell membranes need to be studied. The fundamental properties of proteins in muscle, milk, eggs and other animal products and the interactions of these proteins with each other and with cellular structures should be established.

INTERRELATIONSHIPS BETWEEN ALIMENTARY TRACT MICROBIAL ECOSYSTEMS, DIGESTIVE PROCESSES AND THE HOST ANIMAL

Quantitative analysis and simulation of the dynamics of gut microbial ecosystems of ruminant and nonruminant food-producing animals should include nutrient requirements, end products and interactions between microbes and the host.

The large intestine of all food-producing animals contains extremely high concentrations of microorganisms. They require various nutrients and perform activities which may be advantageous (vitamin synthesis) or deleterious (amine formation) to the animal. Growth-promotion by various antimicrobial agents is exerted primarily via the microbial ecosystem, although the mechanism is unknown. The effects of dietary fiber also appear to be mediated through the microbes.

Microbial and host activities are closely interrelated at all sites of the alimentary tract, except possibly the acid stomach. The relationships between nutrient intake and rate and extent of digestion, rate of passage and metabolic demand require clarification. Effects of hormones, enzymes and other secretions which limit nutrient digestion and absorption need study. Knowledge of hormonal roles in modifying alimentary tract environment is essential in obtaining maximal transformation of feed into animal products.

High fiber feedstuffs unsuitable for human consumption can be useful as animal feed. Their use requires better understanding of the factors limiting digestion of cellulose and hemicellulose. Lignification is the chief limiting factor, but crystallinity and other features within the plant cell wall structure also require elucidation. Microbial agents can unlock this store of carbohydrates; however, the problem is to understand and exploit their actions and to render the carbohydrates in the cell wall suitable for animal utilization. Biological and chemical agents can be applied in various sequences and combinations, along with physical treatments, to explain and alter fiber structure for the purpose of improving bioavailability. These methods also can be used to assess rate and extent of digestion of diverse feedstuffs.

NUTRIENT CONSERVATION IN MEETING ANIMAL NEEDS

For efficient animal production, the conservation of nutrients dictates that nutrient requirements be determined in relation to the changing genetic makeup of animals and to husbandry practices. Requirements have not been established adequately for various stages and levels of performance or productive state (reproduction, growth, maintenance, milk and egg production), or for animals of different sex or age, or for animals exposed to various climatic and environmental stresses and disease states.

To prevent deleterious effects of nutrient deficiencies on animal production, procedures are needed to determine nutritional status. Metabolic profiles, to directly detect nutrient deficiences, should be developed for rapid assessment and correction of problems. Effects of nonnutrient feed additives must be considered in establishing nutrient requirements.

CONTROLLING FEED INTAKE TO MAXIMIZE ANIMAL EFFICIENCY

Feed intake is a major determinant of animal performance and efficiency. Research is needed to define optimum feed intake for desired efficiency during various productive states. The interactions of factors affecting intake need to be described so that feed intake can be predicted and modified more accurately. Physiological factors that control appetite must be defined as they relate to nutritional needs of animals in different productive states and environmental conditions. Acceptability and aversive factors in feeds must be isolated and defined, including nutrient composition, moisture level, texture, physical form, aroma and taste. The role of diet variety and of gastrointestinal factors such as digesta composition, secretions, rate of passage, hormones, enzymes, microbial digestion and end products upon feed intake must be clarified. Since maximum efficiency during certain production states may require feed intakes below those voluntarily consumed by animals, means of controlling satiety must be developed.

INTEGRATED NUTRIENT MANAGEMENT TO ENHANCE PRODUCTION EFFICIENCY

Numerous nutrient sources must be used to attain production goals. Availability and cost of forages, grains, protein sources, nonconventional feedstuffs and other nutrient resources must be integrated for cost-effective animal production. Research is needed to determine replacement values among nutrient resources, particularly in the case of energy management. Studies are needed to determine the economic and energetic consequences of promoting animal tissue storage of nutrients during periods of abundant nutrient supply for use when nutrient supplies are limited. The issue of feed vs fuel, particularly as it relates to livestock environment, is central to efficient energy management. Information relative to alternatives in nutritional management will be needed to develop appropriate models to aid in optimizing economic efficiency.

NUTRITIONAL CHARACTERIZATION OF FEEDSTUFFS

Research characterizing the nutritional composition and availability of feeds must be expanded in order to meet accurately and economically the quantitative nutritional requirements of animals. Increasing emphasis needs to be placed upon relationships between compositional attributes and animal performance. Development of rapid and economical analytical schemes to provide nutritional information for diet formulation prior to feeding is urgently needed. Information on bioavailability of all nutrients, especially those yielding energy, is seriously lacking and must be quantitated. Anti-quality factors, nutrient antagonists and toxins in feeds must be characterized and their effects on animal performance more clearly defined. As nutritional data on feeds are accumulated, improved methods of data capture, cataloging and retrieval must be developed and made readily accessible nationally and internationally.

REFERENCES

Aberle, E.D. and D.P. Doolittle. 1976. Skeletal muscle cellularity in mice selected for large body size and in controls. Growth 40:133.

Allen, C.E., D.C. Beitz, D.A. Cramer and R.G. Kauffman. 1976. Biology of fat in meat animals. North Central Regional Research Pub. No. 234.

Anonymous. 1979. Proc. VI International Symp. on Ruminant Physiology. Clermont-Ferrand, Sept. 4-9, 1979. Ruckebusch and Thivend (Ed.) (In press).

Baile, C.A. 1979. Part II. Regulation of energy balance and control of food intake, p. 291. In: Digestion Physiology and Nutrition of Ruminants, Vol. 2. (D. C. Church, Ed.) O & B Books, Inc., Corvallis, OR.

Baile, C.A. 1980. Control of feed intake in healthy and diseased animals. Distillers Feed Conf. Proc. 35:45.

Baile, C.A. and J.M. Forbes. 1974. Control of feed intake and regulation of energy balance in ruminants. Physiol. Rev. 54:160.

Bassett, J.M. 1978. Endocrine factors in the control of nutrient utilization: ruminants. Proc. Nutr. Soc. 37:273.

Bloom, S.R. 1978. Gut hormones. Proc. Nutr. Soc. 37:259.

Bray, G.A. and L.A. Campfield. 1975. Metabolic factors in the regulation of caloric stores. Metabolism. 24:99.

Cheng, K.J., D.E. Akin and J.W. Costerton. 1977. Rumen bacteria: interaction with particulate dietary components and response to dietary variation. Fed. Proc. 36:193.

Della-Fera, M.A. and C.A. Baile. 1979. Cholecystokinin octapeptide: continous picamole injections into the cerebral ventricles of sheep suppress feeding. Science. 206:471.

Foss, I. and O. Trygstad. 1975. Lipoatrophy produced in mice and rabbits by a fraction prepared from the urine from patients with congenital generalized lipodystrophy. Acta. Endocrinol. 8:398.

Gibbs, J. and G.P. Smith. 1977. Cholecystokinin and satiety in rats and rhesus monkeys. Amer. J. Clin. Nutr. 30:758.

Goering, H.K. and P.J. VanSoest. 1970. Forage fiber analysis: apparatus, reagents, procedures and some applications. Agr. Handbook 379. USDA, Washington, DC.

Goldspink, G. 1972. Postembryonic growth and differentiation of striated muscle, p. 179. In: G.H. Bourne (Ed.) The Structure and Function of Muscle. Vol. I., Part A., 2nd. Ed., Academic Press, Inc., New York.

Hespell, R.B. and M.P. Bryant. 1979. Efficiency of rumen microbial growth: influence of some theoretical and experimental factors on Y_{ATP}. J. Anim. Sci. 49:1640.

Hungate, R.E. 1966. The Rumen and Its Microbes. 533 p. Academic Press, Inc., NY.

Klopfenstein, T.J. 1973. Treatments of crop residues. In: Proc. Sixth Research-Industry Conf., Amer. Forage Grassland Council, Lexington, KY.

Konigsberg, I.R. 1961. Cellular differentiation in colonies derived from single cell plating of freshly isolated chick embryo muscle cells. Proc. Nat. Acad. Sci. 47:1868.

Laurent, G.J. and D.J. Millward. 1980. Protein turnover during skeletal muscle hypertrophy. Fed. Proc. 39:42.

Liptrap, D.O., E.R. Miller, D.E. Ullrey, K.K. Keahey and J.A. Hoefer. 1970. Calcium level for developing boars and gilts. J. Anim. Sci. 31:540.

Loew, F.M. 1972. The veterinarian and intensive livestock production: humane consideration. Can. Vet. J. 13:229.

Luff, A.R. and G. Goldspink. 1967. Large and small muscles. Life Sci. 6:1821.

Mertens, D.R. 1977. Dietary fiber components: relationship to the rate and extent of ruminal digestion. Fed. Proc. 36:187.

Mertens, D.R. 1980. Fiber content and nutrient density in dairy rations. Distillers Feed Conf. Proc. 35:35.

NRC. 1974. Nutrients and Toxic Substances in Water for Livestock and Poultry. National Academy of Sciences - National Research Council, Washington, DC.

NRC. 1975. The Effect of Genetic Variance on Nutritional Requirements of Animals. National Academy of Sciences-National Research Council, Washington, DC.

NRC. 1975. Nutrient Requirements of Domestic Animals, No. 5. Nutrient Requirements of Sheep. National Academy of Sciences - National Research Council, Washington, DC.

NRC. 1976. Nutrient Requirements of Domestic Animals, No. 4. Nutrient Requirements of Beef Cattle. National Academy of Sciences - National Research Council, Washington, DC.

NRC. 1977. Nutrient Requirements of Domestic Animals, No. 1. Nutrient Requirements of Poultry. National Academy of Sciences - National Research Council, Washington, DC.

NRC. 1978a. Plant and Animal Products in the U.S. Food System. National Academy of Sciences-National Research Council, Washington, DC.

NRC. 1978. Nutrient Requirements of Domestic Animals, No. 3. Nutrient Requirements of Dairy Cattle. National Academy of Sciences - National Research Council, Washington, DC.

NRC. 1979. Nutrient Requirements of Domestic Animals, No.2. Nutrient Requirements of Swine. National Academy of Sciences - National Research Council, Washington, DC.

Nir, I., N. Shapira, Z. Nitson and Y. Dror. 1974. Force feeding effects on growth, carcass and blood composition in the young chick. Brit. J. Nutr. 32:229.

Oltjen, R.R. and D.A. Dinius. 1975. Production practices that alter the composition of foods of animal origin. J. Anim. Sci. 41:703.

Owens, F.N. and H. R. Isaacson. 1977. Ruminal microbial yields: factors influencing synthesis and bypass. Fed. Proc. 36:198.

Sclafani, A. 1976. Appetite and hunger in experimental obesity syndromes, p. 281. In: D. Novin, W. Wyrwicke and G. A. Bray (Ed.) Hunger Basic Mechanism and Clinical Implications. Raven Press, NY.

Stacher, G.H., H. Bauer and H. Steinringer. 1979. Cholecystokinin decreases appetite and activation evoked by stimuli arising from the preparation of the meal in man. Physiol. Behav. 23:325.

Swick, R.W. and H. Soong. 1974. Turnover rate of various muscle proteins. J. Anim. Sci. 38:1150.

Underwood, E.J. 1977. Trace Minerals in Human and Animal Nutrition. 4th Ed., Academic Press, Inc. NY.

Warwick, E. J., P.A. Putnam, R.L. Hiner and R.E. Davis. 1970. Effects of castration on performance and carcass characters of monozygotic bovine twins. J. Anim. Sci. 31:296.

Waterlow, J.C., P.J. Garlick and D. J. Millward. 1978, Protein Turnover in Mammalian Tissues and in the Whole Body. North-Holland, NY.

TABLE 5.1. NUTRIENTS PROBABLY REQUIRED BY FOOD-PRODUCING ANIMALS

Water	Cobalt[a]
Energy	Manganese
Nitrogen[a]	Selenium
Arginine	Chromium
Glycine + serine	Fluorine
Histidine	Nickel
Isoleucine	Silicon
Leucine	Vanadium
Lysine	Tin
Methionine + cystine	Arsenic
Phenylalanine + tyrosine	Molybdenum
Threonine	Vitamin A
Trytophan	Vitamin D
Valine	Vitamin E
Essential fatty acids	Vitamin K
Calcium	Thiamin
Phosphorus	Riboflavin
Magnesium	Niacin
Sodium	Vitamin B_6
Chlorine	Pantothenic acid
Potassium	Biotin
Sulfur[a]	Folacin
Iron	Vitamin B_{12}
Copper	Choline
Iodine	Inositol

[a]Used as substrate by ruminoreticular microorganisms for synthesis of nutrients required by tissues of the host. Amino acids, vitamin K and the B-vitamins are some of the substances synthesized.

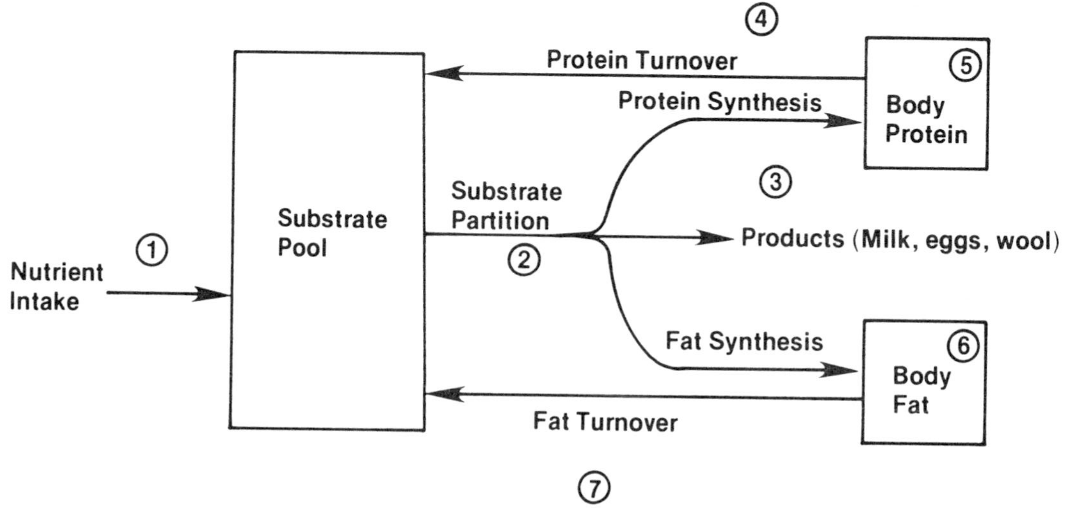

Figure 5.1. Depiction of metabolism of protein and fat for growth and product formation in animals.

6

T. C. Cartwright (Co-Chairman), R. J. Gerrits (Co-Chairman), C. A. Kiddy (Rapporteur)

F. Bazer, G. E. Bradford, K. I. Brown, J. Dendel, G. E. Dickerson, N. L. First, R. H. Foote, J. Gorski, H. D. Hafs, C. Kaltenbach, R. B. Land, E. Lasley, L. D. McGilliard, R. H. Nelson, I. T. Omtvedt, H. Schroeder, R. S. Sechrist, G. Seidel, C. R. Shumway, R. W. Touchberry, and H. A. Tucker

ANIMAL GENETICS AND REPRODUCTION

SUMMARY

Genetics and physiology play a central role in programming and directing reproduction and production of food animals. To be most effective in producing food for humans, the programming and direction must be sensitive to the total economics and ecology of production systems and have capability to respond quickly to ever-changing resource bases, economic conditions and consumer needs. Three major research imperatives were identified.

REPRODUCTIVE EFFICIENCY

Reproduction is the most limiting constraint to efficient production by food animals. Components of reproduction form a chain, in sequence, that results in a complex process. Research on these components is critical to overcoming reproduction far below potentials.

Imperatives for research are:

a) Estrus, ovulation and fertilization.

b) Embryo survival and development.

c) Parturition and perinatal survival.

d) Embryo technology - particularly challenging but promising areas of this component include; 1) harvesting ova, fertilization outside the body and maturation and transfer of ova; 2) frozen embryos, sex control and twinning; 3) microsurgery, chimeras, cell fusion and parthenogenesis; and 4) cloning.

PROTEIN SYNTHESIS AND PRODUCTION

Farm animals vary greatly in muscle growth and other forms of protein production related to lactation and egg production even within breeds. The functional basis for genetic variation in traits, such as growth and lactation, relates to enzymes, hormones and various intracellular processes. Almost nothing is known about genetic cause and effect relationships in these. An understanding of the genetic basis of physiological traits would be used to direct selection and breeding of farm animals that are more efficient in protein production and that deposit fat in proper balance. Research imperatives to accomplish these understandings are:

a) Cell energetics.
b) Growth, lactation and egg production.
c) Neural and endocrine control.
d) Digestive physiology.

GERM PLASM IMPROVEMENT AND UTILIZATION

Established methods of genetics applied to livestock breeding have created improved breeds and strains and will continue to be essential. New scientific knowledge and procedures in physiology, genetics, immunogenetics and systems analysis enhance capabilities of animal breeders to mold livestock for increased productivity, for improved dietary quality of products and for increased immunity to disease. Application of systems analysis to match effectively genetic potential for life cycle with the multifaceted production resources is needed to exploit new knowledge and to keep farm animals economically efficient and in tune with consumer needs.

Research imperatives for translating scientific developments to relevant genetic improvement are:

a) Clarification of objectives.
b) Criteria of merit for selection.
c) Genetic strategy - including systems analysis.
d) Genetic engineering - including recombinant DNA and gene splicing and transfer.

The research imperatives are listed under discrete headings, but stress is on interrelationships among these. Research in each of the physiology subjects may utilize both qualitative and quantitative variation. Benefits of cloning will be realized only if embryo transfer and storage technology are successful and inexpensive, and if consistently high rate of intrauterine survival can be achieved. Application of knowledge of cellular synthesis and tissue deposition of protein to increased supply and quality or decreased cost of human food protein will require many steps. Selection criteria will have to be indentified and used in the development of improved strains. These strains will be multiplied through artificial insemination and other reproduction technologies, and integrated into efficient production programs using methods of systems analysis. That is, basic research is needed to develop knowledge of biological processes; then, an integrated approach is needed for effective utilization of this knowledge for food animals.

INTRODUCTION

Products of animal agriculture used in meeting human needs are a renewable resource. Renewal is accomplished by two functions of animals: 1) reproduction or increasing their numbers; and 2) production or increasing the quantity of their products through body growth, lactation and egg production. These functions are accomplished through physiological processes which are influenced by genes. The efficiency with which livestock reproduce and produce to meet human needs has been increased remarkably during the 20th century by modifying the animal and its environment. Yet production resources and demands of people continue to change at an ever increasing pace.

Continued improvement of the nutritive value of animal products and of the efficiency of reproduction and production are essential if quality of life is to be maintained. Physiology and genetics play key roles for improvement of animals matched to the changing availability of resources. That is, these research areas provide potentials for improvements which have a favorable payoff in economic benefit per cost, as well as relieving such problems as energy, pollution and fat content of diet.

Reproductive inefficiency is one of the most costly and production-limiting problems facing the livestock and poultry industries. Reproductive performance is influenced by genetics, management, nutrition, disease, and environmental factors such as temperature and photoperiod. Each species has physiological limitations for reproductive capacity, but we are far from realizing the reproductive potential of livestock. Although problems differ for each species, the greatest opportunity to improve reproductive efficiency of farm animals lies in the development of basic research information that will permit us to control or alter specific reproductive events and reduce postnatal losses.

The role of genotype in mediation of reproduction and production depends on the nutritional, climatic, disease and managemental environment in which the animal is placed. Modification of genotype by selection or breeding may enhance productivity and profit. Modification of any of the components of environment also may enhance productivity and profit. However, to realize the potential of either a genetic or an environmental improvement, modifications in other components often are required. For example, genetic selection to increase production potential (growth, milk, eggs) does not always translate into increased profit unless the nutrition is sufficient in quality and quantity to support increased requirements to meet the production potential. Also, the market, which may be considered part of the environment of a production enterprise, may place bounds or restrictions on productivity because of market standards (e.g., composition of milk, quality, grade of beef or egg size). These interactions imply that genetic change for increased productivity must be integrated with the production environment for improved genetic capability to be realized fully. And, vice versa, developments in the other components of production may not be effective in increasing productivity unless suitable genotypes are available. For example, physiological manipulation to cause twinning in cows must be in concert with the genotype of the cow and available feed and management inputs.

Manipulation of either genotype or physiology does not imply necessarily that a high technology or intensification is required, but rather a proper matching of genotypes or techniques to production resources. Livestock of the same species are produced in a wide range of ecozones, intensifications, managements, sizes of enterprise, degrees of vertical integration and other aspects of production systems. Relationships among these effects on production systems change across time due to cost/price cycles, technological advances and other trends. The definition of genetic merit is neither completely uniform nor static, though some elements may be nearly so. Genetic improvement is, therefore, a continuous process that should involve a strategy for coping with different goals that change with time.

BEEF CATTLE

Beef cattle are especially important to mankind because as ruminants they can convert crude plant materials into meat. Breeds have evolved over a long time in response to natural and artificial selection that has led to their adaptation to a variety of climatic, nutritional and economic conditions. Differences among these breeds illustrate opportunities for genetic improvement of efficiency of beef production by tailoring cattle to specific human needs. All of the components of production must be integrated to realize the enhancement of genetic potential through selection, breeding and hybridization.

Breeds that have a wide range of genetic capability for mature size, milk production, maturing rate, local adaption and reproductive capacity have been imported. These newly available breeds provide a greatly expanded genetic resource, but their potential contributions to different production systems and efficiencies of production in various regions have been identified and exploited only partially. An evaluation of life-cycle productivity emphasized the importance of reproduction to efficient livestock production. Estimates derived from inventory figures, numbers marketed and calf death loss data as well as experimental data indicate that on a national basis only about 70-75% of cows exposed to bulls wean calves. Delayed puberty of young heifers, insufficient libido and poor semen quality of bulls, failure of cows to become pregnant and calf losses, prior to, at and shortly after birth are problems. Reasons for failure to achieve greater reproductive success are varied and not fully understood.

Production of multiple births from even a part of the cattle herds throughout the world would increase beef production. Hormonal treatment, embryo transfer and selection have been successful in producing multiple births. Predictable superovulation and maintenance of multiple-fetus pregnancies are a major problem.

Postpartum anestrus is highly variable with the result that many cows do not rebreed in time to conceive for calving in consecutive years. Only 55-65% of matings of artificial insemination result in birth of a calf at term. Early embryonic mortality is a major factor, but fertilization failure and abortion or resorption of fetuses are also important. A high percentage of calf losses from birth to weaning is due to deaths within 36 hr of birth. Many deaths are related to dystocia; other causes are largely unknown.

Application of presently available technology likely could result in 88-90% calf crops. This net reproduction would be possible with calving rates of 93-95% and calf death losses of not more than 3-5%. However, more information is needed on basic physiological mechanisms to optimize reproductive rate in specific environmental and nutritional situations.

Understanding physiological processes associated with low reproduction, determining nutritional conditions needed for optimum reproduction, and failure to identify useful and responsive selection objectives constitute the major constraints associated with the production and improvement of beef cattle.

DAIRY CATTLE

During the past decade, milk production per cow increased by over 1 ton per cow (23%), and the increased productivity saved 88 billion liters of corn while meeting the dairy product needs of U.S. consumers. Well over half the increase is attributed to permanent genetic improvement from widespread use of genetically superior bulls available through artificial insemination (AI). Superior bulls are identified by progeny tests (sire summaries) which are part of the nationwide Dairy Herd Improvement program, a management information system. Reasons for extensive use of AI include frozen semen technology, value of a single dairy cow, the way dairy cattle are managed and predominance of selection criteria based on characteristics limited to the female. This system has resulted in genetic potentials for milk production which match or even exceed the ability to provide increased environmental-management inputs (notably more concentrated feeds).

Utilizing research tools similar to those that led to understanding of the genetic influence on milk production, there is increasing interest in determining genetic influences on traits supporting milk yield, such as reproductive rate, feed utilization efficiencies, mastitis resistance, longevity and physical structure as these traits relate to net worth of the cow as a food producer. Genetic opportunities exist to develop breeding programs which emphasize protein and other nonfat components of milk as the industry strives to adjust to new market demand.

Economic and biological implications of intense selection for milk yield are not understood competely because of unknown related responses in other economically important characteristics. Limited research results suggest that health and reproduction may be impaired by selection for milk yield. Nutritional energetics of milk production suggests that genetic gains in milk yield may be at the expense of fertility from negative energy balance at time of insemination.

Reproductive failure is a major cause of reduced annual milk yield and shortened productive life. Even in well-managed herds, 10% of the cows fail to conceive each year, and the average interval between calvings of 13.5 months is longer than the ideal interval of 12 months. The prolonged calving interval results from a pregnancy rate of only 50-55% after first insemination. Cows that do not produce calves after one insemination require repeated inseminations because of failure to detect estrus, defective eggs and sperm, and embryonic death. These losses are due, in part, to the stresses of high milk production, high environmental temperature and humidity and inadequate nutrition that result in physiological irregularities, endocrine imbalances and ovarian and uterine dysfunction. In addition, lack of knowledge with respect to the physiology of milk secretion hampers our ability to maximize milk production and prevent losses due to udder infirmity.

Research on semen physiology is responsible for the rapid growth of AI, the increase in number of offspring per bull, identification of infertility problems in the bull vs the cow, and development of procedures for preservation of sperm through freezing. Studies of reproduction in the cow have provided better understanding of complex interrelationships among embryo, uterus, ovary, pituitary gland and brain, and have provided a means for control of estrus and ovulation. The opportunity now exists to learn more about why early embryos die so often and to develop ways to control the sex of calves and improve procedures for preserving sperm, eggs and embryos for later transfer to cows.

SWINE

Rearing in confinement and crossbreeding are accepted widely by the swine industry. Average backfat thickness has been reduced substantially, but reproductive efficiency has been essentially unchanged during the past 20 years. The number of pigs marketed per litter has remained at about 7.2, and sows farrow an average of 1.8 litters per year. Rearing of breeding stock in confinement has increased environmental stress causing delayed puberty, irregular estrous cycles and reduced conception rates.

Progress has been made in evaluating genetic aspects of most economically important traits in swine, but identification and exploitation of genetically superior breeding stock has not been achieved to a significant degree. Artificial insemination is not being utilized widely.

Opportunities exist for improving reproductive efficiency through additions or manipulation of new breeds, even though most measurements of reproductive performance are influenced only slightly by genetics. Ovulation rate can be increased through selection (about four additional ova in 10 generations), but there is not a resulting increase in the number of live pigs farrowed. Embryonic mortality is a major problem that must be overcome. Most sows ovulate an average of 17 ova. About 95% of the ova are fertilized, but only 9.4 live pigs are farrowed and only 7.2 pigs are marketed per litter. Embryo transfer techniques are available but are not useful to the industry at this point.

Studies on genetic and physiological relationships among traits are needed to establish indexes of optimum performance. Efficiency of growth often is in terms of units of feed required to produce a unit increase in body weight. However, increase of body weight in the form of fat is neither efficient nor desirable, and emphasis on increased efficiency of desposition of lean tissue is desired. Research is needed to determine the biological processes controlling conversion of nutrients consumed into lean as opposed to fat tissue.

Reproductive efficiency in swine is greatly dependent on interactions between developing embryos and their intrauterine environment. There is a loss (pregnancy wastage) of 25-40% of fertilized ova.

Fertilization failure is not a primary factor limiting reproduction in swine. However, refinement of techniques of artificial insemination and means for identifying boars of high genetic merit must be developed before artificial insemination can be employed widely. Likewise, our inability to identify females of high genetic merit prevents effective use of embryo transfer and would limit application of anticipated breakthroughs, cloning and other aspects of ovum in embryo cloning and modification.

A major constraint to improvement of growth and reproductive efficiency is our inability to identify key physiological and endocrinological factors associated with high reproductive efficiency.

Estrous synchronization is essential to effective use of artificial insemination. Although hormones are available for this purpose and have been effective, they have not been approved for use by the Food and Drug Administration.

Physiological and endocrine factors associated with efficient growth and feed requirements per unit of body weight gain are not understood. Consequently, effective criteria for selection for efficiency of growth are not available.

SHEEP

The ability of sheep to produce two or more lambs at a time, to lamb more often than once a year and to produce a desirable carcass on forage alone give them special advantages for utilization of range and forage resources.

The sheep industry in the United States is diverse in types of production systems, but there are broad geographic areas where operations are similar. In general the eastern two-thirds of the country is characterized by small flocks where sheep usually represent a minor part of the farm enterprise. The majority of U.S. sheep are located in the western one-third of the country. Production systems range from untended sheep in fenced pastures in Texas and New Mexico to herded sheep on unfenced ranges in the intermountain states. Along rivers and in farming areas of the west, small flocks exist under more intensive management systems with cultivated pastures, by-products, harvested forages and grain.

The major source of income from sheep in the U.S. is lambs for meat, with wool as a secondary product. Crossbreeding with rams from specialized meat type breeds to produce market lambs is a common practice. Size and growth rate of these sire breeds have been increased markedly by selection. Wool and growth characteristics are used in selection of rams in the range breeds of sheep used as dams in crossing programs. Little effective selection has been practiced as yet for fertility characters, but the importation of the highly prolific Finnsheep has made consistent production of multiple births possible. A few producers are now introducing 25 or 50% of inheritance from this breed to increase prolificacy and decrease age at puberty. Lambing rate can be increased by selection and crossing of domestic breeds and by improved nutrition and management. Lambing once a year is the norm in extensive operations, but there is interest in accelerated lambing where feed supply permits. Technology is improving for ovulation control, artificial insemination, embryo transfer, induced parturition and complete confinement systems, but it is not used, primarily because of economic factors such as labor costs.

Major problems of the U.S. sheep industry include low lambing rate, restricted breeding season, perinatal lamb losses, losses due to predatory animals, and, cost and availability of labor. Carcasses with a high percentage of fat, as a result of both genetic potential and producer decisions on time of marketing, are a problem at times. In addition, decreasing sheep numbers lead to closing of processing facilities, loss of competition in marketing and lack of development of new animal health products and other products useful in sheep production.

POULTRY

Genetic progress in increasing efficiency of egg and poultry meat production has been substantial. In recent years, however, progress has been especially slow for traits that are only slightly influenced by inheritance and influenced to a large extent by environment.

There is a rapidly developing shortage of useful germ plasm from which to develop new and better hybrids and strain crosses. To solve this problem, we need to develop and maintain diverse genetic stocks for adaptation to varying demands in market requirements.

Considerable improvement is needed in selection for disease resistance. Research with viruses and antigen-antibody interrelationships between now and the year 2000 should provide valuable tools for selecting stocks with resistance to disease.

Because the egg is the food product from egg type chickens, reproductive efficiency is of great importance. Currently good egg producing hens lay 250 eggs per year. The ultimate goal should be an egg a day or 365 eggs per year. There is still considerable potential for increasing egg production in optimum environments. However, under current environments used by industry, this 250 eggs per year has not changed for a number of years. Thus, there is a need to develop birds that can increase egg production in the environments that are thought to be most efficient.

While selection still makes progress in growth rate in broilers, responses to selection are becoming smaller and smaller. Artificial insemination of broiler breeding stock in cages offers a potentially significant means to achieve a rapid increase in body weight. By intensive selection on the male line, .27 kg (.12 lb) per generation may be possible. However, selection for body weight alone results in higher percentages of body fat and often results in leg problems. An index of selection needs to be developed that allows continual improvements in feed efficiency and growth rate while providing the ratio of fat to protein demanded by the market.

In broiler and turkey meat production, reproductive efficiency is important not only for increasing the number of chicks or poults per hen housed but also because male and female sex hormones are important growth regulators and determinants of carcass composition and quality. As broiler and turkey breeders have bred birds to mature earlier, grow faster and convert feed into product more efficiently, the apparent negative genetic correlation with reproductive efficiency resulting in limited egg production, fertility and hatchability has become an increasingly important biological barrier to further improvements. More basic knowledge concerning reproduction and growth is needed to improve reproductive efficiency of broiler and turkey hens without reducing growth rate and feed efficiency.

Hatchability of broiler and turkey eggs is much lower than that of eggs from layer chickens (broilers, 82%; turkeys, 80%; layers, 90%). Causes of embryonic mortality must be identified and approaches to improved hatchability developed if we are to improve hatchability of broiler and turkey eggs.

A multidisciplinary approach is needed in the next few years to improve artificial insemination with semen stored for extended time for both broiler chickens and turkeys.

Although much has been done on lighting programs for egg production and growing poultry, we still do not understand the physiological mechanisms and how they are affected by genetics. Intermittent light schemes in which there are 14-16 hr from the first to last light period appear to offer potential for improvements in feed conversion for egg production and growth of both broilers and turkeys while saving considerable electricity.

REPRODUCTIVE EFFICIENCY

Reproductive inefficiency is one of the most costly and production-limiting problems facing the animal industry. Reproduction can be influenced by genetics, management, nutrition, disease, and environment. Although problems are different in each species, the greatest opportunity to improve reproductive efficiency of farm animals lies in development of basic research information that will permit us to control or alter specific reproductive events and reduce postnatal losses. Once defined, the new information must be available to producers and used.

ESTRUS, OVULATION AND FERTILIZATION

Current Status

Animals cannot reproduce until they reach sexual maturity (puberty). Age at puberty is affected by genotype, plane of nutrition, social interaction and season of birth. Fertility usually is limited at first estrus and ovulation. Establishment of cycles appears to be physiologically similar at puberty, in the period after giving birth and in resumption of reproductive activity in seasonal breeders (sheep). Furthermore, factors influencing reproductive activity at these times appear to be similar, with variation being related to differences in requirements for growth and lactation.

Prolonged lack of sexual activity (anestrus) in dairy cattle decreases annual milk production and reduces the number of calves born. Similarly, anestrus decreases the crop of beef calves and lambs. Birth of young at times when forage is poor also has serious economic consequences.

Weight changes, both before and after calving, and body condition interact to affect the interval from calving to first estrus in suckled beef cows.

In sheep, breed, age and lactation appear to be more important than nutrition in determining the interval from lambing to first estrus. Most studies have dealt with effects of undernutrition following lambing. We do not know if undernutrition during late gestation will affect the interval from lambing to first estrus.

Sows do not come in heat while nursing their young. Length of the interval from weaning to estrus is of less importance to swine producers than is failure of sows to show estrus following weaning. Many attempts have been to decrease the interval from weaning to estrus by manipulating feed. Effects of feed level have not been consistent because of the influence of other factors such as age, temperature, day length and breed on the occurrence of estrus following weaning.

Information from allied fields may provide clues about regulation of ovulation in farm animals. In monkeys, a model has been proposed which appears to explain onset of puberty and maintenance of the normal menstrual cycle in primates. This model is based on hourly pulsatile releases (sudden surges) of luteinizing hormone releasing hormone (LHRH) from the hypothalamus (structure at base of the brain) each hour. If hourly release is maintained, follicle development is initiated, estrogen production increases and a positive estrogen effect causes a large ovulatory release of luteinizing hormone (LH). Preliminary evidence in cattle indicates that sensitivity of the cow to LHRH (GnRH) increases during the postcalving period. There also is evidence that progesterone, LH and estrogen patterns must be in harmony before estrous cycle activity is renewed.

In cattle, the lactation-suckling stimulus has a marked effect on the time from calving to resumption of ovarian cycles. Removal of calves at birth resulted in intervals of 25 and 28 days compared to 65 and 76 days, respectively, for suckled controls. Similarly, allowing calves to suckle only once daily significantly shortened the time to first estrus. Beef cows have longer intervals to first estrus and first ovulation than dairy cows because of the repeated suckling stimulus compared to twice daily milking.

Photoperiod (day length) is another factor which affects production and reproduction. In sheep, increasing day length leads to seasonal anestrus by affecting the pineal gland of the brain. The pineal hormones, melatonin and serotonin, appear to influence the hypothalamus and alter LHRH release sufficiently to disrupt the estrous cycle. Research is needed to establish such relationship, and whether systems of management could be developed to allow year-around breeding of sheep and goats.

Photoperiod is important in poultry. Chickens ovulate on a 26-hr cycle. They are stimulated by 14-16 hr of light to release LH and ovulate. Only a few hours of light per day are required providing it is pulsed at correct times. This finding provides an opportunity to decrease energy costs for lighting to maintain egg production by about 75%. Of greater potential significance is the opportunity to select genetic lines of birds to perform well under minimum light, and this could lead to daily egg laying. The present 26-hr cycle in the hen results in periodic absence of laying while the hen resets her biological clock.

Ovulation time is variable during estrus in sheep and swine and occurs after estrus in cattle. A test for time of ovulation or its control would permit insemination at an optimal time. Presently, successful artificial insemination depends on proper detection of estrus, a problem which has increased as dairy herds get larger. It also is a constraint to artificial breeding in range cattle and sheep. Understanding events leading to ovulation could lead to management recommendations which would help to: 1) ensure reinitiation of ovulation at the optimal time; and 2) control multiple ovulations consistent with uterine capabilities. The number of follicles developed to ovulate at a particular estrus may be set during the previous cycle. If this is true in domestic animals, controlled multiple ovulations could become possible (i.e., twinning in beef cattle, multiple births in sheep and cows).

At least 10% of eggs ovulated in livestock are not fertilized, even when mating or insemination is at the proper time. When timing is off, aging sperm and eggs may initiate fertilization, but the resulting embryo often dies.

In competitive fertilization, sperm from certain males compete much more favorably than others in fertilization. Unpublished data on chickens support the idea that there is selective transport of sperm cells by the female. There are unpublished reports that biochemical separation of sperm types may lead to higher fertility.

Proposed Research Approaches

Initiation of ovulation and pregnancy. A 12-month calving interval is important in cattle. Estrous cycles must resume by 40-50 days after calving to allow insemination to a new time. Studies will determine if the concepts derived from the monkey model apply to onset of estrus and ovulation at puberty, after calving and during the nonbreeding season (sheep). Application of these and other findings could lead to procedures for producing estrus and ovulation at the desired time.

Photoperiod alteration of ovulation in poultry. Studies are needed to determine how hens respond to various light-dark periods. Hormone profiles and ovulation patterns in genetically diverse hens exposed to multiple light-dark combinations could result in birds with a 24-hr cycle as in some other avian species.

Day length also regulates the breeding season in sheep and goats, and similar studies of light and endocrine relationships are needed to provide means for eliminating the seasonal breeding problem.

Development of ova (folliculogenesis). A long-range goal will be to pursue concepts now being developed as a result of observations in the rat. One concept emerging is that follicles which are destined to ovulate one cycle later are identified and prevented from undergoing degeneration. As understanding of this process emerges, more reliable control of ovulation rate should develop.

Automated detection of estrus. Large dairy units are becoming more mechanized, and some have microprocessing equipment. This offers an opportunity to resolve the problem of detecting estrus which has caused some dairymen to diminish their use of superior bulls available through artificial insemination. Several approaches to the problem are being studied and need to be continued. Variation in cow activity is closely related to estrus, and small electronic devices should be developed to follow each cow's activity and signal increases associated with estrus. Such equipment also could be adapted to beef cattle and other species to help make artificial insemination practical.

EMBRYO SURVIVAL AND DEVELOPMENT

Current Status

Embryo mortality is a major cause of pregnancy wastage in swine, cattle and sheep. In swine, sows ovulate an average of 17 ova and 95% are fertilized, but an average of only 9.4 live pigs are born and 7.2 piglets are weaned. In cattle and sheep, fertilization rate is also high (90%), but embryonic death claims about 20% of the embryos.

Embryo mortality is poorly understood and deserves considerable attention because of its impact on production of offspring in all animals. Pregnancy still is maintained in swine after loss of 25-40% of embryos, but death of embryos in cows and ewes usually leads to termination of pregnancy and return to estrus. Mechanisms causing embryo death may be similar, but consequences in swine (smaller litters) and cattle and sheep (return to estrus) have different effects.

Hatchability of eggs from broilers (82%), turkeys (80%), and laying hens (90%) indicate that embryo death adversely affects production in poultry as well as mammals.

Embryo mortality may be influenced by: 1) genetic defects; 2) immunological incompatibility between the embryo and dam; 3) elevated body temperature (mammals) or environmental temperature (poultry); 4) failure of embryos to develop adequately and provide appropriate hormonal signals to the dam; and 5) numerous other factors which may be directly or indirectly related to nutrition, social factors or physical stress.

Genetic defects account for 10% of embryo deaths. Gross chromosomal abnormalities are most frequently cited as a cause of this mortality.

Quantitative genetic variation among dams accounts for much more embryo mortality than genetic factors carried by embryos in mice. Although genes responsible for this variation are largely recessive, it is possible by selection to develop strains with high embryo survival. This approach has not been tried with livestock. Embryo transfer experiments on mice with high and low survival rates suggest that the basis for high embryo death rates may be lack of synchrony between uterus and embryo.

Competitive mating systems (mixed semen from two or more males) result in offspring of one male at a higher frequency than would be expected on the basis of random fertilization.

Reproductive performance of swine, sheep and cattle is depressed during hot weather. Part of this effect is on the male (lowered sperm production and increased numbers of abnormal sperm), and part is due to increased embryo death rate. The mechanisms responsible for increased embryo mortality are not known. However, reduced blood flow to the uterus and decreased rate of synthesis of cellular components of "heat stressed" embryos have been implicated. Successful establishment of pregnancy and embryo survival is dependent upon actions of hormones from the very young embryo, corpus luteum and possibly the pitutiary on the uterine lining (endometrium). The endometrium responds by secreting factors which promote embryo survival. Research must be initiated to determine the precise details of embryo development and pregnancy maintenance. Implantation (attachment of placental membranes to the uterine wall) and establishment of a functional fetal-placental unit are essential for continuation of a successful pregnancy. Placental transport of nutrients appears to be influenced by steroid (estrogens and progestins) and protein (placental lactogen) hormones of placental and ovarian origin. Estrogens also stimulate increased uterine blood flow and nutrient delivery. In addition, placental hormones appear to affect mammary development to ensure uninterrupted nutrition of the fetus after birth.

Proposed Research Approaches

Genetic compatability. The importance of incompatability between embryo and uterus as a cause of embryo death needs to be evaluated. This can be approached by competitive mating systems using semen mixed from several males to inseminate females. Genetic markers can be used to study interactions between sire and dam lines affecting embryo survival. If such interactions are significant, cell surface antigens (histocompatibility antigens) should be examined as one approach to explain why.

Maternal recognition of pregnancy signals. Embryonic signals required for establishment of pregnancy require considerable study since failure of embryos to provide these signals results in termination of pregnancy or death of "deficient" embryos. Also an understanding of these phenomena may provide the basis for diagnosis of pregnancy (cattle, sheep and swine), predicting litter size (sheep and swine) or determining placental competence.

Studies with pigs should focus on several points. First, early pig embryos produce forms of estrogen which appear in the blood as early as day 11 of pregnancy. These estrogens (estradiol, estrone and estriol) provide the recognition signal of pregnancy in swine. Measurements should be focused on estrone sulfate of plasma and estriol sulfate. The estrone sulfate is the major estrogen, but estriol sulfate is unique to the pig during pregnancy. Studies are needed to determine when hormones best can be measured in blood to test for pregnancy and to predict litter size. Sows having litters of less than six or seven piglets could be culled. Research should focus on day-11 embryos to determine whether failure to initiate estrogen production is a primary cause of embryo death and whether estrogen or estrogen:progesterone therapy could increase survival of embryos. The role of estrogens of embryo origin in stimulating secretion of uterine proteins and prostaglandins may be critical.

Similar data are not available to indicate how sheep and cow embryos establish pregnancy. Information on this point is critical because embryos that develop normally to day 20-30 usually are carried to term. In pregnant cows, early embryos produce 5 beta-reduced progestins while the endometrium produces only 5 alpha-reduced progestins. These 5 alpha-reduced progestins should be evaluated for their potential as the basis for pregnancy diagnosis.

The pregnant uterus of the cow, sheep and pig appears to be engaged actively in metabolizing steroid hormones. These steroids may serve as immediate precursors for estrogen production by the early embryos. Therefore, we must determine if steroid metabolism by the uterus is essential to establishment of pregnancy.

The potential benefits of this research are to: 1) determine mechanisms whereby pregnancy is established; 2) provide means of developing methods for pregnancy diagnosis; 3) provide methods to assess litter size in pigs and sheep and in cattle with multiple embryos; and 4) determine if hormones can be used effectively to increase embryo survival.

Furthermore, effects of elevated temperature, social stress, adverse nutritional status and other factors leading to increased embryo death may be evaluated directly by measuring effects on the embryo's ability to signal the uterus to establish pregnancy. Therapeutic means of dealing with this problem then may be developed.

Placental function and embryo survival. The placenta is essential for transporting nutrients to the fetus. However, little is known about mechanisms whereby these nutrients are transported, the role of the fetal fluid compartments in nutrient storage, or hormonal regulation of nutrient transport. Research needs to determine effects of estrogens and progestins of placental and ovarian origin and placental lactogen on placental transport of nutrients, uterine blood flow and physiology of the dam's cardiovascular, pulmonary, digestive and endocrine systems. We need to know how and when these hormones act and how they are influenced by nutrition, temperature and other factors that may affect fetal survival. For example, the amount of fetal tissue which can be supported by a female may be limited by her ability to perfuse the uterus with blood because of an upper limit on the percent of cardiac output which can be diverted to the pregnant uterus.

The purpose of this research is to determine the hormones essential to development of the fetal-placental unit. Selection of sire and dam lines producing placentae with the right endocrine activity can be used to enhance fetal survival. The sire and sex of the fetus appear to influence placental endocrine function, fetal weight and mammary gland function. These endocrine reflections of the genotype of the fetus may lead to genetic selection to improve fetal well-being and postnatal survival.

Selection for embryo survival. Genetic approaches to decreasing embryo mortality include: 1) characterization of breeds (and selected or inbred lines when available) for extent and timing of prenatal mortality; 2) evaluation of crosses of these breeds and lines to determine the extent of heterosis and mode of gene action; and 3) selection for increased embryo survival, probably in conjunction with inbreeding. Breeding programs to increase embryo survival could be developed as a result of such studies. These approaches are suggested by results of a few experiments with mice. More studies with laboratory animals should be tried before extensive experiments with livestock are started. However, the success in producing mice with embryo survival rates of 80-90% as opposed to 55-65% in pigs suggests that this approach should be tried. The economic benefit to the swine industry could be enormous. Strains differing genetically in embryo survival also could provide a valuable resource for explaining physiological mechanisms involved in embryo survival in other species.

PARTURITION AND PERINATAL SURVIVAL

Current Status

In beef cows nutrient intake is critical from 60 days before calving until next conception. Inadequate nutrition during this period can delay postcalving estrus, reduce conception rates and inhibit milk production, all of which produce fewer and lighter calves. Reducing nutrient intake during the last trimester of gestation reduces birth weight of calves, but this reduction in weight has no effect on the incidence of calving difficulty.

Some think that overfeeding the dairy cow during the last 60 days of pregnancy causes reproductive problems during the next lactation, but experimental evidence is lacking.

Low birth rates are detrimental to survival in several species including sheep, humans and cattle. Reduction in total nutrient intake during the last third of gestation increases losses of newborn calves. Inadequate protein intake during the last 4 months of gestation increased neonatal death (70% vs 0%) and produced some calves with symptoms of "weak calf syndrome". Incidence of "weak calf syndrome" was related inversely to the amount of crude protein consumed by herds.

Some 72% of anatomically normal calves lost at birth were lost from dystocia (difficult or prolonged parturition). Dystocia usually results from disproportion in size of dam and fetus, with large calves being the most important cause. Other important factors include pelvic area and precalving weight of the dam and sex of the calf. In addition to calf losses, dystocia also decreased conception rates and percentage of cows detected in estrus during the first 45 days of the breeding season. Productivity is decreased by reduction in number of calves and weaning weight of calves in the subsequent year. Precalving nutrition can affect birth weight but has little effect on incidence of dystocia unless overfeeding produces obesity and excessive fat is deposited in the birth canal. Induction of parturition with hormones can reduce birth weight, but contrary to expectation, there is no effect on incidence of dystocia. When all factors are considered, selection for shorter gestation length appears a way to reduce calving difficulty.

Proposed Research Approaches

To minimize incidence of calving difficulty by shortening gestation length would help to reduce calf losses at birth. Gestation length in cattle is affected substantially by genetics. Selection for shorter gestation period could be effective and should be studied. Attention should be directed to relationships between gestation period and other production traits.

Fifteen percent of calf losses are attributed to calfhood diseases, and susceptibility to disease is increased if nutrient intake by the dam is restricted during fetal development in late gestation. Mechanisms by which this affects subsequent suceptibility to disease need to be ascertained.

Stillbirths cause losses of approximately 10% in swine. Incomplete cervical opening appears to be one cause of stillbirth in swine. When farrowing time exceeds 6-8 hr, or when the interval between births is as great as 45 min, the incidence of stillbirth increases. Relaxin from ovaries appears to cause cervical opening, and there is reason to believe that fetal death may be caused by inappropriate timing of prostaglandin or relaxin release. Specific studies to determine the relative role of each of these hormones and mechanisms that control their release are necessary to understand this important problem.

The potential benefits of reducing losses at or near birth are large; in beef cattle alone, calf losses remove at least 3.7 million animals annually.

EMBRYO TECHNOLOGY

Current Status

The development in animal breeding often rated the greatest impact on animal production is artificial insemination. This tool effectively has utilized the genetic contribution of superior sires to increase milk production 50% in the last 25 years. Tools allowing similar genetic reach in female selection have not been available. Embryo production, sexing, storage, multiplication and transfer technology could accelerate genetic improvement of livestock by providing large numbers of genetically superior offspring from individual females.

The base for application has been developed and is being applied in a rapidly growing embryo transfer industry. Approximately 10,000 bovine pregnancies were produced in North America in 1978 by embryo transfer and approximately 18,000, or 80% more, in 1979. The commercial value was about $20 million in 1979. Although these figures are impressive, the cost of this technology is so high that application is limited to a small fraction of the population (about 1 in 10,000 cows). There is need for simpler procedures so that costs can be reduced.

Currently, it is possible to recover six to seven normal embryos per superovulated cow and obtain three to four calves. Each embryo has about 60% chance of developing into a calf. Donors can be superovulated every 2-3 months. Thus, 15-18 calves could result per donor per year. There are three problems. First, there is tremendous variation from donor to donor, and it is impossible to predict which donors will do well; second, there is little information on repeated superovulation, especially after three or four times (some cows may become refractory); and third, cows are out of production when used as donors. Ability to sex embryos and store them for later use is especially important. Freezing kills about half of the embryos.

Proposed Research Approaches

Harvesting ova. Each ovary is filled with hundreds of thousands of potential ova (eggs) during fetal life, but production of new ova ceases before a female is born. Because there is constant degeneration of ova in the ovary, older animals have fewer ova. In some species about 90% of ova degenerate before puberty. Considerably more than 99% of ova degenerate and fewer than 1% ovulate during a female's life. There is tremendous wastage. Wastage of sperm is considerable, but sperm are produced constantly, and the nature of the problem is different.

The most common scheme for increasing the harvest of ova is superovulation. Hormones are injected to increase the number of ovulations by a factor of up to 50, but averaging 3-8, depending on the species.

Although superovulation is generally successful, it is exceptionally unreliable for an individual. In cattle, for example, the number of normal ova produced per cow varies from 0-30 or more. To date, methods of superovulation have been developed primarily by empirically changing the kind, dose and timing of administration of hormones.

It is possible to harvest ova by mincing the ovary and adding enzymes. However, we do not know how to fertilize them and develop them into normal young.

One of the great unresolved processes in biology is that by which the ovary metes out ova in a controlled way over the reproductive life of females.

Future research should be in two areas: 1) understanding the physiology of ovum maturation, development, and ovulation so that more rational superovulation regimens can be developed; 2) study of the feedback relationships among ovary, pituitary and hypothalmus of the brain; and 3) studies to obtain ova from the ovary by enzymatic digestion and to grow and develop such ova in the laboratory.

Embryo recovery and transfer. Although embryo transfer has developed in cattle, there is still much to be done to improve procedures, especially nonsurgical transfer to recipient cows.

Embryo recovery and transfer can be applied successfully to sheep, goats, and swine as well as cattle. A major constraint, however, is that recovery and transfer must be done surgically in these species. This is expensive and may damage animals. It would be appropriate to develop nonsurgical methods for these species, but this is not as important as research to develop much better refinements of techniques for use in cattle.

Maturation, fertilization and culture of ova outside the body. Recovery of large numbers of ova from ovaries of a cow requires that the recovered ova undergo maturation, fertilization and development in the laboratory to a stage where the embryos can be stored or transferred. Present technology allows only slight success in completion of each of these three steps. However, the three have not been applied together in domestic animals, and much improvement is needed in each.

Maturation of ova is arrested from early fetal life until a follicle containing a specific ovum is selected for ovulation. Ova will resume nuclear meiosis and undergo maturation upon removal from the ovarian follicle or within follicles when inhibiting substances are removed by natural or artificial administration of luteinizing hormone to the follicle. Nuclear maturation alone is not sufficient for fertile ova. It must be accompanied by maturation of the cytoplasm and loosening of follicle cells around the ovum. This change in the loosening of follicular cells is induced by follicle stimulating hormone.

While some ova matured outside the body have been fertile, considerable work is needed to determine how maturation is controlled if we are to be able to make a real breakthrough in the harvest and use of thousands of ova from a single female.

Fertilization of ova outside the body. Ova harvested in large numbers and matured in the laboratory should be fertilized there because fertilization in a living female requires surgical deposition and removal. Also, working out mechanisms of fertilization in the laboratory, where the process can be directed, will provide an understanding of the natural process.

Techniques for fertilization outside the body have been developed for the hamster, mouse, rat, rabbit, guinea pig, dog, cat, human and nonhuman primates and recently the cow. These techniques operate with various efficiencies and stages of success in each species. Live offspring have resulted from such fertilization only in the mouse, rat, rabbit and human. In domestic animals, various pieces of the needed technology have been developed.

Much research is needed on fertilization in the laboratory. Cattle experiments failed to determine if the fertilized ova would develop into live offspring. The rate of fertilization was less than perfect, and reactions in the ovum were not entirely normal. Research is needed to: 1) perfect the technology for fertilization of cow ova, including study of normal physiological events such as capacitation of sperm, acrosome reaction of sperm and cortical granule reaction in the fertilized egg; 2) develop fertilization methods for sheep and swine; and 3) develop technology to produce normal young from ova fertilized outside the body.

The tools needed to perfect methods for fertilization outside the body and to develop an understanding of the fertilization process are now available, (e.g., the ability to accomplish fertilization of one egg by one sperm, and the ability to separate and identify several components of the fertilization process). Besides the obvious application of increasing numbers of fertilized ova available for embryo transfer, fertilization under laboratory conditions can be useful in studying such things as variation in fertility among sperm.

Culture and development of embryos. New fertilized ova are not at a developmental stage compatible with present technology for frozen storage, sex determination or nonsurgical transfer. To be useful for embryo transfer, ova fertilized outside the body must undergo development to a stage equivalent to that occurring approximately 8 days after fertilization in cattle. With present techniques the proportion of fertilized ova which develop by culture to the 8-day stage is small. In general, the more stages of development the embryos are cultured through, the lower their survival. This is true whether the stages are early or late in embryo development. Research is needed to develop culture systems which are compatible with maintenance and development of the embryo from fertilized ovum to early embryo stage. Finally, successful development of ovarian ova will require a combination of highly efficient methods for maturation and fertilization of the ova outside the body and maturation of most to early embryo stage.

Cloning and nuclear transfer. Sexual reproduction is a game of chance because eggs and sperm contain only a random half of genes of the respective parents. The nearly infinite number of possible combinations of genes produces such a variety of progeny that some of them are likely to survive and reproduce. Others will die without producing offspring.

Animal breeders try to improve succeeding generations of animals by removing some of the chance associated with sexual reproduction. Examples of such efforts include formation of breed societies, progeny testing schemes, amplification of the reproduction of outstanding individuals by artificial insemination and embryo transfer, and establishment of inbred lines to capitalize on uniformity.

Another consequence of the great variation achieved by sexual reproduction is that certain outstanding individuals are produced that satisfy the needs and desires of people to a much greater degree than the average. The curse of sexual reproduction is that progeny of outstanding animals usually are less outstanding than their parents although they are generally above average. This regression toward the average stems partly from the fact that individuals are a product of genetics and environment and partly from chance when an offspring receives a random half of the genes of each parent.

It is only within the last century that people have had enough information even to dream that they someday could apply technology to produce genetically identical copies of outstanding animals. The ability to clone embryos through several generations of identical clones would provide a tool for accelerating genetic change which is potentially more powerful and effective than artifical insemination. For example, the mating of a proven bull and an oustanding superovulated cow will produce approximately eight embryos. If at the four cell stage individual cells are removed from the embryo and each again cultured to four cells, 16 embryos would result. If each of these could be broken into four cells and this process repeated six times, 4,096 identical embryos would be produced. Culture of these embryos to the stage where they could be sexed and frozen would provide over 4,000 embryos of one sex for transfer to recipients. This and other clonal lines from the same superior mating then could be characterized for a production trait such as milk production by transfer of a number of frozen embryos from each clonal line to recipient cows to produce heifers to be production tested to evaluate the clone. Once characterized, the embryos from superior clonal lines would be available to create entire herds with the capacity for high production. This also would be a powerful tool for research and genetic improvement.

An alternative and perhaps more immediately feasible approach to the cloning of embryos would be separation of embryo cells at the four or eight cell stage before their developmental fate is committed, followed by transfer of their nuclei to fertilized eggs before first cell division. Results of recent research suggest the long range possibility for development and application of cloning and nuclear transfer technology. However, a vast amount of research is needed to develop the necessary techniques and to determine the limitations of the proposed procedures. In the simplest sense cloning research starts with the production of twins.

A final method of producing genetically identical mammals is by inserting the nucleus of a body cell into a fertilized egg and then removing or destroying the original genetic complement of the embryo. This already has been accomplished in the mouse. The procedures were not very different from methods used in amphibia.

Embryos from mice and rabbits have been separated into individual cells at the 2, 4 and 8-cell stages, then cultured as single cells back through 2, 4 and 8 cells to the early embryo stage. In general, more survived to become blastocysts when cells were separated at the 2-cell stage than at later stages. At best, one-third of the cells at the 4-cell stage became embryos. Considerably more research is needed to map out the mechanisms important in embryo manipulation.

Nuclear transplantation. Replacement of male and female pronuclei in the fertilized mouse egg with a nucleus from a cell of a mouse embryo results in a viable embryo which can develop into a live offspring. Nuclei from cells of an early embryo could be used in nuclear transplantation to form about 100 clones. The following questions concerning nuclear transfer need to be answered:
1) Can live offspring be developed in domestic animals, especially cattle, from embryo nuclear transplantation? 2) Is the stage of the cell cycle of the donor cells at the time of nuclear transfer critical for continued development of the newly formed clone? 3) Is the developmental stage of the pronuclear recipient egg critical for future development of the clone?

Several other exciting technologies of relevance are parthenogenesis, production of chimeras and development of inbred lines. Parthenogenesis is the production of young without fertilization. It occurs naturally in all classes of vertebrates except mammals. Of particular interest is a strain of turkeys with this characteristic. Only male turkeys are produced in this way, but in mammals only females would be produced. Although parthenogenetic mammalian embryos develop part way through gestation, they do not go to term. Solving this problem would make parthenogenesis extremely useful for some aspects of genetic engineering.

Chimeras are animals produced by mixing cells from two or more embryos. Chimeras are extremely valuable for certain experiments. For example, some have certain tissues such as liver derived from one embryo and other tissues such as heart from a second embryo. Also, when parthenogenetic embryos were mixed with normal embryos, some of the tissues of the resulting adults were derived from the parthenogenetic embryo. Another example is that certain cancer cells can be mixed with embryos resulting in chimeras with perfectly normal tissue derived from the cancer cells; that is, cancer cells were changed to normal cells.

Production of inbred animals deserves special mention. Such animals are essentially genetically identical with others in the line although they may perform poorly due to inbreeding. If two such lines are crossed, however, hybrid vigor results and the animals from such a cross are also genetically identical. Thus, the chance element of sexual reproduction is controlled resulting in a uniform predictable product. This is of special value for controlling genetic variation in experimental situations.

These are but a few of the questions requiring answers if successful methods for cloning by embryo cell segregation or by nuclear transfer are to be successful. Some tools and basic knowledge to undertake such research are already available.

Storage of eggs, sperm and embryos. It is often necessary to store sperm, eggs or embryos for several minutes to several decades. Usually the objective is to maintain these cells in the same state as they were prior to storage although in some cases it may be desirable to have embryos continue development rather than to remain quiescent. Because of varying requirements, storage may be as simple as placing a test tube of embryos on a laboratory bench or as complicated as sequential dilution of semen in a series of complex fluids followed by freezing to liquid nitrogen temperature in a computer controlled apparatus.

Successful and inexpensive methods are available for storage of reproductive cells of most species up to several hours. In many cases storage also can be successful for a day or two, but techniques are more complex and sometimes less effective than for shorter storage. For storage more than several days, the method of choice is freezing to liquid nitrogen temperature.

Adequate methods are already available for freezing bull and turkey semen. However, there is room for improvement; about half of the sperm are killed in the process, and others damaged so that three times more frozen sperm are required per insemination to obtain fertility comparable to unfrozen semen. Furthermore, semen from some individual animals cannot be frozen successfully at all.

In general, good methods for freezing semen of sheep, goats, swine and chickens are not available. Recent research led to considerable improvements with these species, but much remains to be done.

Relatively little research has been done with freezing ova, but this may be feasible. There has been considerable experimentation on freezing embryos. This is not yet possible with swine, but embryos from sheep, goats and cattle can be frozen although at a cost of killing half of them. Even with 50% loss, there are already certain situations where this technology is being used commercially. There is great demand for frozen bovine embryos, and this technology will be used widely if losses due to freezing can be decreased from 50-20%.

The major constraints in freezing reproductive cells are lack of basic knowledge concerning mechanisms of cell injury, how cryoprotectants work, and why cells from certain individual animals and species are so susceptible to freezing damage. Although strides have been great in developing a theoretical groundwork for freezing technology in recent years, much remains to be done so that we can move from the current phase of trial and error testing of infinite combinations of freezing rates, thawing rates, cryoprotectants, buffers, etc., to a more directable strategy.

Further research should be of two types: 1) basic research on how and why cells respond as they do to cryoprotectants, freezing and thawing; and 2) continuation of the empirical approach using clues from basic research to determine which factors should be studied.

The benefits of storing sperm, oocytes and embryos at liquid nitrogen temperatures are enormous, both from commercial and research perspectives. Such storage greatly increases flexibility because the subjects need not be synchronized precisely in time and space. This leads to tremendous savings in manpower and feed costs.

<u>Sex selection: sperm fractionation</u>. There is no repeatably successful procedure in any species for separating sperm that determine male offspring (Y-bearing sperm) from those that determine female offspring (X-bearing sperm). Efforts have appeared promising in the past (for review, see Sex Ratio at Birth - Prospects for Control, Kiddy and Hafs, 1970) and will continue to excite scientific inquiry. However, in the most recent review of literature on sex determination and differentiation R. E. Short (Hawk, 1979) concludes, "And the tantalizing goal of artificial regulation of the primary sex ratio remains a mirage on the horizon."

In many birds and mammals there is a distinct difference between the sexes in morphology of chromosomes. In mammals, females have two X chromosomes and males one X and one Y chromosome. Theoretically, one needs to observe the chromosomes in only one cell of an animal or embryo to determine its sex. A review of this methodology was provided by Hare and Betteridge (1978).

The current state of the art is such that embryos can be divided with greater than 95% accuracy, into three approximately equal groups by sex chromosomes: male, female and unknown. Pregnancy rates following transfer of sexed embryos is reduced slightly. Nearly a full day is consumed from the time an embryo is biopsied until sex has been determined. Before this technology can be applied commercially, it must be made simple, fast,inexpensive, reliable, and safe for embryos.

A more sophisticated method of sexing is to observe whether male- or female-specific proteins (or RNA) are being produced. One such protein produced by all male cells and not by female cells is the H-Y antigen (Ohno, 1979). This protein appears to be coded by DNA on the Y chromosome, and thousands of copies are inserted into each male cell membrane. Because so many molecules are in a cell, detection of the protein might be a more sensitive assay than direct observation of the DNA, of which there is only one copy. The presence of the protein is detected by antibodies. A method for sexing embryos with anti-H-Y antibody and a fluorescence technique is in preliminary stages of development. It may enable sex determination without the necessity for killing or biopsying the embryo although death may occur to a few cells as a result.

There are only hints of successful approaches to separating X- and Y-bearing sperm of domestic animals. The situation for sexing embryos is much more promising, but the technology is still too clumsy for routine commercial application.

The attractiveness of sex control by separation of X- and Y-bearing sperm dictates that further research effort should be devoted to this problem as new leads become available. Experiments should continue on improving the sexing of embryos by sex chromosome morphology. Probably the major emphasis in experimentation on sexing should go to detection of gene products of the Y chromosome such as H-Y antigen. Production of monoclonal antibodies to H-Y and use of such antibodies to sex embryos are examples of what might be done.

Twinning by embryo transfer. Nearly 80% of the cows in North America are beef cows with sole function to produce a calf each year. Approximately 70% of the nutrients consumed by each beef cow are for her own maintenance, whereas only about 30% are for growth and maintenance of the calf during pregnancy and lactation.

The greatest limitation in converting forage to meat is the low reproductive rate of cattle. The conversion of feed into meat would be more efficient if cows had twins. The approach offering the most hope for greatly improved meat producing efficiency in cattle is the induction of twinning by embryo transfer, and its successful development and implementation would be a major breakthrough in the technology of food production.

Currently, embryo transfer is the most effective method of inducing twin pregnancies in cattle. When twinning is induced, pregnancy rates range between 67 and 91%. Between 27 and 75% of pregnant cows deliver twins. The most cost-efficient method is to transfer an embryo to a heifer that was inseminated at estrus. On the other hand, pregnancy and twinning rates were higher when two embryos were transferred to cows that had not been inseminated. This was attributed to low fertility following artificial insemination in the group that received a single embryo. Hormonal induction of twinning, a modification of superovulation, has not been developed into a reliable procedure. Even if the technology were available, it would not be the best method since both ovulations might be on the same ovary.

Impact of Research

The potential impacts and long range payoff for the research described as embryo technology is enormous. With current technology and success rates, embryo transfer could lead to an increase in rate of genetic improvement of from 10% to nearly 100% depending principally on current intensity of sire selection and on the trait(s) in question. Van Vleck (1980) estimated possible effects of alternative technologies on rate of genetic improvement in milk yield in dairy cattle in kilograms/cow/year: regular AI 100; sexed semen 115; embryo transfer 158; sexed semen and embryo transfer 166. Van Vleck (1980) also showed that cloning, if available, could be used to increase accuracy of genetic evaluation of dairy cattle over that possible from sister and progeny tests. The accuracy is doubled under conditions giving maximum advantage to cloning and about one-third to one-half higher under more realistic assumptions.

These technologies have important applications in addition to increasing rate of genetic improvement. Sexed semen has economic implications. For example, based on the estimate that a male calf produces 10% more lean beef than a female on a given amount of feed, a reduction in the proportion of females from the present 50% to the approximately 20% needed to maintain the beef cattle breeding herd, would increase total beef production per unit of feed by 3%.

Embryo transfer also has important economic applications in multiplying rare breeds, strains, or individuals, and, in conjunction with use of frozen embryos, in avoiding the high costs of animal shipment and quarantine in transferring livestock breeds to new environments for either production or research.

Some of the most important applications of these technologies are to research in other disciplines of animal science. Identical twins have been used to provide genetic uniformity for nutrition and other research, but their use is severely limited by their scarcity and the cost of proving identity. The availability of clones would reduce the numbers of animals needed for specified precision by an estimated 20-50% in many livestock experiments.

Embryo transfer is a powerful tool for separating contributions of maternal and offspring genes to prenatal survival and growth in mammals. It has been used effectively to measure uterine capacity and genetic control of the initiation of parturition. The technique has many potential applications to development of new knowledge of uterine function and interactions between the embryo and uterus.

POTENTIAL ANNUAL BENEFITS

Economic benefits from increasing reproductive efficiency in food producing animals are potentially enormous. Benefits by species from application of technology in 1) estrus, ovulation and fertilization; 2) embryo survival and development; and 3) parturition and perinatal survival have been calculated:

Beef cattle Calf crop, as a percentage of cows exposed to bulls, could be increased from 75-90% by increased pregnancy rate and reduced perinatal mortality. For the U.S. cow herd, 1980, of 39 million head, to increase the beef calf crop from 75-90% with maintenance requirements per cow of 1,140 kg total digestible nutrients (TDN), feed costs of $80/ton TDN (i.e., $60/ton hay for 4 months and $30/ton pasture dry basis for 8 months), and nonfeed costs for cow maintenance $50/cow; benefits could be a reduced cow herd to 6.5 million cows and reduced energy requirements with savings per cow of $150. Total savings could be $975 million per year.

Dairy cattle. The dairy cow population is about 10.8 million head. By controlling onset of estrus and ovulation earlier in the postpartum period through approaches described and by improving automation of estrus detection so that effectiveness of detection is maintained with increasing herd size, calving interval could be reduced from 13.5 to 12 months. The poorest 10% of cows could be marketed, and the remaining 90% would provide the current milk supply at an annual savings in estimated feed costs of $332 million from reduced energy required and in labor costs of $108 million, total $440 million per year.

Swine. Application of new technology to increase embryo survival, perinatal survival of piglets, and frequency of farrowing should increase pigs weaned per sow per year from about 13-20. With this increase in efficiency, present pork production could be maintained at an estimated savings in feed and management of about $273 million and required energy reduced if the number of sows were reduced from 6.4 million to 4.7 million head. This is for $120/sow/year for feed and $40/sow/year in nonfeed costs. Potential benefits from development of more fertile sire and dam lines, early culling of females predicted to have small litters, and increased use of estrous synchronization and artificial insemination cannot be assessed at this time.

Sheep. Improvements in lambing rate by increasing both number of lambs per birth and lambing frequency could increase lambing rate by 24%. This is in line with the USDA National Research Program for Sheep which estimated that application of findings from proposed research would increase total gain in efficiency 13% for the sheep industry. Total dollar value of a 13% increase could approximate $8.5 million annually.

Poultry (egg production). Hens ovulate on a light sensitive 26-hr time clock. By altering ratios of light to dark it should be possible to increase egg production by 10%. Thus, the present production of eggs could be maintained with a 10% reduction in number of hens at an estimated annual savings of $132 million. This derives from 292 million laying hens with 10% increase in productivity per hen and maintenance costs of $5 per hen providing a reduced number of laying hens to 26.5 million and reduced energy requirements for artificial light.

Thus, the total estimated annual savings in nutrient supply and energy available for redistribution is substantial. If, in the next decade, only half of these goals were achieved, savings annually at present prices are nearly $900,000,000.

PROTEIN PRODUCTION AND SYNTHESIS

GROWTH, LACTATION, AND EGG AND WOOL PRODUCTION

Current Status

Research has produced marked increases in both the rate and efficiency of production of animal protein for human food and fiber. In the 30 years from 1945 to 1975 the average milk production per dairy cow per year increased from approximately 2,000-4,700 kilograms. In 1945 it required approximately 26.6 million dairy cows to produce 54 billion kg of milk. In 1975 only 11.6 million cows produced 55 billion kg of milk. The feed that would have been required for maintenance of the 15 million fewer cows not needed in 1975 is largely feed saved. If all of the nutrients required for maintenance of the 15 million cows came from corn, it would require approximately 40.5 billion liters of corn per year. This is an enormous amount of feed that would be consumed without producing any edible product. In addition there are large savings in labor and facilities required. The protein production per cow per year increased from approximately 65 kg in 1945 to 150 kg in 1975.

Growth and development of muscle likewise produce protein. In 1978, approximately 40 million cattle were slaughtered. The average length of time to produce a 454-kg steer or heifer has decreased by approximately 200 days during the last 30 years. The feed required for maintenance of 30.6 million 360-kg beef cattle for 200 days is equivalent to 29 billion liters of corn. There would likewise be a large saving from producing less fat and more lean meat in the beef carcass.

In 1975 pigs reached market weight (100 kg) at approximately 160 days while approximately 200 days were required in 1945. In addition, a carcass of a 100-kg pig in 1980 contains 14 kg less lard than that of a comparable pig in 1945. Approximately 70 million barrows and gilts are slaughtered in the U.S. each year. The maintenance requirement for 70 million 82-kg pigs for 40 days is equivalent to 3.7 billion liters of corn per year. In addition, the feed required to produce 95 million kg of fat is now being converted into edible meat.

Advances in genetics, nutrition and management in poultry have enabled the broiler industry to produce a 1.7-kg live broiler in 7 weeks instead of 12-14 weeks 20 years ago. The kilograms of feed required per kilogram of live broiler is near 2 compared to 4 kg in 1940.

The genetic increases in rates and efficiencies of protein production have resulted from selection and breeding programs based upon physical traits such as yield of milk per cow per lactation, weight at a given age, percentage of lean meat in the carcass, backfat thickness and number of eggs produced per hen per year. Crossbreeding and the crossing of inbred lines have been used in many species to enhance further the expression of many traits. Other parts of the increases have resulted from marked improvement in reproductive efficiency and mostly improved diets formulated specifically for different species and for different ages within the same species.

With current knowledge we have described genetically our breeds and groups in terms of physical measurements. However, even though animals of a given breed group vary greatly in expression of a trait such as milk production, we know essentially nothing about the functional basis of genetic variation in physiological traits such as enzymes, hormones and various intracellular functions that play an important role in causing such enormous variation among animals for the economically important traits, meat, egg and milk production. We must initiate research to define our farm animals genetically and phenotypically for both physical and physiological traits and explore relationships among such traits.

Proposed Research Approach

There have been marked improvements in instrumentation and techniques for quantitatively measuring substances in biological tissues and fluid, and systems have been developed to study various intracellular functions. To define our animals genetically and physiologically will require scientific efforts of individual scientists and teams of scientists representing several disciplines of biology. For examples, the genetics and nutrition of systemic diseases may be as important as description of the disease, or genetics of endocrinology as related to reproduction, lactation and egg production may be more important than the quantitative genetics of weight at a given age or amount of milk per lactation.

To regulate meat, milk and egg production, we must initiate basic research to understand, quantitate and manipulate cellular and subcellular events responsible for controlling protein production. These needs are described in the following sections and list of priorities.

CELLULAR ENERGETICS AS RELATED TO PROTEIN PRODUCTION

Current Status

Production of meat, milk and eggs requires a complex protein-synthesizing system and major expenditures of energy plus precursor materials by the producer tissue. In many cases the whole cell's economy is devoted to production of a specific protein product; for example, egg white proteins by glandular cells of the chicken oviduct, muscle proteins by muscle cells and casein synthesis by mammary epithelial cells. This necessitates coordination of protein-synthesizing and energy-producing systems. An understanding of these cellular and molecular processes in producer tissues, as well as their systemic or organismic and genetic controls, is essential to increasing the total output and efficiency of production of animal protein.

Protein metabolism. Protein production involves net accretion of body proteins and secretion of egg and milk proteins. Structural proteins, such as those of bone and muscle, as well as cellular enzymes are constantly being degraded and replaced. This important aspect of protein metabolism is called turnover. Thus, protein production is the difference between synthesis, on the one hand, and turnover on the other. Egg and milk proteins are not turned over during production. Therefore, egg and milk protein production is much more efficient than that of muscle proteins (meat). In a growing steer for instance, about 2.3-3.6 kg (5-8 lb) of muscle protein are replaced each day. The mechanisms modulating protein turnover have not yet been elucidated; however, understanding of these modulating mechanisms maybe as, or more, important than those of controlling protein synthesis.

Proposed Research Approaches

Protein synthesis. This is one of the most important considerations in protein production. More biochemical details are available for protein synthesis in microorganisms, particularly bacteria, than in laboratory animals and more is known in the latter animals than in meat-producing animals. Studies of protein synthesis in muscle of meat-producing animals or mammary tissue of cattle have been minimal. A notable exception is the large research effort devoted to synthesis of egg protein. The chicken has become a favored experimental animal for modern molecular biologists concerned with control of gene expression in animal cells.

Concentration of messenger RNA apparently plays a critical role in regulation of protein synthesis in production of protein of egg white and egg yolk. Messenger RNA availability in turn is from synthesis, processing and turnover of RNA. Information on these three processes is limited, and all three have to be considered as rate limiting for protein synthesis in farm animals. Protein synthesis also can be regulated at the translational level where messenger RNA availability is not concerned. In most growth systems (generally in rodents or experimental cell culture systems) ribosome number is correlated closely with changes in growth rate. Little is known about this in protein-producing systems of farm animals. Other translational control mechanisms include modification of the activity of factors which initiate protein synthesis. This is important in red blood cells and could be important in other tissues.

First priority in the study of protein-synthesis regulation in animal cells probably should be determination of whether messenger RNA or ribosomal RNA production are rate limiting as they seem now.

Protein secretion. Production of both egg and milk protein involves movement of proteins from their site of synthesis on ribosomes associated with cytoplasmic membranes to the exterior of the cell. While the biochemical steps of this process have been described, its regulation and relationship to efficient milk and egg production need to be studied.

The relationship of secretion to synthesis also deserves further study. In the same tissues protein synthesis is blocked if the protein is not secreted. Thus, if a cow is not milked, involution of mammary gland cells occurs.

One crucial overriding constraint on protein production concerns the question of cell number in protein producing tissues. Animals differing in muscle mass, for example, may have the same protein content per muscle cell nucleus but differ in the total number of nuclei per muscle. Most of the nuclei ultimately residing in muscle cells accumulate during postnatal life, even though muscle cell number does not change significantly. Consequently, the postnatal increase in muscle cell nuclei provides the genetic material required for additional protein accretion resulting in hypertrophy of the cell. Similarly, the number of producing cells of milk or egg constituents is as critical as production per cell. It may be that different physiological and genetic strategies are involved in increased cell number and increased cell size. Genetics may be more important than nutrition in determining cell number while the reverse may be true for cell size. The genetic and nutritional effects on protein producing cells may also apply to fat cells. A thorough understanding of the biological mechanisms controlling proliferation and differentiation of protein and fat producing cells is imperative.

Energy metabolism. This term is used to bring together all those cellular processes that combine to provide the energy required for protein synthesis in the form of ATP. Anabolic processes, including protein synthesis and fat synthesis, require large amounts of energy as well as amino acids and other precursors.

In farm animals, two major sources of cellular energy must be considered, glucose and short chain fatty acids produced by rumen microflora. Under adverse nutritional conditions, energy as well as precursor amino acids control the rate of protein formation. It is much less clear how high planes of nutrition affect rates of protein formation. In ruminants this may be even more complex because of the utilization of short chain fatty acids for energy and fat synthesis.

For both energy sources there exists a type of cellular competition that is of great importance in animal production. Fat cells use the same energy inputs to produce fat as are used to produce protein. Some of the regulators of protein anabolism such as insulin cause both increased fat deposition and muscle protein synthesis. We need to learn more about the physiology of both protein-producing and fat-producing cells and the manner in which energy and resources are partitioned between tissues.

Another aspect of energetics is transport of energy substrates into cells. Glucose transport and fatty acid uptake by protein-producing tissues under different physiological states could be important. Also, intracellular transport of short chain fatty acids and their utilization in mitochondria need to be examined, particularly in muscle, liver and mammary gland of farm animals.

In all of the metabolic steps discussed above, the crucial questions concerning the production of protein for food from animals are: 1) What limits the rate of protein and fat formation? 2) What regulatory systems control the steps that lead to protein and fat formation? and 3) What is the genetic variation in these? If a step is not rate limiting, it is not likely to be regulated and is not amenable to manipulations. However, as one rate limiting step is modified so that it is no longer limiting, it can only follow that another step becomes limiting.

While emphasis in this area is cellular and molecular, studies of systemic and organismic energetics particularly as related to nutrition should continue.

Control of the rate limiting steps is likely to involve neural and endocrine factors discussed in the following section. The rate limiting steps identified above will become sites where progress by either classical genetics or genetic engineering must take place. Genetic engineering technologies have progressed at much faster rates than believed possible. Their practical value is likely to be limited not by the technique but by the problem of identifying rate limiting steps in the specific genes to be engineered. Similarly, classical genetic and animal breeding strategies outlined in another section will depend more and more on finding additional selection criteria. Multiple gene control probably will continue to require quantitative genetic approaches; however, where specific genes that effect performance can be identified, appropriate genetic strategies can be applied.

ENDOCRINE AND NEURAL CONTROL OF PROTEIN PRODUCTION

Current Status and Research Approaches

Peptide hormones. In laboratory species a number of peptide hormones increase protein production. These peptides include growth hormone (GH), somatomedin, growth factors such as epididymal growth factor, prolactin and insulin. Limited supplies of these peptides constitute the major limiting constraint in testing these materials in domestic animals, but with the recent rapid development of recombinant DNA techniques, virtually unlimited quantities of peptide hormones probably could be available in the future from "genetic engineering companies" if this led to more efficient production of protein for human diets.

Initial research should be designed to identify those peptides which most likely will enhance production of animal protein. Development of appropriate model screening test systems will be needed. For example, the ability of peptides to stimulate protein production could be tested outside the body; induction of protein synthesis in mammary tissue and in muscle tissue cultures has been demonstrated. Another approach would be to choose an in vivo animal model in which protein production could be tested without reusing large quantities of the limited quantities of peptides. Initially doses and routes of administration of peptides which maintain increased concentration of the hormones in the model animal will have to be established.

Once a peptide is identified which may have promise as a stimulator of protein production, large quantities of the peptides will be needed to test their long-term efficacy in increasing protein production. In other words, do animals become resistant to continuous administration of the hormone? The effects of physiological, nutritional, genetic and environmental states on responsiveness to the peptide must be evaluated. Delivery systems for the peptides must be devised which will take into account differing animal management systems, (e.g., range vs feedlot). In addition, peptides should be examined for their ability to increase feed efficiency and to increase ratios of protein to fat. The most promising peptides for increased protein production or efficiency must be tested for efficacy on commercial farms. To ensure a supply of safe food, data on the human safety impact of possibly increased endogenous hormones may be needed. Current sensitivity of method (SOM) dogma followed by FDA is a major limiting constraint in the use of biological and endogenous hormone manipulations. Other approaches to regulate protein production should include development of methods to alter endogenous secretion of peptide hormones thought to be important for protein synthesis. Such methods may include manipulation of physiological, nutritional, genetic or environmental factors. Use of drugs may be included. The impact of these factors on concentration, metabolic clearance rate, secretion rates, episodic patterns of release or other metabolic transformation of appropriate anabolic peptides may provide clues as to which peptides are of greatest importance for protein production. Also, insight into the mechanisms whereby peptides regulate protein synthesis may be established. A peptide per se will not evoke a biological response unless specific receptors for that peptide are in the tissue. Thus, receptor numbers are a rate limiting step in the mechanism whereby peptides regulate protein production. Studies should measure changes in appropriate peptide receptor numbers with changes in target-tissue protein production induced by physiological, nutritional, genetic, pharmacological or environmental factors known to affect protein production in farm animals. Such studies may lead to eludication of a portion of the mechanism whereby peptides regulate protein production.

Appetite and liver function play key roles in regulation of supplies of nutrients which limit protein production. Thus, it seems logical that studies of endocrine control of these factors may lead to new means whereby protein production may be stimulated. Such studies could be designed as integral parts of some of the experimental protocols previously outlined.

Steroid hormones. Synthetic estrogens have been used successfully to stimulate protein production in several domestic species. In view of theoretical interrelationships between estrogens and cancer in humans, it seems likely that FDA approval of use of these compounds is not likely. As an alternative, means should be devised to increase secretion of endogenous estrogens, reduce their rate of breakdown or stimulate receptor numbers in the appropriate target tissues. Again, approaches used to study peptide hormones seem appropriate, namely, attempts should be made to alter these variables through manipulation of physiological, nutritional, genetic, pharmaceutical or environmental systems.

Neural control. Animals constantly receive signals from their environment via their sensory systems (eyes, ears, etc.) and adjust their productive capabilities to maintain homeostasis. A typical example is increased ambient temperatures decrease milk, meat and egg production; during cooling (above freezing temperatures) the reverse occurs. Thus, studies should be implemented to elucidate mechanisms (probably neuroendocrine) whereby environmental variables affect protein production. Such variables could include temperature (solar and thermal radiation) photoperiod, humidity, rainfall, wind, barometric pressure and gaseous content of the animal's microenvironment.

Since the hypothalmus plays a key role in regulation of hormone secretion and appetite, it seems logical that studies of this area of the brain could lead to new methods of altering protein production. Several hypothalmic neurotransmitters and endogenous opiates affect a variety of anabolic hormones in laboratory species. Similar evidence in domestic species is limited. Thus, experiments should be designed to study the effects of hypothalmic factors on secretion of appropriate anabolic hormones, protein production and appetite.

Severance of a nerve will cause atrophy of muscle. Thus, the nervous system directly can regulate muscle size. For this reason, it seems likely that studies of direct neural control of muscle could lead to additional procedures whereby accumulation of muscle mass could be stimulated.

Biological supplies. There is real need for a continuing supply of biologicals such as animal hormones, antibodies, monoclonol antibodies, cell culture lines, cloned DNA's, etc. A nationally coordinated program including a collecting and distribution system appears to be the most efficient operation. A continuing supply of items identified is essential to permit scientists to pursue new basic research approaches that have considerable potential to increase efficiency of producing animal protein.

DIGESTIVE PHYSIOLOGY

Current Status

There is considerable variation among breeds of animals in their capacity to utilize selected feeds or nutrients. Thus, selection for animals with this unique genetic ability could have significant impact on efficiency of protein as meat, milk or eggs.

Proposed Research Approaches

In light of the potential there is a need for research with present animal populations to develop new lines to identify variation in their ability to consume, digest, absorb and assimilate feed and selected nutrients. It may be possible to select for, or through genetic engineering to develop, microbes with increased ability to assimilate nonprotein nitrogen and to digest unique types of plant cells.

There is a need to identify gut hormones, enzymes and other secretions and their interactions which limit digestion and absorption. In addition, more study is needed of the relationship between intake and extent of digestion, rate of digesta flow and metabolic demand. As these factors become better understood, it should be possible to select animals that have increased genetic capacity to digest and absorb nutrients. If at the same time animals could be selected for improved ability to assimilate amino acids into muscle protein, the efficient conversion of plant protein to animal protein could be enhanced greatly.

Research also should be on the possible genetic relationship of disease to gut function. For example, pale-bird syndrome markedly reduces lipid absorption from the gut. Other disease-causing microbes that cause inflammation of the gut lining may adversely affect gut function.

Feeding high fiber diets to broiler pullets restricts their body size, markedly increases gut size and improves subsequent hatching egg production. This larger gut size improves conversion to eggs when the hens are placed on a low fiber laying hen diet. This suggests that selection for increased gut size in chickens and turkeys may speed growth and improve feed conversion by more efficient digestion and absorption of nutrients in the gut.

GERM PLASM IMPROVEMENT AND UTILIZATION

STATE OF THE ART AND SCIENCE

Achievements are shaped by germ plasm farm animals inherit and by their environments. Together these prescribe the biological characteristics of each animal, which contribute to its commercial value to the producer and ultimately and most importantly, to the consumer. Selective breeding for valuable biological components will lead to improvement. Genetics enables efficient breeding programs to take into account the ease with which different aspects of commercial value such as efficiency of food conversion and reproductive performance may be changed. Thus, an appropriate balance between components can be developed in an index of merit that breeders can use in selection programs wherein efficiency of the entire production system, as by ratio of output to input, is emphasized. Improvement of germ plasm then is complemented with improvement in environment for production to advance performance as has been so spectacular over the last quarter century.

CONSTRAINTS

Contributions of genetics to advances in performance depend on the potential rate of genetic change and its achievement. Constraints to the application of genetics are unclear biological objectives and limitations in rate of genetic change. Research is needed to specify changes in production over lifetime which will reduce costs of the animal product.

The rate of genetic change in life-cycle efficiency is affected by genetic variation and by the ability of animal breeders to identify superior animals for an environment. Research is needed to develop ways to choose superior animals more surely.

Objectives are determined by circumstances of social and economic factors, such as costs, prices and consumer demand that themselves are subject to change. Further, it is expected that the rate of such changes will increase in the future. Keynote to providing useful food animals will be the development of capability for rapid genetic adaptation. Inferior strategies for altering populations of animals may limit change for human needs.

PRIORITIES FOR RESEARCH

Preferred research must be to develop methods to increase rates of improving food animals genetically. The knowledge and techniques derived from sections - Efficiency of Reproduction and Protein Synthesis - contribute to maximize effects of selection, but genetic improvement requires that objectives be defined clearly for specific production resources.

Clarification of Objectives

Contribution of potential genetic change to lower cost of production. Historically, both breeding resources and time have been underutilized because of incomplete, unbalanced or irrelevant biological objectives in livestock breeding programs. To maximize future genetic improvement, potential genetic changes in components of animal performance need to be evaluated for their expected impact on efficiency (output/input) of full life-cycle systems of animal production. Biological efficiency is the animal product output per feed energy input (e.g., output of meat protein, milk solids, egg mass, etc.). Biological efficiency can be converted to economic efficiency by adding nonfeed inputs, costs and prices for production-marketing situations.

Environmental effects on genetic improvement. Effects of genetic changes in animal activity (e.g., in reproduction, growth, body composition, lactation, heat tolerance or disease resistance) on efficiency of output to input will depend on environment to supply required additional inputs, such as nutrition, shelter, labor or disease control, for full expression of genotypic capability. Environments vary from adverse arid roughland or inferior care to intensive systems with abundant feed, caring labor and appropriate shelter. Genotypes optimal then will differ widely among such environments.

Potential benefits from genetic and environmental changes. Analyses of production systems by species are needed to project benefits from genetic changes and from fitting environments for efficiency of output to input. Under intensive management preliminary predictions suggest that increased reproductive rate is more important than increased growth rate in beef and dairy cattle and sheep but less important for more prolific swine and poultry; reduced fat content is most important in lamb, pork and beef, less in chicken and least in turkey. Potential genetic-environmental gains in efficiency of producing lean meat from speeded reproductive rate and growth and from reduced fatness are 45-55% for beef and lamb, 40% for pork, 30% for turkey and 20% for chicken. Rigorous analyses can ascertain more accurately the genetic objectives and potential benefits under the range of existing and anticipated production-marketing systems.

Genetic Strategies

The aim is to match genotype with environment. Four principal strategies may be used to meet this aim.

Selection among existing populations. Comparison of existing material is an essential first step. Research should widen comparisons with exotic breeds and consider species substitutions.

Direct selection is the most effective route of improvement, once suitable populations are chosen and objectives are defined properly.

Hybridization is beneficial when progress is slow, superior lines are available, specialized lines are advantageous or traits show heterosis.

Biologically diverse lines. Considerable genetic progress has accrued from introduction of exotic breeds; large beef cattle and prolific sheep are examples. These diverse stocks were developed under geographic isolation to the American breeder and became available only recently. This approach to improvement could be extended by the planned development of expanded genetic diversity among populations. Such diverse genetic stocks would facilitate rapid genetic changes (by crossing) to meet changing circumstances. Examples are ability to utilize unusual roughages, cattle with higher twinning rates, and pigs with increased litter sizes. This strategy also provides a positive approach to germ plasm conservation in which the potential merits of genetic stocks could be increased further.

Methodology. Merits of new genetic material must be evaluated in environments for which it potentially is adapted. This evaluation requires innovation in experimental design and quantitative-genetic methodology. In particular, quantitative-genetic methods are needed to cope with awkward traits such as reproductive performance in cattle.

Execution. Advances in knowledge related to reproductive efficiency and protein synthesis (first two sections) are incorporated into breeding technology. Real progress in production technology is dependent upon the blending of such knowledge with genetic strategies.

Recognition of Merit

Development of improved measures of genetic merit is of immediate relevance. We are only able to act on what we can see of the animals, yet for breeding decisions many characteristics we wish to change cannot be measured at young ages, in live animals or in both sexes. Amount of lean tissue cannot be measured readily on the live animal; female reproduction, egg laying and lactation cannot be measured in males. Research to understand variation in the physiology and biochemistry of animals should be targeted to overcome these limitations in identification of superior germ plasm. A basis for such research is that males carry genes for female traits and that these may be measurable in some form in the male. Identification of the multitude of individual genes by their primary biochemical products would have limited advantages. However, it may be possible to measure the collective expression of many genes integrated into a physiological change that is associated genetically with the trait to be improved. For example, testis activity in males is correlated with ovarian activity in females and, further, energy metabolism in males may be associated with lactation in females. Such research also should include measurement of biological efficiency with which an animal's feed intake is utilized for producing protein and fat, either in body tissue or milk. The average half-life of protein in the mammal is only about 14 days; about half of the energy consumed by livestock is used merely to resynthesize their body protein. The development of methods to identify variation in the requirements for protein maintenance among animals and selection of those with slower rates of protein breakdown could markedly affect biological efficiency without presupposing that it might be possible to change biochemical efficiency.

Another approach to recognition of superiority is use of environmental stimuli to reveal variation; measurement of useful egg laying characteristics under modified patterns of day length is an example.

While recognition of merit through physiological measures is more likely to be applied to selection for traits which are not expressed before breeding decisions must be made, it is possible that such criteria may aid in selection for traits which are a function of unfavorable environments. An important example is knowledge of the physiology of disease resistance which may facilitate selection of resistant animals in the absence of disease agents. Despite the generalization that measurements of integrated physiological systems are more likely to correlate with performance than are measurements of individual gene action, there are examples of important effects of single genes such as dwarfism, double muscling, susceptibility to stress and resistance to viral infection.

Genetic Engineering

Recent advances in molecular genetics have opened a new dimension in agricultural sciences. This technology falls into two general categories: 1) recombinant DNA in which mammalian DNA is transferred to bacterial systems. This transfer can cause the gene to produce in the bacteria. Thus, bacteria potentially can produce growth hormone, insulin and interferon, examples which may impinge on agricultural production. 2) Gene transfer involves isolating genetic material (perhaps multiplying it by use of recombinant DNA) and transferring it to another mammalian cell. Viral transfer of genes has been accomplished. Under special conditions even naked DNA can be transferred into animal cells to increase enzyme activity. This technology coupled with nuclear transfer and cloning of embryonic cells presents a potentially exciting system.

Recipient cell strategies need to be developed. Can we transfer genes into sperm? Can we transfer "synthetic" X chromosomes into sperm? The undifferentiated embryonic cell might make an ideal recipient because it replicates into all the cells of the organism and would carry the transferred gene to all cells of the organism. On the other hand, it might be more desirable to transfer genes to only some cells and, therefore, cells at a later differentiated state might be more desirable recipients.

Gene transfer is an example of science rapidly being applied to solving problems of human need. Animal agriculture is an area where this application can have immediate as well as long-term benefits. Research is needed to develop and adopt these techniques for food animals.

REFERENCES

Anderson, G.B., P.T. Cupps and M. Drost. 1979. Induction of twins in cattle with bilateral and unilateral embryo transfer. J. Anim. Sci. 49:1037.

Betteridge, K.J. (Ed.) 1977. Embryo transfer in farm animals. Monogr. 16, Agriculture Canada, Ottawa. 92 p.

Elliott, K. and J. Knight (Ed.) 1974. Size at Birth. CIBA Foundation Symp. 27. Amer. Elsevier. New York. 353 p.

Elliott, K. and J. Whelan (Ed.) 1977. The Freezing of Mamalian Embryos. CIBA Foundation Symp. 52. Elsevier, Excerpta Medica, Amsterdam 330 p.

Hare,W.C.D. and K.J. Betteridge. 1978. Relationship of embryo sexing to other methods of prenatal sex determination in farm animals: a review. Theriogenology 9:27.

Hawk,H. W. (Ed.) 1979. 3. Animal Reproduction. Beltsville Symp. in Agr. Res. Allanheld, Osmun, and Co., Montclair, NJ.

Ohno,S. 1979. Major Sex Determining Genes. p. 140. Springer Verlag, Berlin.

Kiddy, C.A. and H.D. Hafs (Ed.) 1971. Sex ratio at birth-prospects for control. A Symp. Amer. Soc. Anim. Sci. 104 p.

VanVleck, L.D. 1980. Cornell Univ., Ithaca, NY.

"The meat-type hog...is a dramatic success story in livestock research and development."

Anson Bertrand

"The average annual yield per milking cow has increased 2.3-2.4 times since 1945...(this) revolution in dairy farming (has been) a triumph of applied research..."

G.W. Salisbury

7

D. P. Anderson (Co-Chairman), W. R. Pritchard (Co-Chairman), J. J. Stockton (Rapporteur)

W. G. Bickert, L. Bohl, W. B. Buck, J. Callis, R. Cypess, J. Egan, D. Gustafson, D. Halvorson, B. Hawkins, A. Holt, A. D. Leman, S. W. Martin, T. D. Njaka, B. I. Osburn, G. Purchase, W. W. Thatcher, H. F. Troutt, J. Williams, and R. G. Zimbelman

ANIMAL HEALTH

SUMMARY

Diseases of food animals cause major losses and are important causes of reduced productivity throughout the world. In less developed countries epidemic infectious and parasitic diseases, many of which kill large numbers of animals, are responsible for the largest proportion of disease loss. Some exotic infections, such as African swine fever and foot-and-mouth disease, are serious threats to the United States' livestock industries. In developed areas, such as the United States, Europe, Australia, and the U.S.S.R., the majority of the highly contagious epidemic diseases have been controlled, but production diseases, many of which cause no recognizable clinical signs, are the most important contributors to reduced productivity. Toxic chemicals from industrial, agricultural, energy-related, transportation, household and other sources are increasingly important causes of reduced productivity in food animals. Important human health hazards result from contact with animal products contaminated by toxic chemicals and from animal pathogens infective to man.

Production inefficiencies which result from factors that impair animal health can be eliminated in the foreseeable future if 1) present knowledge is applied, 2) animal health delivery systems are improved, and 3) new technology is developed through basic and applied research where suitable measures are not presently available. There are good prospects for major breakthroughs in animal health within the next decade which will have implications for animal production in the 21st century.

CRITICAL CONSTRAINTS

The most important constraints to improved animal health are:

1. Deficiencies in specific diagnostic tests and immunologic approaches for many of the important infectious animal diseases. Of particular importance is the lack of an effective immunizing agent for African swine fever and a long lasting vaccine for foot-and-mouth disease. To correct these deficiencies much basic research will be required on disease agents, pathogenesis of disease and host-agent interactions.
2. Lack of methods to interpret causal relationships of the multiple etiological factors responsible for production diseases and strategies for the control of these diseases under farm conditions.
3. Incomplete understanding of health effects of many toxic chemicals that are getting into the environment, and reliable means of detecting, quantitating and monitoring these effects in food animals.
4. Lack of adequate systems for the definitive diagnosis and reporting of animal diseases and means of collecting and managing animal health data. An improved mechanism of animal identification and more effective means of providing support to the animal health system are essential.
5. A deficiency of means to eliminate consequences to human health of animal pathogens, toxic chemicals and drug residues in animal products and to individuals in contact with the production environment.

RESEARCH IMPERATIVES

Integrated Food Animal Health Systems.

Develop integrated (interdisciplinary multifactorial) methods to identify and quantify factors and interactions responsible for losses and inefficiencies in the production environment, and develop analytic systems to construct practical strategies for the control of production diseases. Examine under controlled conditions causative interactions to substantiate on-the-farm assessment of control strategies and construct computer simulation models to evaluate alternative control strategies. Develop a national system of animal disease reporting, animal identification and standardization of diagnostic methods and systems.

Genetic Engineering for Disease Resistance and Immunity.

Identify and enhance disease resistance in animals. Apply recombinant DNA techniques to infectious agents and characterize the means by which animals develop protective immunity. Devise techniques to accelerate protective immunity in the newborn and strategies to produce maximum protection against disease. Special attention should be directed to the importance of developing effective technology for control of the most destructive foreign animal diseases.

Hazards to Food Animals from Chemical Agents.

Develop rapid economical procedures for quantitative identification of a broad spectrum of chemicals (contaminants, naturally occurring toxins and drugs) and their metabolites in animal tissues and feeds for use in the diagnosis and monitoring of animal contamination. Define cellular mechanisms of drug and chemical actions and interactions including how they are affected by stress and disease. Develop new pharmaceuticals and pesticides which are innately less dangerous through fundamental research taking into account mechanisms of drug action which bypass animal and human biological mechanisms.

Elimination of Human Health Hazards.

Develop procedures to eliminate animal pathogens from food animals, animal products and the environment of human beings. Eliminate potentially toxic chemicals and their metabolites from food products. Develop antimicrobial agents, growth promotants and permittants, and production improvers that do not have the potential for adversely affecting human health through transfer of resistance factors.

EPIDEMIC INFECTIOUS DISEASES

STATE OF THE ART

During the past century the U.S. has dealt successfully with many of the most important epidemic infectious livestock diseases. The most devastating such as contagious bovine pleuropneumonia, foot-and-mouth disease, bovine babesiosis, Asiatic Newcastle disease and hog cholera have been eradicated. Those that are being controlled are tuberculosis, coccidiosis, Marek's disease and pullorum disease, to name but a few. Salmonellosis in turkeys is an example of a disease requiring new strategies to effect control. Methods for control of the epidemic infectious diseases of livestock are established and consist of identification of the causative agent, development of diagnostic tests to identify infected animals, definition of reservoirs and vectors and means of spread, institution of sanitary procedures including quarantine to prevent spread of infection, development of immunization procedures and slaughter of infected animals to reduce or eliminate the infectious agent. In a few, such as coccidiosis of poultry, chemotherapeutic means have proved to be the most effective control. These procedures have been successful and generally practicable to apply.

A large number of important infectious diseases remain as serious constraints to efficient livestock production (e.g., Brucellosis, enteric, and respiratory diseases of swine and cattle). Some still exist because present technology has not been applied adequately (e.g., bovine leukemia), whereas others are present because effective means of control have not been developed (e.g.,pseudorabies). Still others have emerged as new problems because changes in environment or in production practices have created new conditions that have permitted new diseases to occur or old diseases to assume new roles. Table 7.1 lists a few examples of research needs for the more important diseases and parasites of livestock.

Certain epidemic infectious diseases of livestock, mainly in developing countries, constitute an ever present and serious threat to U.S. livestock industries. Should any one of these infections gain entrance into the country and establish a foothold in the livestock population, it could cause serious, and in some cases crippling, losses to domestic livestock production. Examples are African swine fever, foot-and-mouth disease and Rift Valley fever.

Laboratory support is essential for definitive diagnosis of diseases in animals. A wide variety of immunological techniques is used for detection of antigens or antibody including complement fixation, virus neutralization, hemagglutination inhibition, agglutination and agar-gel precipitation tests. Tests for antibody may be in small laboratory animals or in tissue culture systems. Newer diagnostic techniques such as the enzyme-linked immunosorbent assay test are being developed. This test requires smaller quantities of diagnostic reagents and is more sensitive than conventional methods. It also lends itself to automation and is useful for examining large numbers of samples as in surveys to determine disease prevalence.

For exotic diseases, programs have been conducted for training veterinarians in the recognition of certain foreign animal diseases. Protocols for diagnosis of these diseases are being described, and quantities of diagnostic reagents have been produced, safety tested and stored for use in diagnostic laboratories in the geographic area where there are outbreaks of foreign animal diseases.

The U.S. Department of Agriculture, in cooperation with the states and certain universities, has established a Regional Emergency Animal Disease Eradication Organization for controlling foreign animal diseases should they appear. A limited number of teams of laboratory workers has been identified and trained in the conduct of diagnostic tests for the exotic diseases. Only a few U.S. nationals have research experience with important exotic diseases.

The most common approach to control of domestic animal viral and bacterial infections is through immunization with inactivated or attenuated vaccines. Some products in use today are not optimally effective, but most are being improved, and new ones are being developed such as has been done recently for pseudorabies in swine. Interferon is being used increasingly for treatment of certain diseases in humans. It will be used soon in veterinary medicine as well. Similarly, antiviral chemotherapy appears to have great potential, perhaps equaling that of antibiotics which have been used with so much success for controlling bacterial infections.

A number of infectious livestock disease agents has been studied at the molecular level generating detailed information on physical and chemical properties of these agents. This storehouse of information now is being utilized in development of genetically engineered bacteria which will produce specific proteins which can be used as vaccines. Better diagnostic methods with particular references to detection of strain differences are being developed through nucleotide analysis. This technique is especially helpful for detecting viral strain relatedness or cross reactions and allows mixing of appropriate viral strains in fabrication of more efficacious vaccines.

CRITICAL CONSTRAINTS

1. Hybridoma derived monoclonal antibody banks are not yet available as reference reagents in serotyping of important animal viruses.

2. Little is known about the function or location of viral genes with regard to their pathogenic activity. Such voids prevent application of techniques to alter the viral genetic structure.

3. Little information is available on nutrition-immunocompetence-infection interactions. In some diseases it appears that nutritional modulation of immunity may be an important determinant of morbidity and even mortality. Immunization procedures may be applied more effectively if such relationships can be assessed.

4. Latent infections are poorly understood. For example, origin, duration, stimuli for recrudescence and return to latency for the more than forty important herpes viruses of animals are in large measure unknown. This is the major reason for difficulties in coping with this important aspect of infectious disease control.

5. Almost nothing is known about specific kinds of cells in animals that a microbial agent will infect and cause altered functions or death. There is a dearth of technology to determine why disease may, for example, occur in only 1-2% of those animals that become infected.

RESEARCH IMPERATIVES

1. Hybridoma derived monoclonal antibody technology needs to be developed to investigate pathogenesis, immunology and taxonomy of microbial agents.

2. Investigations into the genetics of viruses, bacteria, rickettsia, chlamydia, fungi and mycoplasma must be made for development of immunizing agents of many kinds (subunit vaccines, plasmid antigens and hybridized microbial units) in the development of knowledge of gene specification of pathogenetic characters and application of recombinant DNA technology for control or elimination of infections.

3. Investigations are needed to bring into focus the impact of nutrition on immunocompetence and infection.

4. Investigations are needed to determine the impact of environmental, climatic and geographic factors on distribution and character of microbial and multicellular parasite infections.

5. Problems of latent infections (e.g., the forty-odd herpesviral diseases of animals) need to be understood much more completely to permit development of effective strategies for coping with them.

6. It is important in control programs for infectious diseases to be able to distinguish virus strains even though they are serologically identical. Nucleotide analysis is one possible method of accomplishing this.

7. Neonatal losses of pigs and calves due to subclinical infections with swine cytomegalovirus and blue tongue in cows need to be investigated since there is a high prevalence of infection accompanied by a low rate of observable disease.

PARASITIC DISEASES

STATE OF THE ART

Subclinical infestations with both endo- and ectoparasites occur widely among all food animal species in the U.S. It is generally agreed that parasites contribute significantly to production losses. However, quantitative assessments of these effects are not generally available.

Considerable effort is expended in the United States to prevent introduction of a variety of ectoparasitic agents affecting food animals. Some of these serve as vectors of infectious disease agents (e.g., ticks) while others are primary pathogens (e.g., screwworms). The broad range of geoclimatic conditions within the U.S. makes it likely that a number of important arthropod parasites could proliferate and spread once introduced.

Most endoparasitic conditions are endemic in the U.S., and special preventive efforts are undertaken for those associated with specific vectors (e.g., Babesia sp. in Boophilus sp. ticks). However, meat importation regulations have been developed, in certain instances, to preclude introduction of parasitic infections (e.g., cysticercosis in sheep). Regulatory efforts are made within the country to limit transmission of ecto- and endoparasitic diseases (e.g., scabies, trichinellosis), but, for the most part, parasite disease prevention and control in food animals is addressed through private veterinary services and producer efforts involving a combination of chemotherapeutic agents and management adjustments. No biological products are available which are relevant to these efforts.

Infections of food animals with zoonotically important parasites persist in the U.S.; and although the significance of some of these has been decreasing because of modern husbandry techniques and meat hygiene practices (e.g., trichinellosis), others are widely prevalent with uncertain or even unknown impacts on consumer health (e.g., toxoplasmosis, sarcocystosis).

Resistance to parasitic diseases is recognized in some strains of food animals, but no practical selection systems have been devised. Intensive use of chemical agents in treatment and prevention of parasite disease has resulted in appearance of ecto- and endoparasites resistant to many of the readily available insecticides and anthelmintics. Wildlife hosts serve to maintain and disseminate certain important parasite pathogens of food animals in North America (e.g., Fascioloides liver flukes).

CRITICAL CONSTRAINTS

Detection systems for exotic and zoonotic endoparasites (e.g., Theileria sp., Babesia cysticercus) are not adequate for confidence in identification and surveillance.

Excessive dependence upon chemical agents for treatment and prevention of arthropods and endoparasites is complicated by selection pressures for resistant organisms and is at odds with increasing consumer concern over tissue and milk residues and environmental contamination. Biologic agents suitable for preventive strategies or environmental application do not exist, and in vitro cultivation methods are not available to pursue characterization of parasite products for immunogenicity or other important activities.

Combinations of husbandry practices, and climatic, genetic and nutritional factors which influence the spread and extent of parasitism and its manifestations in food animals, economically or clinically, are defined poorly in the U.S.

RESEARCH IMPERATIVES

1. The influence of widespread subclinical parasitism and their economic effects quantified so that appropriate rationales can be constructed for intervention through either management, breeding, nutrition, chemical or biological agents and combinations of these to improve productivity.
2. Modern immunobiological approaches must be used to investigate genetic regulation of resistance to parasitic disease.
3. Specific reagents for standardization of diagnostic methods and for use in immunologic probes and assays must be developed.
4. Epidemiologic investigations are needed to define mechanisms of parasite transmission and population expansion so that appropriate management strategies can be devised.
5. Immunologic consequences of parasitism must be explored, not only as a basis for immunopreventive agent development, rational management, or breeding countermeasures, but also as a means of understanding effects on immune responses to other infectious agents or biologic products.
6. Molecular genetic technology should be applied to the in vitro production of parasite antigens which can act antagonistically in complex biological systems. This could include live bacterial immunizing or sterilizing agents for colonization of specific host sites or dissemination in vector populations.
7. Biological and environmentally-sound chemical approaches should be pursued to influence propagation of parasites as vectors of disease and as primary pathogens affecting livestock production. These should include exploitation of genetic incompatibilities among anthropod strains and invertebrate pheromone attractants.

8. Vector agent interactions must be investigated in relation to the transmission pattern of pathogens, enhancement of infectivity through vector passage and survival through adverse environment episodes and vector insusceptibility.

METABOLIC DISEASES

STATE OF THE ART

Metabolic diseases are those maladies that stem from the inability of the animal's metabolic regulatory systems to adapt to nutritional and environmental conditions. These disorders are of growing importance as livestock production becomes more intensive and as growing use is made of industrial wastes and crop and animal by-products for animal feeds. Losses from metabolic diseases are substantial and appear to be increasing as the productive efficiency per animal unit, feed unit and labor input unit is increased.

Some examples of metabolic dysfunctions that cause major losses include the following: 1) loss of regulation of mineral ion levels in body fluid compartments as in parturient hypocalcemia (milk fever) and hypomagnesemia (grass tetany); 2) faulty regulation of energy metabolism leading to loss of milk production and ketosis; 3) failure of the initiation of milk production following parturition as in agalactia of swine; 4) breakdown of microbial population ecology in the digestive tract under certain environmental and nutritional circumstances as in lactic indigestion and polioencephalomalacia; 5) deranged motor function of the digestive tract under certain nutritional and biophysical stresses as in left displacement of the abomasum and bloat; 6) failure of the neuroendocrine regulation of reproduction with resulting impairment of fertility; 7) disturbances of metabolic regulation attributable to deficiencies or imbalances of trace minerals, e.g., nutritional myodegeneration (white muscle disease); 8) inadequate coordination of development of the locomotor system in rapidly growing animals, particularly in poultry and swine, leading to limb stress and syndromes such as leg weakness, lameness and kinky back; 9) deranged regulation of myocardial development leading to cardiomyopathy (round heart disease); 10) faulty regulation of muscle development leading to susceptibility to conditions causing losses through carcass condemnation, e.g., pale, soft, exudative pork, malignant hypothermia of swine and excessive susceptibility to handling stress involving muscle tearing and hemorrhage in turkeys; and 11) failure of growth-generating processes leading to low productivity and unthriftiness.

Although sophisticated methods have been developed and applied to many of the basic systems involved in metabolic disorders, there have been too few studies conducted on actual cases. There are many unanswered questions relating to pathogenesis and control.

CRITICAL CONSTRAINTS

Sporadic and unpredictable instances of many metabolic disorders make it difficult to arrange for studies of actual cases. Satisfactory experimental disease models have not been developed for most conditions of this type.

Neuroendocrine aspects of metabolic regulation are complex and present problems of methodology. Similar difficulties are encountered in estimating the dynamic fluxes of metabolites, ions and water.

There is a lack of an adequate data base of the population dynamics, species characterization and biochemical properties of rumen and enteric microbes as they relate to metabolic disease.

The regulatory factors in food intake and the physiology of the digestive tract are still poorly characterized, both at the basic level and in dysfunctions.

The study of emerging problems associated with growth and the locomotive system is at a very primitive stage. Development of new methodology is required in this area.

RESEARCH IMPERATIVES

1. Quantitatively characterize the factors contributing to susceptibility of defective neuroendocrine and tissue level metabolic disorders, including ketosis, milk fever, agalactia and grass tetany. This work must be carried out on animals that become affected with these diseases.
2. Determine the pathogenesis and control of metabolic diseases involving dysfunction of the microbial, motor and secretory aspects of the digestive tract as they relate to diseases having a nutritional component in their etiology. This should encompass studies of the effects of new feeds.
3. Characterize the nature of susceptibility to musculoskeletal and cardiac disorders of rapidly growing animals, especially poultry and swine. This should extend to evaluation of susceptibility to stress and trauma, including factors leading to condemnation and slaughter.
4. Identify and quantify the factors contributing to impaired reproductive performance as it occurs in field situations.
5. Manipulate the regulatory processes to maximize productive efficiency and to minimize the incidence of metabolic diseases.

PRODUCTION DISEASES

STATE OF THE ART

Despite the fact that most developed countries including the U.S. have been successful in controlling many serious epidemic infectious diseases, losses to productivity from other livestock diseases remain high. Most of these losses are caused by what will be called for purposes of this discussion "production diseases." As a class they are complex epidemiologically and have infectious, toxic, genetic, metabolic and nutritional etiologies or combinations thereof. They generally result from several etiologic factors acting in concert with environmental and production factors. This group includes: 1) infectious complexes, such as mastitis, neonatal disease, multiple gastrointestinal and other parasitisms; 2) reproductive disorders; 3) stress related syndromes such as the bovine respiratory disease complex, salmonellosis and transport tetany; 4) metabolic imbalances such as ketosis and hypomagnesemic tetany; 5) digestive system disorders, such as bloat and lactic acidosis; 6) nutritional disorders and marginal malnutrition-infectious disease complexes provoked by deficiencies in minerals and other essential nutrients; 7) toxicoses caused by industrial and agricultural chemicals and wastes, poisonous plants and mycotoxins; 8) infestations with ticks, fleas, lice and biting insects; and 9) combinations of two or more of the above.

Some production diseases are capable of killing large numbers of animals, and others may produce high morbidity in affected herds. However, as a class they cause their greatest reduction of productivity of livestock enterprises often without producing clinical signs of disease.

Much is known about some of the important production diseases and surprisingly little about others. Reliable means of control are available for a few, whereas prevention and control procedures are disturbingly inadequate for most of the others. For example, despite all that has been learned about calf diarrhea since the classical studies of over a half century ago, it continues to be of one of the most important causes of neonatal losses in animals throughout the world. Similarly, a great deal has been learned about mastitis through research during the past 50 years; yet it is still one of the world's most costly diseases. Essentially, the same situation applies to reproductive inefficiencies, mixed parasite infections and many other insidious disease complexes.

CRITICAL CONSTRAINTS

The majority of production diseases are different from the less complex epidemic infectious diseases that veterinary research scientists have been largely concerned with during the past century. Consequently, suitable methods of determining their causes are not available. Research methods capable of assessing causal influences and interactions of multiple etiologic factors are needed. The chief constraint for production diseases is methodological. There also are no disease and production monitoring systems to collect data on disease factors, disease determinants, production and production environment essential to the effort.

There also is a lack of information on biological aspects of many production diseases. It is likely that mathematical methods will help to identify many important gaps in understanding production diseases. These gaps must be filled through traditional veterinary research approaches (microbiological, immunological, physiological, pathological).

There is a critical need to develop and implement methods that detect production losses and inefficiencies and assess causal influences of important factors contributing to production diseases. This research will require an interdisciplinary approach involving veterinary and other scientists in team efforts. A wide range of factors must be examined and quantitatively related to productivity. These factors include infectious and toxic agents, housing systems, ventilation (temperature, humidity, gases, aerosols), climate, diet (content, character, storage and delivery) and animal factors including behavior, genotype, phenotype and immunologic capacity. Whenever possible, these factors must be analyzed simultaneously and related to states of production disease.

RESEARCH IMPERATIVES

1. Develop combined disease and production monitoring systems which interface data with diseases, disease determinants and production practices in the production environment.
2. Develop integrated (interdisciplinary-multifactorial) methods to determine and quantify various factors and interactions responsible for loss and inefficiencies in the production environment.
3. Devise analytical systems to develop practical strategies for effective control of production disease.
4. Examine under controlled conditions the causative interactions and their effects to substantiate on-the-farm assessments to improve strategies of disease control.
5. Develop computer simulation models for production diseases to assist in the development of control procedure and to predict biologic and economic consequences of particular strategies for disease control.

CHEMICAL AGENTS

STATE OF THE ART

We live in an era in which a variety of chemicals generated from industrial, agricultural, pharmaceutical, energy-related, household and other sources is expelled into the environment. Many are overtly toxic to animals, plants and man; others produce subtle effects that may be manifested by reduced fertility, growth, productivity and decreased resistance to infectious diseases. Equally important is the multitude of fungal and plant toxins that may be in animal feeds. Ionizing radiation, microwaves and ultraviolet light also can have acute and chronic effects.

Industrial effluents that are hazardous to food animals are represented by fluoride, molybdenum, lead, arsenic, polychlorinated and polybrominated biphenyls, dibenzodioxins and others. Agricultural chemicals include insecticides, herbicides, rodenticides, fungicides, soil sterilants and fertilizers. Energy related substances include crude oils, petroleum distillates, chemicals associated with oil drilling and mining operations including sodium chloride, acids and heavy metals, and by-products of combustion such as carbon monoxide, nitrogen oxides and ozone as well as much chemically complex particulate matter. There are numerous chemicals classed as feed additives or potential feed substitutes which, when given in excess or fed to the wrong species, may be toxic. These include antibiotics, growth promotants, wormers, insecticides, nonprotein nitrogen, animal wastes, industrial wastes and stillage (from alcohol production). Associated with increased confinement of livestock and energy conservation through building design or heating systems is the accumulation of toxic gases such as hydrogen sulfide, ammonia, methane and carbon monoxide. Some therapeutic drugs also become toxicants under certain conditions and may cause adverse reactions or interactions with other drugs or chemicals.

Toxicology involves understanding poisons, their chemical properties, identification, biologic effects and treatment of disease conditions caused by them. To be most effective it should include biodynamics of uptake and elimination, mechanisms of action, dose-response relationships and principles of treatment. Available expertise in food-animal toxicology in the U.S. is located predominantly in a relatively few state and federally supported laboratories which have toxicologists on their staffs.

Although there has been considerable research done on acute toxicity of chemicals in food animals, there are large gaps in our knowledge of:

a) absorption, distribution, metabolism and excretion of these agents (toxicokinetics) as they affect animal health and productivity;
b) effects of low exposures to chemicals, toxic gases and natural toxins on animal productivity;
c) limits to which toxicologic data can be extrapolated appropriately across species lines;
d) limits to which data from high exposure studies can be used to predict effects of prolonged low exposure on animal health;
e) methods to decontaminate animals once they have been exposed to chemicals that bioaccumulate in the tissue;
f) effects of toxicants on animals' resistance to infections and production diseases;
g) interactions of pesticides and therapeutic agents in food animals;
h) effects of stress factors such as heavy production, unfavorable climate, crowding and transportation on drug metabolism and interactions; and
i) effects of disease states on drug and chemical metabolism.

Animals as Sentinels of Environmental Quality

Domesticated animals and many wildlife species are excellent sentinels of environmental quality. In event of accidental or intentional release of toxicants or organisms into the environment, animals are often the first exposed. A comprehensive program of chemical monitoring and disease reporting on a state or regional basis which could have major responsibility for defining the magnitude of environmental contamination does not now exist. If a better understanding of animal effects were known, a potential for significant benefits to humans would exist. This could be of great significance if a foreign adversary were to release a toxicant or biologic agent into our environment to impair agricultural production in the United States.

Pharmaceuticals

One of the areas of greatest potential for improvement of the efficiency of livestock production is the manipulation of basic biological mechanisms with specific pharmaceutical manipulation. This applies to both the basic physiological processes and the competing biological agents that cause disease, thereby obstructing attainment of the animal's biological potential. Although progress has been made and some significant products have been developed, remarkably little research has been done in this area that accounts for a high proportion of the producer's animal health expenditure.

Currently, animal pharmaceuticals are available for a variety of animal health problems. These involve nutritional substances, reproduction regulatory compounds, antimicrobial drugs, anti-inflammatory compounds and antiparasitic drugs. Some pharmaceuticals are narrow in spectrum of activity while others have broad effects. There is a serious lack of information on mechanisms of action of many drugs in many food animals.

CRITICAL CONSTRAINTS

There is within the food animal establishment only limited analytical capability to screen rapidly animal tissues and feeds for chemicals and natural toxins. When contrasted to human health, there is also a lack of multidisciplinary approaches to animal drug and toxicant research; specifically lacking is involvement of sufficient numbers of trained persons in molecular biology, pharmacology and toxicology. There is a paucity of information concerning the molecular and cellular mechanisms of action of chemicals, including mutagenicity, teratogenicity and carcinogenicity. It is not possible to extrapolate directly toxicologic data across species lines and from high exposures to low exposures to predict effects on animal health and productivity.

There has been failure to recognize the importance of pharmaceutical developments and, conversely, detrimental effects of chemical contamination and natural toxins in food animal production. This failure has resulted in a paucity of public support for training food animal pharmacologists and toxicologists, and similarly there has been limited support for animal toxicology and pharmacology laboratory facilities.

State or federal systems are not available to permit rapid retrospective detection of feed and animal tissue contamination. Further, there are no centralized systems for reporting animal diseases and toxicoses. The absence of animal poison control centers with capabilities of receiving, screening, categorizing and transmitting information to users is a further constraint.

The search for new pharmaceutical agents has been limited by inability to relate cellular mechanisms to clinical effects. Further, drug delivery systems that provide optimum drug concentrations at the appropriate time have not been developed. The latter results in unnecessary drug exposure in some instances while providing insufficient concentrations in others.

Constraints to pharmaceutical development include the high costs of new product development because of the regulatory requirements of efficacy and human safety. This factor has had a substantial inhibitory effect upon industrial research. Research in this field in universities and government laboratories is extraordinarily little. Lack of quantitative research on efficacy in relevant disease models and clinical situations is lamentable. There are still many diseases for which there is no adequate pharmaceutical control, either prophylactic or therapeutic. A major thrust is required in development of new products to meet this need. There is a serious shortage of qualified pharmacologists for essential clinical efficacy trials and pharmacokinetic studies.

RESEARCH IMPERATIVES

1. Develop economical and rapid screening procedures to identify and quantify a broad spectrum of chemicals in animal tissues and feeds.
2. Determine basic cellular mechanisms of drug and chemical actions and interactions with the goal of predicting the liklihood of these interactions causing adverse reactions. Such studies also could provide information which would lead to the detoxification of chemically poisoned animals.
3. Determine effects of chronic low exposure to chemicals and natural toxins (mycotoxins, plant toxins) on animal reproduction, growth, productivity and resistance to infectious agents.
4. Understand drug action so that desirable effects can be achieved more often and undesirable effects can be minimized or avoided. This understanding would be achieved more frequently by incorporation of molecular biologic and pharmacokinetic approaches to drug research. Needed studies include: a) Research to seek new agents and approaches that will allow manipulation of the animal's physiology to achieve productive advantages; and b) Research to seek new agents and approaches for control of epidemic and endemic infectious agents (bacterial, viral, fungal and parasitic) that threaten survival or impair productivity. This should be addressed to diseases for which there is no adequate control.

5. Explore means of stimulating host defense mechanisms to lead to pharmaceutical products for more effective disease control, including improved drug efficacy and lowered capacity for resistance in microorganisms.
6. Determine bioavailability of drug and chemical metabolites in animal food products to assess the significance of their hazard to humans.
7. Develop integrated model approaches to assess the safety of waste products for animal feeds, including animal wastes, agricultural and industrial by-products and stillage (from alcohol production).
8. Develop computerized system(s) to meet needs of animal poison control centers. These centers would receive, categorize and transmit information as well as assist in the interpretation of clinical and laboratory data.

HUMAN HEALTH HAZARDS

STATE OF THE ART

Diseases of animals transmissible to people (zoonoses) have been of concern throughout history. As long as animal products constitute an element in the human diet and as long as animals and people share the same environment, there will be concern for transmission of hazardous agents one to the other.

New zoonotic disease problems constantly are arising. For example, new human influenza strains may arise from genetic recombination of existing human and animal strains. Thus, animal influenza viruses have potentially high significance to human health. Also, recent characterization of babesia infections in people have placed this infection among the zoonoses. The possibility of new zoonoses arising is increased as environmental and ecological changes continue to occur. In the last decade, swine vesicular disease caused by one of the swine enteroviruses, spread throughout Europe causing economic losses to swine producers, and problems of differential diagnosis in those countries where foot-and-mouth disease is endemic. In comparative studies of enteroviruses of people, strong cross reaction with Coxackie B-5 virus has been demonstrated, indicating that swine vesicular disease may be a new zoonotic disease. Research workers infected with swine virus developed illnesses similar to those in Coxsackie B-5 infections.

Occupational hazards exist for slaughter house workers, veterinarians, research and diagnostic laboratory personnel, livestock producers and many others through their contacts with animals and animal products. Concerns for biocontainment are surfacing as problems of general public interest. Human exposure to hazardous agents may occur even though there is no overt disease in animals sharing the same environment. For example, ornithosis infection in people may arise through contact with normal appearing turkeys in processing plants. Infection with antibiotic-resistant bacteria as well as drug residue exposure may occur through contact with and consumption of healthy animals fed antibiotics or drugs. Salmonella infections, particularly in poultry, create important public health problems, and currently this is the most important zoonosis in the United States.

Many naturally occurring toxic substances (e.g., mycotoxins) which find their way into food products of animal origin are recognized as potential problems to people.

The animal production, processing and merchandising industries constantly must be alert to consumer attitudes and consumer acceptance of products of animal origin. Confidence in safety of food will become an increasingly important consideration for food animal industries.

RESEARCH IMPERATIVES

1. Determine the bioavailability of potentially toxic compounds and their metabolites in animal products in the food chain.
2. Determine the existence and distribution of potentially hazardous agents in the food animal population and the consequences of dissemination resulting from human consumption of a wide variety of infectious agents and chemicals.
3. Assess the impact of the real or perceived effect of human health from contaminated or potentially contaminated products on consumer acceptance of livestock products.

ANIMAL HEALTH SERVICES SUPPORT

STATE OF THE ART AND CRITICAL CONSTRAINTS

Food animal health services in the U.S. are provided by veterinarians engaged in private practice and state and federal preventive medical programs. Both are supported by extension agents, feed and pharmaceutical industries, academic institutions and state and federal diagnostic laboratories. A few private diagnostic laboratories are in operation. There is no national system for reporting animal disease data and, thus, no adequate information base upon which to establish national animal disease control strategies. Diagnostic Laboratory support for food animals is inadequate in many geographic areas. Kinds of services offered, operating policies and methods of reporting vary among laboratories. In a majority of instances only limited services are available. Only one full-service federal laboratory for mammalian species (National Veterinary Services Laboratory, Ames, IA) and one reference laboratory devoted to exotic diseases (Plum Island National Animal Disease Center) are available.

Diagnostic laboratories could be an invaluable resource for collection of animal health data, disease surveillance, epidemiologic investigation and integrated health care delivery support. Among presently operating laboratories data collection, recording, reporting and storage are not standardized. Many tests are not standardized, reagents are not always available and their quality is not regulated. Automated testing systems are being developed and evaluated, but their use is limited and they are not certified for regulatory programs. The U.S. animal health support services system is not capable of coping effectively with the demands that will be made on it to serve needs of the 21st century.

RESEARCH IMPERATIVES

1. A national animal disease reporting system is needed to collect epidemiological data on animal diseases, maintain a data bank, and devise means best to utilize it in support of the food animal health services delivery system. An economical, convenient and permanent means of identifying individual animals is required.
2. Operational research is required to evaluate alternative systems for providing diagnostic laboratory and epidemiological support services to the food animal industries.
3. There is need to develop a system for standardization of diagnostic methods and certification of the competency of diagnostic laboratories to perform services for the food animal industries.

IMMUNOLOGY AND IMMUNOGENETICS

STATE OF THE ART

Domestic food animals are susceptible to an array of viral, bacterial, protozoan and metazoan assaults which may decrease production efficiency or cause overt disease. Prevention of disease often is approached through immunologic intervention and protective vaccines which have been developed for most of the epidemic and some of the endemic problems. However, these have come about primarily through empirical approaches, and there is room for improvement in development of immunopreventive agents and strategies in the future. Immunological techniques are used widely in diagnostic procedures both clinical and epidemiological for infectious and noninfectious agents (e.g., hormones, drugs), but the specificity of the reagents and their standardization are problematic.

Immunologic investigations of the persistent production diseases such as mastitis and neonatal diarrhea are at the stage where preliminary work shows success of induction of enteric or secretory resistance (e.g., oral transmissible gastroenteritis vaccines, nasal vaccines). Progress has been slow toward practical immunologic measures for control of either endo- or ectoparasites. There is recognition from results in human beings, laboratory animals and chickens that immune responsiveness is regulated genetically, but no comparable definition has occurred in mammalian food animals. There is accumulating evidence of heterogeneity and characterization of immunoglobulins, immunocompetent cells and paraimmunological defense mechanisms in food animals, but this is not yet exploitable in ways to enhance productivity by preventing or treating diseases.

The genetic base for animals narrows from selection for specific traits. There are stocks that have traits of potential value for improvements in resistance to disease. There should be a means of identifying and storing such stocks.

CRITICAL CONSTRAINTS

1. A lack of understanding of development of the immune system in food animals limits ability to enhance immunopreventive measures by proper utilization of vaccines in the optimal dose, form, vehicle, route and regimen. All neonatal animals have only partially developed immune systems.
2. Primitive state of knowledge of immunogenetic traits and regulation of the immune response. This makes it impossible to use genetic markers as an instrument for selection of those food animals best suited to resist infectious or noninfectious disease.
3. Lack of identification of elements of infectious agents which lead to protective responses. This affects our ability to improve ways in which food animals can be made to respond optimally.
4. Lack of understanding of the nature of influences of environmental and nutritional factors on immune responsiveness. This limits the extent to which management, housing and nutrition can be adjusted to maximize disease resistance in food animals.

RESEARCH IMPERATIVES

1. There is great need for early definition of genetic and molecular bases for disease resistance; for example, genetic resistance of cattle to pink eye or of poultry to leukosis. This will require characterization of the major histocompatibility systems and other polymorphic blood traits of each food animal species for precise investigation. Identification of histocompatibility systems also will contribute to specific identification of animals both for genetic studies and for disease surveillance.
2. There is need for development of new and improved vaccines containing known protective antigens in quantifiable and stable forms so that optimal immunizing schedules giving reproducible results can be applied.
3. There is need for ways in which different elements of the immune system, such as macrophages and lymphocytes, interact with one another, especially as this relates to the nonspecific innate, or paraimmunological mechanisms of resistance. Substances such as complement, transfer factor, or interferon must be examined. Expanded production of some of these factors may provide means of passively protecting animals during critical times (e.g., neonatal and stress situations) or treating infections which are currently incurable.

4. There is need to develop by modern immunogiological methods highly specific immunochemical reagents which can be used to enhance the sensitivity and specificity of diagnostic procedures for infectious agents and for immunoassay of hormones, drugs and chemicals. The purity, homogeneity, specificity and reproducibility of in vitro produced reagents will overcome difficulties in standardization of methods.
5. There is need for comprehensive understanding of development of immune responsiveness in all food-animal species with the specific objective of enhancing the effectiveness of maternally derived protection for the offspring. Studies should emphasize resistance mechanisms which cooperate to establish protective responses at the mucosal surfaces such as in gut, respiratory and genital tracts that are prone to assault by infectious agents.
6. There is need to find ways to obtain, evaluate and conserve germ plasm of domestic animals so that needed traits can be introduced into food animals to reduce their vulnerability.

GENETIC ENGINEERING

STATE OF THE ART

Microbes are being altered genetically by biochemical procedures to produce human hormones. Thus, the genetic information for production of insulin and growth hormone has been introduced into Escherichia coli, a common bacterium. These hormones soon will be produced in large amounts in a highly purified form by bacteria growing in large cultures. Work has begun on introducing genetic ability to produce antigens of foot-and-mouth disease virus into E. coli so that safe vaccines can be produced against this important disease.

Bacteria that produce diphtheria are harmless unless they are infected with a virus. Virus-infected bacteria produce a highly potent toxin. How most other disease organisms cause disease is not known. This is particularly true of organisms which, unlike diphtheria organisms, live within the cell and affect normal functioning of the cell. Virus infections are examples of intracellular parasites which interfere with normal metabolism and may cause disease.

CRITICAL CONSTRAINTS

1. Techniques for genetic engineering of microbes of medical importance have just become available. They are primitive and need refinement. Also, researchers in animal health need to acquire and use the techniques since there are many animal health problems to which the techniques should be applied.
2. There is a societal concern that microbes may be engineered so that they become more dangerous and aggrevate rather than help animal health problems.

3. Because techniques are new, many reagents are not available commercially. Thus, researchers must prepare the reagents themselves, which is costly of time and resources.
4. Little is known about the molecular genetics of microbes important to animal health. Such information is necessary before genetic engineering can be started.

RESEARCH IMPERATIVES

1. Molecular and biochemical mechanisms of pathogeneity of disease-causing organisms need to be determined so that pathogenicity can be blocked.
2. Antigens against which the animal's immunity is directed should be identified, isolated, and their molecular structure determined. Through recombinant DNA techniques the genetic information for these antigens should be introduced into harmless bacteria so that large quantities of purified antigens can be made. These antigens must be tested for their safety and for their ability to induce a protective response against disease. The antigens also may be useful in diagnostic tests for disease.
3. Bacteria, viruses, and other parasitic organisms need to be modified genetically so that they do not cause disease but retain their immunizing ability. They also may prevent disease by establishing an infection and interfering with the pathogenic mechanism of the disease-causing organism or by parasitizing the animal, organism or arthropod. For example, some organisms growing in the gut will exclude competitively others from growing there too.
4. Research is needed on the promising area of genetic surgery, i.e., removal of genetic information from one organism and inserting it into another. This technique can be applied to almost all living organisms from viruses to mammals. Animals could be engineered genetically to be more resistant to disease, to be more productive and to have many other desirable characteristics.

ANTIMICROBIALS FOR LIVESTOCK

STATE OF THE ART

Utilization of antimicrobials has revolutionized food animal production and health. Along with control of important bacterial diseases, use of antimicrobials has been associated with more rapid growth rates and feed efficiency of meat producing animals. Current governmental restrictions which derive from human health related problems threaten to limit use of antibiotics in animals. Without new knowledge and a systematic analysis of their effectiveness for systems of animal production and their impact on human health and medical practice, use of antibiotics could be lost for food-producing animals.

CRITICAL CONSTRAINTS

These constraints affect availability of antibiotics for use in animals for production of food.

1. Increased number and frequency of antibiotic-resistant strains of bacteria that cause disease in humans has led to possibility of governmental regulation of drug availability.
2. Of concern to those attempting to control important infectious diseases is the increased number of antibiotic-resistant strains of bacteria associated with fulminating disease in food producing animals.
3. There is a lack of knowledge of animal to animal or animal to human transfer of resistance factors. Design of experiments to test these interactions is difficult in light of the widespread use of antibiotics for therapeutic and prophylactic purposes in animals and humans.

RESEARCH IMPERATIVES

1. Better detection and screening techniques for residues of antimicrobials and their metabolites in animal products are necessary for assuring human safety.
2. Bioavailability and potential activity of drug metabolites in animal products destined for human consumption warrant investigation.
3. The role that antimicrobials or their residues in food play in sensitization and subsequent hypersensitive reactions in humans should be ascertained.
4. Associations and frequency of antibiotic resistances and their relationship to frequency of animal diseases need to be evaluated.
5. Biomolecular techniques need to be applied to studies on the mechanism of action of antimicrobials as well as their metabolic pathways in species of animals and in humans.
6. Continual efforts need to be directed towards broadening the spectrum of antimicrobials for control of diseases, internal parasites, external parasites and for potential use as growth promotants.
7. Further definition of effective additive or compromising therapies resulting from the use of combinations of antibiotics is needed (e.g., certain combinations of antibiotics compromise the therapeutic benefits of each of the antibiotics in the combination).
8. The influence of antibiotics on microbial populations or metabolism of mammalian systems leading to adverse effects (e.g., tumors) needs clarification.

ANIMAL WELFARE

As we approach and plan for the 21st century, giving attention to the availability, wholesomeness and nutrient quality of food products of animal origin, let us not lose sight of the fact that what is desired is obtainable only through life . . . animal life. The comfort and well-being of animals in production units should not be in conflict with production efficiencies, as an optimum physiological state should enhance the opportunity to reach the objective of food-animal agriculture.

"...the great time compressor for productive change is scientific research as the scouting party for what ought to be done next."
Glenn Salisbury

"...The ruminant is the key to using vast amounts of land that would otherwise go to waste as a food-producing resource -- land that is too steep, too arid or otherwise unsuitable for crop production."
Anson Bertrand

TABLE 7.1 STATUS OF THE TECHNOLOGY FOR CONTROL OF SELECTED IMPORTANT INFECTIOUS DISEASES AND PARASITES OF FOOD ANIMALS

Disease	Cause	Research needs					
		Diagnoses	Epidemiology	Vector	Vaccine	Prophylaxis/ therapy	Control methods
Exotic diseases and parasites							
Foot and mouth disease	Virus				X		
African swine fever	Virus		X	X	X		
East coast fever	Protozoan			X		X	
Contagious bovine pleuro-pneumonia	Bacteria					X	
Screwworm	Insect		X				X
Heartworm	Richettsia	X		X			
Domestic diseases and parasites							
Anaplasmosis	Bacteria			X	X		
Brucellosis	Bacteria	X			X		
Bluetongue	Virus		X	X	X		
Scrapie	Unknown	X					X
Pseudorabies	Virus	X			X		
Influenza	Virus		X		X		
Coccidiosis	Protozoa					X	X
Roundworms	Helminth					X	X
Mange	Arthropod	X				X	X
Ringworm	Fungus					X	

8

W. F. Wedin (Co-Chairman), H. J. Hodgson (Co-Chairman), J. E. Oldfied (Rapporteur), K. J. Frey (Rapporteur)

C. W. Deyoe, R. S. Emery, L. Hahn, V. W. Hays, C. H. Herbel, J. S. Hillman, T. J. Klopfenstein, W. E. Larson, V. L. Lechtenberg, G. C. Marten, P. W. Moe, D. Polin, J. M. Sweeten, W. C. Templeton, P. J. Van Soest, R. L. Vetter, and W. J. Waldrip

FEED PRODUCTION

SUMMARY

One of the primary constraints to animal production is an adequate supply of feed. As the world human population increases, with consequently increased demand for animal products (especially proteins) in the human diet, there must be increases and improvements in the supply of feed for animal production.

Much of the improvement in animal productivity of meat and milk that has taken place in the U.S. during the past 2 or 3 decades has resulted from extensive use of feed grains. Attention should continue to be given to feed grain production because of its importance, with emphasis on the potential for genetic improvement by plant breeders. There is evidence, however, which suggests that the importance of grain as a feed base may be declining and that its place, especially for ruminant animals, will be taken, to an increasing degree, by forage or alternate feed sources. Other uses of grains, including their export as an international trade item and their domestic use as a fermentation base for liquid fuel production, as well as decreased demand for lean meat, all contribute to this trend.

There is also an increasing emphasis on conservation of resources, which will increase the cost of land and water and the inputs required for feed grain and forage production. Excessive soil loss from mismanaged cropping is coming under increasing criticism and regulation.

Given that animal production systems in the 21st century will be subject to higher energy costs and that instability will result from alternate demands for a limited supply of grain crops, the exact type of enterprise for the future cannot be predicted. Yet the forces impinging on feed production for animals could lead to a new era in which the animal production systems are part of the total resource conservation for nutrients, water, soil and energy, while producing the feed needed.

The need is apparent for establishment of an extended base of animal feedstuffs. Although forage has served as feed for centuries, the technology of its production and processing has not developed to the extent of that applied to the production of grains and other concentrate feeds. Alternate feed sources include both materials already studied and utilized to some degree, as well as materials not currently thought of as animal feed sources.

Given likely constraints of the future, and a history of past research breakthroughs, we can delineate imperative research areas for the future. The imperatives in three major program areas are:

I. Increase utilization efficiency of ligno-cellulosic plant materials.

Ligno-cellulosic (plant fiber) materials comprise one of the most abundant organic compounds on earth. They are renewable but, unfortunately, they are utilized with relatively low efficiency. Through research, their efficiency of utilization can be increased. There are five research areas which require major research attention.

1. Develop basic information on biochemistry of synthesis and degradation of ligno-cellulosic forage materials.
2. Improve procedures for measuring forage quality.
3. Improve harvesting, storing and processing technologies to preserve and enhance feed quality.
4. Enhance forage quality through plant breeding.
5. Optimize feed quality through improved crop management.

II. Increase quantity of animal feed produced.

Increased plant productivity may be accomplished with the inputs presently available, provided a more nutritious plant product can be developed. But because feed for animals, particularly forage, of necessity may be produced on marginal lands that have mineral or water stress, there are both genetic and environmental (and their interactions) factors which must be researched. Four research areas are of urgent importance.

1. Improve forage and grain species by plant breeding.
2. Integrate new crop cultivars into feed production strategies that optimize and preserve natural resources, yet maximize efficiency of energy use.
3. Improve crop management, harvesting and storage methods to minimize losses of plant material.
4. Increase exploitation of legumes that fix their own nitrogen and that have higher feed intake potential and overall forage quality than do grasses.

III. Evaluate alternate feed sources.

Identification of new and different feed materials should be pursued. Particular attention should be given to the recycling of by-products and wastes from agricultural, domestic and industry sources, in the interests of conservation of resources and economy. Favorable yield characteristics of unconventional feed sources, such as certain types of trees (alder, aspen, poplar), may be exploited to further extend the feed base. Feed production areas can be increased through investigation of the nutrient value of aquatic materials, both plants and animals (fish). Three research areas require prompt consideration, such that early progress can be made.

1. Develop efficient methods for delignifying carbohydrate lignin complexes in woody and vegetative tissues of plants.
2. Improve methods used for handling, processing and storing residues to preserve and improve their nutritional quality and safety for use as feed.
3. Identify and evaluate new sources of concentrate feeds.

All of these research imperatives, by increasing the supplies of available nutrients for animal feed, can effectively support a continued productive animal agriculture in this country and the world and, therefore, represent wise investment in the future welfare of mankind.

INTRODUCTION

Feed supply is the basic resource for the animal industry. Adequate feed quantities with appropriate quality for the various classes and ages of livestock are essential for sustained, economic production of human foods from animal origin.

The past several decades have been typified by large increases in the amounts of concentrate feed available, usually at reasonable prices, for livestock. This has brought about efficient production of animal products. Even ruminant livestock have become large consumers of grain and other concentrates.

Feed production, as we presently know it in the U.S., will change in the decades ahead. Factors which are fundamentally agronomic, environmental, social, technical and economic will influence these changes.

Because of anticipated changes in our understanding of what feed shall be in the future, in this chapter we define feed, production and feed production as follows:

Feed - Any material ingested by animals from which they obtain one or more of the major nutrients (carbohydrates, protein, lipids, minerals, vitamins);

Production - Procurement of a product from growing, harvesting or salvaging;

Feed production - The husbanding of inputs (land, nutrients, cardinal growth factors, plants) to procure feed for animals.

Americans consume about 645 kg of food per capita annually, of which some 273 kg are foods of animal origin. If animal production is to sustain this consumption level in the future, especially with anticipated population growth, it is obvious that an expanded feed resource for animals must be provided. Presently, basic components of animal diets in the U.S. are chiefly plant materials, i.e., either concentrates (mostly cereal grains and their by-products) or roughages (vegetative portions of plants in fresh or preserved form) (table 8.1).

Provision of needed animal feeds for the future likely will involve more than simple expansion of the present feed categories. A number of forces are at work which suggest significant departures from the traditional feed supply. In particular, cereal grains for livestock feeding may be limited for several reasons. Grains are a major export commodity. Their export is likely to continue and to grow. Domestically, large quantities of grains may be used to produce liquid fuels. These forces almost certainly will cause substitution of forages for grains in the animal diet, a factor which could change production to patterns using less fossil energy in animal agriculture. Changes in American dietary patterns are demanding less fat in food products of animal origin which argues for less use of grain and greater dependence on forages in livestock finishing diets.

Plans for animal feed production in the future, therefore, must call for a greater variety of feedstuffs. As such, a technology that utilizes agricultural and industrial by-products and wastes as feeds is attractive because of its dual potential to provide nutrients for animal production and to reduce environmental contamination. Also needed are efforts to improve feed materials already used, by increasing their yields of available nutrients, improving their utilization by livestock, preserving their values over prolonged storage periods and removing inhibitory or hazardous factors they may contain. Such efforts, as detailed below, will provide for continued production of foods from animal origin to enhance the health and well-being of the citizens of this country.

BACKGROUND

FEED--RESOURCES, PRODUCTION

Land and Water Resources

Feed production in the U.S. is dependent on soil and water resources. On the other hand, production of animals is a major, or sometimes the only, avenue through which these material resources can be utilized for contributing to our human food supply. That is, much land in this country is marginal and only via animals can products of value be generated from it. In this section, a brief description of land and water use is outlined.

Total land use in the U.S. approximates 918 million hectares (2,264 acres). In 1974 (table 8.2) 17% was in cropland (excluding cropland used only for pasture), 30% in available grassland pasture and range, 32% in forest and woodland, 8% in special uses and 13% in unclassified uses. Urban uses now claim more than twice as much land as in 1950, and highway and airport use is growing. Some land classified as forest and woodland is grazed by livestock. Part of the agricultural land that has been shifted from cropland to other uses since 1950 could be put back into production, even though it has small field size, poor fertility and steep terrain.

Water and related natural resource phenomena also impinge on the animal feed resource base. Water quantity and quality, as well as water utilization, are major factors affecting potential feed and forage production. Other factors in addition to soil and water involve the aggregate productivity: wetlands, riparian vegetation, wildlife habitat, windbreakers, recreation, wilderness areas, landscape resources and others.

Our land and water resource base presently appears sufficient to provide for intermediate-term feed and forage production. Several constraints to be detailed later in this chapter, are mentioned here to further describe the land base. Quality of the land base is dropping because Class, I, II and III lands are diminishing in amount. Wind and water erosion, salt intrusion and other destructive forces are prime obstacles to increasing production. Millions of hectares have been, or are being, converted to urban and other uses. Ground water depletion is serious, and is critical in some areas.

Grain, Hay, Silage and Pasture and Rangeland Production

The plant resources for production of animal feeds consist of forage and silage crops, feed and food grains and crop residues, together with pasture and rangelands and by-products of crops processed for other primary uses.

Areas of land devoted to production of grain, hay and silage crops together with estimated 1979 production, are presented in table 8.3. Nonruminant animals depend heavily on grain and soybean producing lands, while ruminants utilize production from not only cropland shown in table 8.2, but also from pasture, rangeland and other grazing land. Major grain crops occupied 66.8 million hectares and produced some 293 million metric tons of grain in 1979. Corn and wheat production increased from 73% of the total area producing grain crops in 1973 to 81% in 1979.

The last decade represented the culmination of the post World War II period during which grain production increased dramatically as a result of the application of research-based technology. This technology included improved crop cultivars and soil management practices, such as earlier planting, higher plant populations, better pest control and increased use of fertilizer and irrigation. Generally, further increases in yield from these management practices will be difficult, and most additional gains must come from genetic improvement or new, innovative management techniques.

During most of the last decade, the increased grain production, except for wheat, was utilized as livestock feed. Certainly, large quantities of low-priced grain have played a major role in increased animal productivity during this period. Table 8.4 indicates the percentages of the leading grain crops that are used for various purposes. Note that the portion of grain production used as feed declined between 1971 and 1977, while the amount exported rose sharply. Further growth in world population probably will increase the portion of grain production that is exported.

Industrial usage of grain remained relatively constant between 1971 and 1977. Now emerging, however, are new industrial uses such as production of sweeteners and liquid fuels from corn, which may demand substantial portions of our grain production. Some estimates have been as high as 100 to 150 million metric tons of grain annually being converted to liquid fuel production by the year 2000.

Liquid fuel production and export markets will compete with livestock for grain. Unless production is greatly increased, the availability of grain for livestock, particularly ruminants, is likely to be substantially reduced.

Hay and silage production approximates 29 million hectares (table 8.3). Also more than 30 million hectares of cropland are devoted to pasture. Thus, the cropland area devoted to forage production nearly equals the area devoted to grain production. Additionally, permanent pasture and rangeland used some 275 million hectares, most of which is unsuitable for cropland use. Production from these lands, plus crop residues, provides 60% of the feed units for all livestock and 80% for ruminants.

Permanent pasture and rangeland occupy large areas of land unsuitable for the cultivation of crops (too steep, arid, wet, rocky). Alternate uses are restricted, yet major improvement is possible from control of unwanted plants, water-management, revegetation and fertilization, and through management systems with one or more of these improvements.

Breeding herds of sheep and beef cattle receive a substantial portion of their nutrient needs from pasture and range forage. On range, particularly, production per unit area is inherently low, but production per unit of input (capital, labor, energy, etc.) is high.

In contrast to grain production, technology application to forage production is generally low. Thus, the potentials for increasing livestock feed from forage producing lands is large. Application of improved technology to cultivated forages, pastures and rangeland could increase ruminant livestock production several-fold. Increases in yield of dry matter and improvements in dry matter digestibility could occur from fertilizer and lime application, increased use of legumes, more timely harvest and grazing, and improved harvest and storage technologies.

Almost all the forage produced in the U.S. is consumed by ruminant livestock, and there are few competitive uses. Recent interest in the development of technology for the economic conversion of cellulosic materials to liquid fuels, however, have given rise to estimates of several hundred million metric tons of ligno-cellulosic materials being used annually for their new development. Whether such levels are reached by the year 2000 is problematical, but the potentials exist. Should these technologies develop, the impacts on ruminant livestock production would be substantial. Feed production in animal agriculture would remain important, but provisions for research and development to offset this new demand would be necessary. Increased yields resulting from application of presently available technology could make it possible to meet the needs without reducing the amounts of forage currently available to animal agriculture. Additionally, the alternative use would provide an economic market for low quality forage. There would be an incentive for increasing quality of forage fed to animals.

The use of feed grains and concentrates in general is diminishing as more forages are being used for production of ruminant animals (table 8.5). The proportion of the animal diet supplied by forages (corn equivalent basis) increased by 1978 for four animal-type categories. The proportion of the available grains that were actually used for feed in 1971 vs 1977 was given in table 8.4. The alternative uses for grain mentioned earlier may accentuate this trend in the future.

Agricultural Residues

Crop and animal residues provide an animal feed resource that may play a significant role in the future. The crop and animal residues can substitute for grains, or they can increase production through an expanded feed supply. A reasonable level of residue utilization can be compatible with soil conservation practices. Agricultural residues available by regions in the U.S. are summarized in table 8.6. Almost 60% of all residue available is found in the Corn Belt (eight states) and Plains (eight states).

For the U.S. as a whole, stovers of corn and small grain offer the best potential as feed. Furthermore, of the total available crop residue, about 78% occurs in the Corn Belt and Plains regions (table 8.6). Corn, sorghum, barley, oats, wheat and soybeans contribute 95% of these residues, with corn and wheat comprising 54% and 20%, respectively.

Residues of fruit and vegetable crops are local in nature and are not easy to use efficiently. Forestry residues make up a substantial portion of total available agricultural residues (table 8.6), but these residues probably will have the most economical use in the fabrication of new wood products or as fuel. Manure residues comprise about 6% of all agricultural residues (table 8.6). Some of these animal wastes can be fed. It can be concluded that the stovers of corn and small grain, plus cattle and poultry wastes, offers the most promise as feed. In least-cost diets, they often appeared to be potential substitutes for traditional feeds.

Energy Implications

Grain and forage crops have traditionally constituted the major feed source for the U.S. livestock industry. Competitive uses for feed grains, such as exports, food and industrial uses have long existed, but these uses have represented a relatively small proportion of the U.S. feed grain crop. Since 1977 exports of feed grains have increased and now represent approximately one-third of the total U.S. grain production. Exports are expected to represent an even greater proportion of production in the future.

In addition to increased exports, the industrial use of grain crops could increase substantially in the near future. A recent report by the Office of Technology Assessment (OTA) of the U.S. Congress (1980) estimates that the potential exists to produce as much as 11 billion liters of liquid fuel from grain and sugar crops annually by 1985.

Whether this potential is reached depends on economic developments and policy decisions at the national level. Beyond the year 2000, the use of U.S. grain for fuel production appears less certain. The development of other energy alternatives could then reduce the use of grain for liquid fuels. Even if the conversion of grain to liquid fuel does not reach the upper level cited, it is likely that the combination of new industrial uses and exports will result in greater demand for grain in the future.

Shifts in Usage

Expected increase in demand for grain does not imply that feed grains will be unavailable for animal feeding. Likely, they will be expensive and subject to volatile price fluctuations. Animal production systems of the future will need to be flexible enough to effectively use alternative feeds, e.g., high quality forages and by-products from the industrial uses of grain.

If a national policy was adopted to develop renewable energy resources in the U.S., it would not only affect the supply and price of grain for livestock, but would also have a far-reaching impact on the production and use of fibrous crops that have traditionally been used for ruminant livestock feed. If technology currently under development for converting cellulosic crop materials to fuels becomes economically feasible, a competitive nonfeed use will exist for traditional forage crops. The OTA Report (1980) estimates that from 91 to 233 million metric tons of forage and 55 to 78 million tons of crop residues could potentially be used for fuel purposes annually before the year 2000. The diversion of this quantity of cellulosic crop materials to fuel use, without decreasing animal production, could occur only by an increase in annual forage production of 50 to 100% in the Eastern U.S. While this magnitude of increase appears possible technically, challenging problems would arise in integrating such a system with present livestock feed production.

The development of an industrial forage utilization system would create a cash market for forage crops that would greatly stimulate their improved production and management. This could have several important effects on crop and livestock production. Conservation practices which involve forages in rotation could become profitable.

While this development could increase the cost of forage feeds for animals by creating an alternative demand and use, it could also stimulate the production and feeding of high quality forages by providing a stable market for low quality forages, rained-on hay and forages otherwise mismanaged. Higher quality forages could be reserved for animal feeding systems requiring their use.

Protein Sources

It is generally acknowledged that animal protein is a major contributor to the human diet and, because of its amino acid composition in relation to human dietary needs, its continued involvement in human diets is advocated. There is substantial evidence that proteins from meat and milk played a significant role in the history of human development and that strong and stable cultures and nations have emerged through association with animal agriculture.

The significance of this historical development in relation to feed production is that, in order to produce the protein needed for human use, supplies of protein, or of nitrogen in the case of ruminants, must be found for animal diets. The energetic efficiency of the animal diet depends upon a correct mix of a number of dietary essentials, including protein. It is not enough, therefore, merely to identify sources of energy, but adequate supplies of protein and other nutrients must be available also.

There are two major avenues by which improved supplies of feed proteins can be approached: 1) Existing protein sources can be improved in some way so as to improve their efficiency of utilization by animals; and 2) new and different forms of protein can be identified and tested in animal-feeding experiments.

To assess these approaches, it is necessary to differentiate between the protein requirements of various animal species, particularly between ruminant and nonruminant animals. It has been demonstrated clearly that ruminant animals cannot only exist but also can grow and reproduce on diets containing no preformed protein, if sufficient nitrogen is present. On the other hand, nonruminant species such as the pig require preformed protein of high quality and of high biological availability. The usefulness of dietary protein may be impaired by the presence of indigestible materials.

Plant fiber occurs in close association with protein in many feed sources as, for example, legume and grass forages. Attention has been directed worldwide to the production of "leaf protein concentrates" (LPC) which would effectively free protein from the undesirable association. Several strategies have been proposed. Simple physical separation of leaf from stem in crops such as alfalfa can enhance the protein value of the leaf fraction considerably. Most efforts, however, have been directed toward coagulation of protein from juices expressed from forage plants under pressure. Such processing has reached the point of limited commercial application. The LPC produced has been prosposed for direct consumption by humans; if such a market develops, the residual stemmy materials could find useful outlets in diets for ruminant livestock.

Another example of improving existing protein resources lies in the animal product area. Traditionally, residual materials from slaughterhouses, including blood, hair, viscera and trimmings have been recycled into the animal feed trade as meat and bone meals and tankage. In addition to their recognized nutritional qualities, use of such products is encouraged by their consequent removal as potential pollutants of the environment. The products are frequently less than ideal protein supplements, however, due either to characteristics of their base composition (excess bone gives too high a calcium content, for example) or to problems caused by the processing they have undergone - too high a temperature may damage essential amino acids, such as lysine. Improvement may be obtained by fractionation procedures.

Identification of new sources of protein is one of the most exciting and potentially most valuable research areas in the whole animal feed production picture. It can involve variations on traditional feedstuffs such as high-lysine corn, high-lysine barley and the more recent hybrid cereal, triticale. Other new variants of recognized protein sources have occurred among the residual meals produced in the processing of oil-bearing subcrops. Rapeseed meal, for example, is being produced in great quantity in Canada as a result of encouragement toward crop diversification of wheat lands by the Canadian government. Early experiences with rapeseed meal as a protein source were unsatisfactory due to presence in the meal of glucosinolates with goitrogenic properties. These problems have been overcome through the efforts of Canadian plant breeders who have successfully produced low-glucosinolate cultivars, such as Tower.

An area of great promise in the development of new protein sources is the investigation of microbial cultures (Single-Cell Protein). The ability of microorganisms to synthesize protein and other nutrients, plus their characteristics as "biological filters" in removing or modifying substances that might otherwise result in environmental insult, make their exploitation as protein providers potentially valuable. Single-cell protein production has application in the disposal or recycling of animal manures and of certain industrial wastes that are produced in large quantities in some areas, such as the sulphite liquor effluents from paper manufacturing plants. In some cases, too, microorganisms can be produced from hitherto-unstrapped substrate materials such as hydrocarbons or petrochemical residues. Microbial activity also can be used to enhance the nutrient qualities of feeds, as exemplified by the production of "activated sludges" from domestic and industrial wastes and the generation of needed supplementary nutrient sources, including vitamins and amino acids, through fermentative processes.

FEED--QUALITY, PROCESSING AND STORAGE, UTILIZATION

Forage and Feed Grain Quality

Forage and feed grain quality is crucial to efficient production of food products by animals. High levels of intake of highly digestible and nutritious feeds run hand in hand with efficient production. If intake is high, animals then consume protein, energy, minerals and other nutrients in excess of those needed for body maintenance. These extra nutrients are available for meat, milk and egg production. While nonruminants such as poultry and swine consume most of the protein, energy and other nutrients from feeds that could be used for human food, ruminants such as cattle and sheep consume forage legumes and grasses, as well as other cellulosic feeds, which provide the basic dietary needs. Presently, additional concentrate feeds are fed in substantial quantities to ruminants to insure high levels of meat and milk production in American animal agriculture. The relationship of feed quality to animal needs in different production categories (e.g., brood cows that require lower levels of nutrients than do high-producing milking cows) has often been overlooked.

Management schemes have been devised to increase forage quality. Among these are frequent cutting of immature plants when at high quality, improved fertilization practices, use of legumes with both high intake potential and high protein and use of legume-grass mixtures where they are adapted. Quality of a few selected forage species has been markedly improved by breeding for increased digestibility (e.g., bermudagrass and pearl millet) or for reduced anti-quality constituents (e.g., indole alkaloids in reed canarygrass). Factors that cause bloat in grazing animals on high-quality legumes have been partially identified. Heritability of some of the quality and anti-quality factors and the influence of the environment on their presence have been established. Yet potential for additional quality improvements is great. Through searches for other means of identifying and quantifying quality parameters or anti-quality substances, progress can be made. It is commonly accepted that environmental or genetic control of these quality factors is much greater than that which has already been realized.

In practice, dairy and beef farmers can shift from excessive dependence on feed grains requiring substantial fossil fuel energy input. If high-quality forage is substituted for low-quality forages, high yields of milk and meat can be realized. Top-grade alfalfa hay production is possible on many more farms. The potential for consistent procurement of such high-quality forage can be increased through more research. For grazing, management schemes have included an oversupply of forage early in the pasture season and deficiencies during the "summer slump" and late-autumn periods. More research is needed to determine the potential for integration of alternative high-quality plant species into season-long grazing systems.

In most instances, plant breeders of grain crops have developed cultivars that are high yielding, lodging and disease resistant, but generally they have not bred or monitored for changes in feed quality of grain per se or straw and stover. Some past objectives in breeding grains may have caused indirect reduction in feed quality of cereal residues, e.g., stiff, lodging-resistant corn stalks, which contain a high concentration of indigestible lignin. Certainly, increasing protein and other nutrients in grain crops has not been a primary objective of breeders. In fact, corn in the U.S. today averages about one percentage unit less protein than corn of 40 years ago. Plant breeding research might alleviate this situation.

Processing and Storage

For efficient storage of feed, the harvested nutrients must be conserved. Changes in composition and quality of stored feeds is associated with losses from respiration and fermentation. The alteration in composition results from a replacement of metabolizable components by the products of microorganisms. There is often loss in net mass through respiration. This includes selective loss of the more metabolizable and valuable components, i.e., protein and energy.

Molds and aerobic processes can cause greater loss of organic matter, and can produce more toxins and heat, than is the case with anaerobic fermentation. The heat, in turn, can promote nonbiological degradation of the protein and carbohydrate by means of the nonenzymatic browning reaction.

The losses attendant upon such biological and nonbiological processes are very substantial and unpredictable. The technology controlling and releasing these factors is inadequate, and the mechanisms are incompletely understood.

On most farms, forages are not produced as cash crops. Therefore, forage harvest, storage and processing are not given first priority. Because of inclement weather and other demands for equipment and labor, forages are too often harvested at a stage of maturity beyond maximum quality. Processing and feeding of hay crops is dependent on equipment, fuel and labor. Processing may aid voluntary intake and may be necessary for proper diet mixing in some cases. Self-feeding of livestock from packages may cause feed losses due to trampling and fouling.

Harvesting losses are high in haymaking, especially of alfalfa which comprises over 60% of the hay tonnage. The recent trend has been to reduce labor and further mechanize haymaking by use of large packages (stacks and round bales). Leaf loss in packaging can be appreciable. In humid areas, rain may further decrease yield and quality. Storage losses have also increased with the advent of large package systems because the packages are often not protected from the environment. This reduces storage-building cost, but it increases feed losses.

Alfalfa and grasses stored as silage or haylage are not as subject to harvesting loss problems. However, storage losses of dry matter and quality can be appreciable. Direct cutting of wet forages leads to nutrient seepage in the silo and to fermentation and animal intake problems. Field wilting can overcome these problems, but it increases equipment, fuel and labor costs. There is an inverse relationship between cost and storage efficiency of storage structures for silage.

Other silage crops such as corn, forage sorghum and small grain forages supply large yields. However, storage losses may be large and variable.

Grains are traditionally dried and transported for livestock feeding or for export. Most grains are processed by grinding, rolling or steam flaking before feeding to animals. Energy costs for drying and processing continue to increase and require research on alternatives. Grains scheduled for livestock feeding can be stored and processed as high-moisture feeds to eliminate drying costs and reduce the potential for field harvest losses. Also, chemical processing of grains can enhance storage characteristics.

Feed grains are stored both on the farm and commercially. Farm storage has increased rapidly in many grain producing areas following changes in production, transportation and marketing plans, and as a result of a lack of commercial storage capacity at harvest times. Losses in grains occur as a result of (a) insect infestation, (b) microbiological activity and physical damage and (c) losses during handling. Grain inspections indicate increased occurrence of insect fragments, infestation, damaged kernels and lower grain grades in the trade. These losses are variable and may represent substantial losses of feed grains available for livestock production.

Harvest, storage and feeding of crop residues can be quite expensive relative to their forage quality and economic value. Corn and sorghum residues can be grazed by ruminants with little or no cost, no fuel use and with minimum reduction in soil tilth. The practice is highly weather-dependent and understandably confined to regions where cattle and residues are both produced.

Most crop residue harvesting has been conducted with large hay package equipment. This equipment is expensive unless used for alternative forage harvest on the same farm. Cost of grinding and feeding residues is high and similar to that of hay. The cost of energy for mechanical harvest of forage crops and crop residues may necessitate increased harvest by grazing.

Treatments of residues with sodium hydroxide, calcium hydroxide and ammonia have increased their digestibility by 10 to 20 percentage units. The increased digestibility may be sufficient to cover the cost of harvest and storage.

CONSTRAINTS INFLUENCING FUTURE DEVELOPMENTS IN ANIMAL AGRICULTURE

There are a number of constraints to increased feed production for animals in the next century. These constraints include the limitation of our natural resources (land, water, climate, energy) and the inability of animals to effectively metabolize cellulose.

LIGNO-CELLULOSE METABOLISM

A major factor limiting the supply of feed nutrients for animals is the inefficiency with which food-producing animals utilize ligno-cellulosic feedstuffs.

It is generally agreed that feed grains will be less available for U.S. animal production in the 21st century than at the present time. Alternative feed supplies, therefore, will be needed to sustain high levels of animal production. Cellulosic plant materials are one of the most abundantly produced materials, with production in nearly every climatic region on earth. The energy content of cellulosic plants is equal to that of starchy crops; however, this energy is not efficiently utilized by animals. The major barriers limiting their efficient utilization are the highly ordered structure of cellulose and the association of cellulose with lignin.

A major breakthrough could be achieved if the energy in ligno-cellulose materials could be made readily and efficiently available to animals. This would make possible the use of many alternative feeds in animal production. Crop residues, cellulosic industrial wastes, animal wastes and a wide array of other materials could be utilized.

Low bioavailability of ligno-cellulose is a constraint of such magnitude and far-reaching implication that a major, high priority research effort should be martialed to develop and obtain the information necessary to "unlock" this resource. Success in this area would not only make available an almost inexhaustible supply of animal feed, but also would have profound effects on the ability to produce fuel from cellulosic biomass.

SOILS AND LAND USE

While it might appear that our natural resource base is sufficient for adequate forage and feed production, several current developments impinging upon agricultural land use will greatly tax our soil and water resources. At present we are losing annually from .4-1.2 million hectares (1-3 million acres) to roads, airports, plus industrial and urban development. Some estimates suggest that we may need to convert from 20-28 million hectares (50-70 million acres) to crops being grown for energy purposes only. In addition, many acres are being seriously damaged by wind and water erosion, salt intrusion and other destructive forces. These losses of cultivated land, plus possible major shifts in agricultural land resulting from the high costs of energy, will cause changes in the type and amount of livestock feed produced, as well as where it is grown. For example, USDA estimates that we have about .4 million hectares (100 million acres) of land not now in crops that could be converted to cropland with low to medium risk. Approximately another 12 million hectares (30 million acres) could be converted, but at high risk from erosion and other hazards. While the additional land is scattered and needs development, much of it is located in the Southeast and Lake States. If it becomes necessary to use the additional land, much of it may be best suited for cultivated forage crops to produce feed for ruminants.

The contemporary concern for the soil resource follows a period in which public agencies infused substantial support to reverse trends in soil exploitation. But, because cropping practices have changed markedly in the past 3 decades, today's patterns differ greatly from those of 3 decades ago (table 8.7).

The reliability of the plant resource from which the bulk of the feed in this country is derived is dependent on the continued productivity of the soils themselves, and the environment for plants which grow on those soils. To consider what feed production will entail in the future, it is necessary to assess this land base as a constraint.

Soil erosion resulting from changes in cropping is discussed thoroughly in Special Publication No. 25 of the Soil Conservation Society of America (1979). In general, all of the reports state that the soil loss from cropland in the U.S. is at alarming levels. As a basis for several studies reported therein, crop areas and yields from 1972 to 1977 were compiled for major land resource areas (MLRA) and used in computer models. Soil loss from cropland in all MLRA's is about 2.7 billion metric tons a year (about 40 metric tons per hectare per year), but estimated tolerance level is about 50% of this amount.

It is evident that our present soil degradation should not continue, particularly with the mandated controls related to environmental quality (e.g., Section 208 of Public Law 92-500 concerning nonpoint water pollution). Crop residue management is one of our best means to control soil erosion. The first need for residues is to maintain soil productivity. Amount of crop residue in excess of those needed to maintain adequate soil erosion control were calculated and included in the Soil Conservation Society Special Report (1979). About 49 million metric tons in the Corn Belt or some 36% of the total crop residue produced was available for removal from the land. The nation's corn crop accounts for 65% of the total residue. Two-thirds of all of the residue which can be removed would come from only four of the 14 MLRA's in the Corn Belt, those soils which are deep and level. In the Great Plains, only 14% could be removed; this is 20% of the residues produced.

Implications to feed production when more restrictive cropping practices are used have been studied in the Center for Agriculture and Rural Development (CARD) at Iowa State University. In one study, four conservation practices (straight row cropping, contouring, strip cropping, terracing) in various crop management systems, and three tillage practices (conventional tillage with the residue removed, conventional tillage with residue left, reduced tillage) as they affect crop production and least-cost diet for livestock were compared. As the level of soil conservation rose, hay acreage showed a steady and substantial increase. Hay acreage expanded because it was economical as a soil conservation measure relative to such alternatives as additional terracing. A further consequence of more hay was a substitution of hay for silage in livestock diets. With increasing soil conservation, a decline in acreage of corn and sorghum silage occurred. Corn production was forced to shift to less productive land, e.g., the Great Plains, and there the amount of fertilizer and pesticide required to raise a bushel of corn increased.

High and low export of feed grains have been examined, also. In a projection to 1985, one study reported that with adequate soil erosion measures, row crops become costly. As a result, the use of corn in livestock feeding declines. The major cost of soil loss to farmers is the loss of future productivity, but future productivity loss is discounted in many cases. In western Iowa, soil loss dropped 30% between 1952 and 1957, but increased 24% between 1957 and 1975, as farmers relaxed use of production practices which control erosion. Shifting to conservation tillage and modifying cropping systems to fully utilize all of the advantages of forages would decrease erosion losses.

CLIMATE

Most of the major feed producing areas of the U.S. are within the temperate zone. Only the deep South and Southwest are outside the temperate region. Within these areas, climate undergoes cyclical changes of a diurnal, seasonal, interannual or periodic nature (e.g., 20-22-year shifts associated with sunspot activity), with considerable variability in each cycle over time at a given location. There is also considerable variability in the weather that makes up climate, whether it is for a small area of a few farms, a few counties or several states. Such climatic diversity has necessitated development of plant cultivars of important forage and grain crop species that can withstand a variety of stresses.

Climatic constraints which influence feed production and availability are temperature, precipitation, thermal radiation, humidity and wind. Lack of historical, current or forecast knowledge of these parameters is also a constraint in decision making (e.g., purchase of an irrigation system, timely use of the irrigation system or scheduling the planting of the crop).

Air temperature is a principal factor which sets the limits where both forage and feed crops can be grown, and in what quantity. It is a dominant factor also in plant composition, which in turn influences feed quality (e.g., lignification). Air temperature also influences the length of the growing season and the time for harvest of forages and grains, which, in turn, constrains the selection of high-producing cultivars.

Precipitation is the primary climatic constraint in feed production. Both the total yearly precipitation and the seasonal distribution markedly affect plant growth. Plant growth in the western range areas in the Plains States is limited by precipitation, unless irrigated. Extended periods of no precipitation during the growing season can drastically reduce yields and interrupt grazing systems.

High intensity precipitation may lead to soil erosion and damage to plants, both resulting in reduced production. Precipitation influences soil salinity, water-containing capacity, or in cases of drought, the threshold value for plant wilting. Precipitation enhancement in dry areas from cloud seeding may impact on areas from which such rain has been shunted to influence crop output. It cannot be assumed that enhanced precipitation will benefit a crop or that a decline in precipitation would be deleterious to feed production (table 8.8). Also, the magnitude of the change is not always predictable (table 8.8). Shifts in the hydrologic cycle and amounts of precipitation may result in large-scale disruptions in current plant productivity and of species distribution.

Precipitation also influences the air moisture content. High air moisture content has a strong adverse impact on natural air drying of harvested forages and grains, and on the long-term storage and quality of those reserves.

Thermal radiation (solar and long-wave) is often a rate-limiting parameter in photosynthesis and plant development. Low intensities of solar radiation, resulting from cloud cover or time of day or season, inhibit photosynthesis. High intensities of solar or long-wave radiation can cause stomatal closure, likewise inhibiting photosynthesis. Excessive outgoing long-wave radiation on clear nights can cause frost damage to plants, even when the air temperature is above freezing.

Wind is usually perceived as a less dramatic constraint in forage and grain crop production. It is an important aspect of evaporation, however, from soil and plant surfaces, and can lead to inhibited plant development. In addition, wind can cause mechanical damage to plants (lodging or blowing soil particles) and can reduce feed production by eroding soil necessary for maximum production.

WATER

Water is essential to plant growth, and maximum growth is favored by an ample water supply. Thus, this resource will be of increasing importance in the future. Although the U.S. is one of the "best-watered countries" in the world, substantial amounts are effectively lost from crop production because of runoff. It is likely that taxpayers increasingly will question the subsidization of high-priced, inefficiently produced feed in irrigated areas. The limited availability of water for private development will make the high cost of water used in irrigation for agricultural production most economical only on crops such as vegetables, fruit, rice, cotton and corn, all which bring a high return per unit area of land. There are areas where irrigation is economical for alfalfa. Projected ahead, however, there will be shortages of water in many areas.

Each year about 76 cm (30 in) of water falls to the surface of the conterminous United States. The average annual precipitation ranges from less than 10 cm (4 in) in parts of the Great Basin and Lower Colorado Regions to more than 508 cm (200 in) in the coastal area of the Pacific Northwest. In addition to surface water, the nation has large supplies of groundwater. A general overview of those supplies is shown in table 8.9. Regional or local shortages of water occur because of uneven distribution of precipitation. Water shortages are generally associated with the arid West, but many humid eastern areas also have periodic water supply problems.

The dependability of surface water is as important as its location. Both yearly and seasonal variations must be overcome. Reservoirs are used to partly offset the natural variation. Our current capability to offset drought may decline considerably unless groundwater supplies are augmented with surface waters or transfers of interbasic water aquifers is practiced.

Agricultural irrigation, much of which is for feed grain and alfalfa production, has a profound effect on regional water budgets because it diverts and uses large volumes of water. Hence, it is increasingly important that increases in irrigation efficiencies be achieved in a variety of ways, principally through better water management and by delivering water only in those amounts needed by crops.

Summing up, all types of water (i.e., precipitation, surface water, ground water) are absolutely necessary for a productive feed, forage and grazing economy. In fact, water in all its forms and locations may be the resource most limiting to a reasonably priced feed supply in the U.S. Because of the demand for water, the greatest potential for increasing feed production, with existing water resources, is in the eastern half of the U.S. Water is not as serious a constraint to increasing crop production in this part of the country as in the Great Plains and Western States.

PLANT SPECIES

Forage and grain at the time of harvest represent the cumulative results of plant growth over time. Both the genotype of the plant and the environment (i.e., land, climate, water, biotic factors and management practices) in which the plants are grown influence the quantity and quality of forage and grain harvested. The environmental factors can, in some instances, be economically amended so as to modify the quantity and quality of the harvested plant material but, generally, the environment amendment must be repeated with each growth cycle. Genotype modification, on the other hand, once achieved, does not require continued amendment input as long as the genotype is continued in use.

Genotype consideration for contributing to grain and forage productivity and quality begins with a choice of species. Grain productivity has been materially increased via plant breeding, e.g., between 1930 and 1970, yield increases attributable to genetic improvement were 50-60% for corn, 40-50% for wheat, 30-40% for soybeans and 60% for rice. Likewise, yield potential for certain forage species has been phenomenal, e.g., bermudagrass yield potential has been increased at least 100% in the southeastern U.S. For certain forage species, such as alfalfa, however, the increase in genetic potential for yield has not been good, and this lack of a large response in genetic potential for forage yield represents one constraint to alfalfa production in certain midwestern areas.

Future production of forage and grain for animal feed may be more concentrated on lands that are marginal because of severe mineral deficiencies and toxicities. Amending these land areas to be even minimally productive may not be economically feasible, but there is evidence that plant breeding can evolve grain and forage cultivars that could be productive there.

Plant breeding has great potential for improving forage and grain quality also. The primary deficiency of grain crops as a feed for nonruminant animals is their generally low quality protein. Mutant types of endosperm are known in corn, barley and sorghum which will produce grain with protein of quality adequate to sustain good growth in nonruminants. With continued research effort to improve the productivity of these mutants, grains can become more adequate as energy and protein sources in a diet.

Forage quality has several aspects which are subject to substantial genetic manipulation. They are nutrient availability, palatability and antimetabolite concentration. Nutrient availability (digestibility) in feedstuffs has two components, the proportion of the dry matter that is available to the animals, and how rapidly the available nutrients are released. A good, rapid in vitro procedure is widely used in the laboratory to predict forage digestibility. The cellular content is highly digestible, whereas the real problem in nutrient availability is associated with the cell wall constituents including cellulose, hemicellulose, lignin and silica. One successful example of genetic manipulation of nutrient availability is the brown-midrib gene in corn, in which the substitution of this allele reduced lignin percentage by 40%. Feeding of this low lignin stover resulted in a 30% increase in dry matter consumption and .45 kg per day greater gain than when normal corn stover was fed to beef cattle. A second example relates to the $\underline{d}_2$ gene, which reduces stem length in pearl millet by 50% and is associated with increased leaf percentage of the forage from 54 to 81%. Resultant average daily gains and live weight gains per unit land area for dairy heifers were 33% greater. Other factors such as the presence of phenolic compounds and alkaloids can affect nutrient availability also. Regarding palatability, plant traits that affect voluntary intake potential of a forage are leafiness, cell wall constituents and factors associated with pest resistance, such as tannins. Generally, factors that affect palatability are not well understood and, therefore, they must be lumped together as "taste related" factors. Good examples of increased potential for intake through breeding are Tiflate-1 pearl millet, Kenhy tall fescue and Morpa weeping lovegrass.

Antimetabolites and their general detrimental role in animal metabolism are not well understood, but prominant groups of such compounds include: a) alkaloids in reed canarygrass; b) saponins in alfalfa, tannins in sorghum and lespedeza; c) cyanogenic glycosides in sorghum, sudangrass and white clover; and d) coumarin in sweetclover. Two examples will suffice to show how breeding can reduce or eliminate antimetabolites in forage. Genes at two loci governed the presence or absence of coumarin in sweetclover, and by substituting the appropriate alleles at these two loci, this compound could be eliminated from sweetclover. Antimetabolites in reed canarygrass are the indole alkaloids, of which there are 10 species. The indole alkaloid series of compounds is controlled by alleles at two loci. A line of reed canarygrass, MN76, with only 1/3 as much indole alkaloid concentration as a commercial check has been developed at the University of Minnesota. Lambs fed these reed canarygrass lines gained weight at twice the daily rate of those fed the commercial check.

Insects and diseases that attack forage plants can and do affect the quantity, and especially the quality of forage produced. Resistance to such pests can be bred into alfalfa varieties.

In addition to breeding programs aimed at improving the quality and yield of individual species, potential exists to improve the soil efficiency of nutrient utilization by plants. This may require that breeding be done simultaneously on various species to ascertain more closely the benefits derived when growing in common and mixed culture. This may increase productivity throughout the growing season and provide better quality of forage.

Thus, we see that plant breeding research can contribute in several ways to the productivity and quality of animal feedstuffs. There are, however, several major constraints to plant breeding making the numerous contributions for which there are potential. These are: 1) all of the factors that contribute to forage quality (from the level of the animal) have not been elucidated; 2) rapid in vitro tests for assaying plant genotypes for many quality factors are generally nonexistent; 3) germ plasm collections of most forage species generally are too small for breeders' needs; 4) available germ plasm collections have not been assayed for forage quality factors; and (5) the inheritance patterns of quality factors have not been elucidated adequately.

ENERGY

The full impacts of energy in the future will likely be immense in relation to feed production. Costs for gasoline and diesel oil have already exceeded those projected and used in studies of the future. It is fast becoming necessary to look closely at the efficiencies of plants, their production and their utilization by animals. Alternative ways in which agriculture can make changes as energy becomes more expensive have been studied recently. Crop rotations can change, forages have a major role and nitrogen fixation by legumes figure prominently in these alternatives. In a 5-year rotation - corn, corn, oats, alfalfa, alfalfa - the use of energy was only 60% as much as with coninuous corn, yet the crop yield in total dry matter produced was only 6% less (Heichel,

1978). Nitrogen is a high energy input into grain crop production systems. The amount of fuel energy used in producing crops undoubtedly can be reduced if some cropping systems which involve legumes were used to provide biologically fixed nitrogen. It is important, however, to recognize that these cropping systems will be voluntarily adopted by farmers on a large scale only if the profitability of the total rotational cropping system is equal to or superior to present continuous grain cropping systems. It cannot be accurately predicted if the relative energy cost of nitrogen will preclude its use in continuous grain cropping systems. Rotations or modifications thereof could become a reality if profitable systems are developed for utilizing greater amounts of the harvested legume herbage.

With respect to the energy in feeds fed to animals, one substitution that has been firmly advocated for more than a decade is the development of the capacity to replace energy from grain with energy from forages, plant materials entirely noncompetitive with food for man. Leading in this advocacy have been Hodgson (1978) and others. Reid et al. (1975) calculated the digestible energy (DE) outputs per unit of fossil energy (FE) inputs in feed production. Cited were DE to FE ratios as follows: pasture herbage, 30-115; grass silage, 8.2; hay, 7.5; corn silage, 4.1; soybean, 2.0; corn grain, 2.5. Energy relationships when ruminant cattle graze have all been studied in situations where considerable energy is expended in the process of obtaining the feed. Tradeoffs occur, obviously, in whether the forage is machine-harvested or grazed. Relative importance of these factors may change in the future.

LOSSES DURING GROWTH, HARVESTING AND STORAGE

Losses during growth, harvesting and storage of feed crops reduce efficiency of animal production, especially from forage-based feeds. Quantity and quality may be lost, lowering feed availability and reducing acceptability, intake and digestibility of the forages by animals. Losses can result from pests, poor management and inclement weather. Some estimated losses from forage pests are shown in table 8.10. Reducing these losses could greatly increase the proportions of available herbage consumed by livestock.

Forage quality decreases markedly with advancing maturity. Weather frequently delays harvesting beyond optimal maturity stages, resulting in losses which may reach 20-25% of the dry matter with even greater decreases in nutritive value. Harvesting during inclement weather increases losses and cost of field operations. Reduction in drying time of hays and wilted silages, which could be accomplished through research on biological, mechanical and chemical procedures, might significantly reduce weather-related losses. Additional research also could reduce mechanical losses.

Losses in dry matter and nutritive value from heat, molds and other causes during storage represent other serious losses in feed value. The extent of these losses in quality are shown in table 8.11. The potential increases in animal production which could be achieved by harvesting earlier, and by reducing postharvest forage losses are presented in table 8.12.

INADEQUATE DATA BASE

Improved methods for acquiring and assembling more accurate statistics should be developed for forages and grains. The need is critical. This is particularly so for forage production where present data are limited and of questionable value. As a result, the importance of and contribution to U.S. agriculture from forages is underestimated. There are limitations also in currently available information on weather forecasting, climatological data and weather modification which, if improved, would facilitate on-farm strategic and tactical decision-making.

Other types of data which are needed relate to land availability and class, crop production estimates, germ plasm information and grading standards for hay.

PUBLIC POLICIES

The feed producer historically has provided an abundant feed supply for animals in American agriculture. The use of a wider variety of inputs (fertilizers and pesticides, etc.) which have increased feed production efficiency has been questioned, in some cases, by the public. In turn, this has led to public policies which in total make feed production more difficult. Consumer advocacy also has influenced the perception of healthful foods. The most evident of policies which constrain feed production in one way or another are briefly described.

The concept of zero tolerance for feed additives that may have carcinogenic effects as exemplified in the Delaney clause of the Food and Drug Act is unworkable in the light of analytical technology presently available. The clause should be superseded by criteria based on dose-response, and the risk-benefit ratios should be established through appropriate controlled research.

Our dwindling land and water resources must be protected from further degradation and loss so that food, fiber and energy production for the future generations is assured. Animal agriculture could be one of the first segments of agriculture to suffer if our soil and water resources become limiting. The need for protection of our natural resources is urgent.

Because export of feed grains and conversion of feed grains and cellulosic crops to fuel are presently on the increase, some policies may be necessary to insure that an adequate feed resource base be available in the future.

A deterrent to efficient range production has been the profusion of some regulatory constaints. These regulations by government agencies, particularly federal, in many instances, have caused significant increases in costs for both producer and consumer. Regulatory decisions should be supported by sound scientific data.

RESEARCH IMPERATIVES

The structure of animal production enterprises in the 21st century is unpredictable Most likely energy will be costly; there will be ultimate demands for a limited supply of grain crops and conservation of natural resources will be a paramount objective. These forces could lead to a new era in which livestock production systems will play a paramount role in a total conservation plan for nutrients, water, soil and energy. For certain, livestock enterprises will continue to play a major role in human nutrition and well-being.

Even though the nature and prominence of livestock production systems and strategies in the 21st century are not predictable now, it has been possible, by studying both research history and the present and predictable future constraints on animal production, to delineate research imperatives.

INCREASE UTILIZATION EFFICIENCY OF LIGNO-CELLULOSIC PLANT MATERIALS

There is growing evidence that cereal grains will be less available as feed in future decades. Ligno-cellulosic (plant fiber) materials are the most likely candidates to replace grain for ruminants and possibly for nonruminant animals. Ligno-cellulosic compounds are the most abundant organic compounds in the world. They are produced in every agricultural system, in every climate and on every soil type. Forages and crop residues comprise a large portion of the herbaceous ligno-cellulosic reservoir. They are utilized by animals, particularly ruminants, but with relatively low efficiency. Great opportunity exists to increase efficiency of their utilization or, in other words, to improve their quality. The most opportunistic means are listed below. (Note that the same research advances which improve utilization of these materials by animals also will render them more useful as biomass resources for liquid-fuel production).

Develop Basic Information on Biochemistry of Synthesis and Degradation of Ligno-cellulosic Forage Materials

This information is fundamental to achieving extensive progress in the subsequent research. Unlocking the ligno-cellulosic bonding by efficient and effective methods is of far-reaching importance.

Improve Procedures for Measuring Forage Quality

Rapid, inexpensive, accurate and precise procedures are needed to identify important attributes of feeds for poultry and livestock.

Procedures to measure specific nutrients and potential digestibility are available. New procedures are needed to identify bioavailability of nutrients and characteristics of feeds which influence the environment within the gut of animals and thereby affect availability of nutrients of other diet components. Techniques are needed to predict rates of fiber, starch and protein degradation. The ability to accurately characterize nutrient content and availability will provide the basis for improvement in feed quality and also allow feedstuff selection and diet balancing for a variety of animals and production levels.

Improve Harvesting, Storing and Processing Technologies to Preserve and Enhance Feed Quality

Losses during and following harvest often are severe and costly. A standing crop of alfalfa at optimum harvest date has 67-70% digestible dry matter. This is often reduced to 55% or less by delayed harvest, and by harvest and storage losses. Dry matter losses commonly reach 25%. Concomitant losses in available protein may also occur. These losses reduce animal production. Research in this area is limited. Payoff potential is very high. Improved methodology here will be beneficial in offsetting reduced availability of grain in ruminant production. Lower quality forage and crop residues can be rendered more valuable by certain postharvest treatments or storage procedures. The potential for increasing nutritive value should be extended and more fully exploited.

Ensiling of forages or high moisture grains reduces harvesting losses but can be associated with considerable storage losses and quality changes. Factors (including additives) which affect nitrogen solubilization, carbohydrate digestion, starch solubilization, feed intake reduction and energy metabolism should be studied further.

Enhance Forage Quality Through Breeding

Genetic improvement of forage quality by breeding and selection includes efforts to develop increased digestion of plant cellulose, increased palatibility to animals, elimination of anti-quality substances in some species and reduction of lignification. This will require understanding of the inheritance of quality factors. Germ plasm resources must be expanded by plant exploration and evaluated for quality factors, and superior cultivars must be developed. Breeding of most forage species is complicated by their polyploid and perennial nature. Furthermore, forage species come from genera in two major plant families, the grasses and legumes.

More plant breeding efforts are needed to improve the content and quality of protein in grains to more nearly fulfill the nutritive requirements of animals, particularly nonruminants. Also, plant breeding may be able to reduce lignification and silicification of cereal crop residues to improve their nutritive value.

Optimize Feed Quality Through Improved Crop Management

Quality varies with crop species, cultivars and maturity. Concurrently, manipulation of management alternatives of forages and livestock offers opportunity to better integrate forage quantity and quality with the needs of animals of differing classes, ages and production levels.

INCREASE QUANTITY OF ANIMAL FEED PRODUCED

Research imperatives for increasing feed supplies for animals are genetic- and management-oriented with respect to plants. They will have the greatest payoff if both research areas are fully integrated.

Improve Forage and Grain Species by Plant Breeding

Genetic improvements of forage species can result in more animal product per unit of land or economic input by increasing plant productivity and nutritive value. Because much animal feed, especially forage, is apt to be produced on marginal land, more emphasis will need to be given to breeding plant varieties that will be productive in stress environments. Furthermore, new varieties will need to be optimally suited to management systems used for animal feed production and have high inherent resistance to diseases and pests.

For range livestock, quantity of forage is often more important than forage quality. Management systems to integrate soil, plant and animal relationships provide opportunity to increase forage production and conserve resources.

Integrate New Crop Cultivars to Feed Production Strategies that Optimize and Preserve Natural Resources, Yet Maximize Efficiency of Energy Use

Maximum production of feed may not be as necessary as optimal production, especially when conservation of natural resources becomes increasingly critical. Production of animal feed is uniquely compatible with conservation practices, but much research is needed to determine the best combinations of plant genotype(s), crop management system(s), and energy and labor inputs to give the optimal results in conservation and animal production in regional and local environments.

Improve Crop Management and Harvesting and Storage Methods to Minimize Losses of Plant Material

The biotic factors (insects, diseases and weeds), management and harvesting methods and storage techniques combine to cause crop and nutrient losses as high as 25% of production. Plant breeding research can provide inherent insect and disease resistances which are nonpolluting, and good field husbandry can reduce weed competition. Research is needed, however, to develop crop management, harvesting and storage strategies that will reduce plant material losses and maintain quality in feed grain and forage.

Increase Exploitation of Legumes in Animal Production

Forage legumes are used alone and in a mixed culture with grass species for pasture, hay and silage. They are unique in their ability to fix atmospheric nitrogen and, when grown with adapted grasses, can replace up to 300 kg per hectare per year of commercial nitrogen. Moreover, animal production from legumes and legume-grass mixtures is superior to that of pure grasses. Breeding and management research on forage legumes is needed to increase their nitrogen-fixing capacity, to improve their persistence and competitiveness, to alleviate seed production problems and to improve their productivity under grazing and stress environments.

EVALUATE ALTERNATE FEED SOURCES

Identification of new and different feed materials should be pursued. Particular attention should be given to the recycling of by-products and wastes from agricultural, domestic and industry sources, in the interests of conservation of resources and economy. Favorable yield characteristics of unconventional feed sources such as certain types of trees (alder, aspen, poplar) may be exploited to further extend the feed base. Feed production areas can be increased through investigation of the nutrient value of aquatic materials, both plants and animals (fish).

Develop Efficient Methods for Delignifying Carbohydrate-Lignin Complexes in Woody and Vegetative Tissues of Plants

Base and acid treatments are used for this currently, but their use is limited by chemical residue in the delignified product. Biological and new chemical approaches should be investigated to overcome this limitation.

Improve Methods Used for Handling, Processing and Storing Animal Residues to Preserve and Improve Their Nutritional Quality and Safety for Use as Feed

Many animal residues, such as slaughterhouse by-products and manure, have considerable value as animal feeds, but are presently underutilized.

Identify and Evaluate New Sources of Concentrate Feeds

Technology is needed to utilize by-products from new industries that use agricultural products as raw materials. Currently a large industry is developing to use corn as a raw material for liquid fuel production. The distillery residue that remains after the liquid fuel is produced is a protein concentrate that may be useful as a livestock feed. By-products from food production and processing industries, forest and wood processing industries, aquatic plants and other new crops need to be evaluated.

REFERENCES

Amer.Soc. Agron. 1976. Biological Fixation in Forage-Livestock Systems. Special. Pub. No. 28. Madison, WI.

Axtell, J.D. 1980. Breeding for improved nutritional quality. In: Plant Breeding II. K.J. Frey (Ed.). Iowa State Univ. Press (In press).

Barnes, R.F. and G.C. Marten. 1979. Recent developments in predicting forage quality. J. Anim. Sci. 48:1554.

Brink, R.A., J.W. Densmore and G.A. Hill. 1977. Soil-deterioration and the growing world demand for food. Science-197:625.

Brown, A.W.A., T. C. Byerly, M. Gibbs and A. SanPietro. 1975. Crop Productivity - Research Imperatives. Michigan State University Agricultural Experiment Station and Charles F. Kittering Found.

Bula, R.J., V.L. Lechtenberg and D.A. Holt. 1977. Potential of temperate zone cultivated forages, P. 7-28. In: Potential of the World's Forages for Ruminant Animal Production, Winrock Rep., Morrilton, AR.

CAST. 1975. Utilization of Animal Manures and Sewage Sludges in Food and Fiber Production. Rep. No. 41. Iowa State Univ., Ames, IA.

CAST. 1975. Ruminants as Food Producers: Now and for the future. Special Pub. No. 4, Iowa Sate Univ., Ames, IA.

CAST. 1980. Foods from Animals: Quantity, Quality and Safety. Rep. No. 82. Iowa State Univ., Ames, IA.

Center for Agriculture and Rural Development (CARD). 1977. A Multigoal Linear Programming Analysis. . . in U.S. Agriculture. Rep. No. 76, Iowa Agr. Expt. Sta., Ames, IA.

Evans, H.J. and L.E. Barber. 1977. Biological nitrogen fixation for food and fiber production. Science 197:332.

Frey, K.J. 1970. Improving crop yields through plant breeding, p. 15. In: Moving Off the Yield Plateau. R. Munson (Ed.). Spec. Pub. 20. Amer. Soc. Agron., Madison, WI.

Heichel, G.H. 1978. Stabilizing agricultural energy needs: Role of forages, rotations, and nitrogen fixation. J. Soil and Water Conserv. 33:79.

Hodgson, H.J. 1978. Food from plant products - forage, p. 56-74. Plant and animal products in the U.S. food system. Proc. Symp. on Complementary Role of Plant and Animal Products in the U.S. Food System. Nov. 29-30, 1977, National Academy of Sciences, Washington, DC.

Hodgson, H.J. 1976. Forages, ruminant livestock, and food. Bio. Sci. 26:625.

Iowa State Univ. 1977. Proc. World Food Conf. 1976. Iowa State Univ. Press, Ames, IA.

Johnson, W.M. 1973. Basic natural resource considerations in land use, p. 55-67. In: National Land Use Policy: Objectives, Components, Implementation. Proc. Special Conf., Soil Conserv. Soc. of Amer., Des Moines, IA. Nov. 1972.

Lechtenberg, V.L., R.M. Peart, S.A. Barber, W.E. Tezeur and O.C. Doering. III. 1980. Potential for fuel from agriculture, p. 125-138. In: Proc. 1980 Forage and Grassland Conf., Louisville, KY.

NRC. 1973. Alternative Source of Protein for Animal Production. Proc. Symp., Blacksburg, VA, 1972.
NRC. 1975. Enhancement of Food Production for the U.S. National Academy of Science-National Research Council, Washington, DC.
NRC. 1976. Climate and Food: Climatic Fluctuation and U.S. Agricultural Production. National Academy of Science-National Research Council, Washington, DC.
NRC. 1977. Crop Productivity (Study Team I) and Animal Productivity (Study Team II) Vol. I, Supporting Papers: World Food and Nutrition Study. National Academy of Science-National Research Council, Washington, DC.
NRC. 1978. Plant and Animal Products in the U.S. Food System. Proc. Symp. Nov. 29-30, 1977.
NRC. 1976. Genetic Improvement of Food Proteins. Proc. Workshop, Nov. 18-20, 1974. National Academy of Science-National Research Council, Washington, DC.
Office of Technological Assessment. 1980. Energy from Biological Processes. U.S. Govt. Printing Office (In press).
Pessek, J. 1980. Presidental Address-Unified Resource Conservation in Agriculture: Role for Agronomy. Agron. J. 72:1.
Pimentel, D., E.C. Terhune, R. Dipon-Hudson, S. Rochereau, R. Samis, E.A. Smith, D. Denman, D. Reifschneider and M. Shepard. 1976. Land degradation: Effect on food and energy resources. Science 194:149.
Reid, J.T., K.L. Turk and R. Anrique. 1975. Comparative efficiency of animals in conversion of feedstuffs to human foods. Symp. of the Rockefeller Foundation. Future Role of Animals in Food Production. New York.
Reid, R.L. 1979. Forage quality as seen by an animal scientist. Proc. 3rd Eastern Forage Improvement Conf. Ottawa, Ontario, Can. July 10-12, 1979, p. 9.
Timmons, J.F. 1980. Protecting agriculture's natural resource base. J. Soil and Water Conserv. 35:5.
USDA. 1971. National Inventory of Soil and Water Conservation Needs. 1967. Stat. Bull. 461. Washington, DC.
Van Soest, P.J. and J.B. Robertson. 1976. Composition and nutritive value of uncommon feedstuffs. Proc. Cornell Nutr. Conf., Ithaca, NY. P. 103.
Van Soest, P.J. 1979. Laboratory techniques for forage quality. Proc. 3rd Eastern Forage Improvement Conf., p. 9. Ottawa, Ontario, Can. July 10-12, 1979.
Vetter, R. L. and M. Boehlje. 1978. Alternative feed resources for animal production, p. 95. In: Plant and Animal Products in the U.S. Food System. National Academy of Science, Washington, DC.
Wedin, W.F., H.J. Hodgson and N.L. Jacobson. 1975. Utilizing plant and animal resources in producing human food. J. Anim. Sci. 41:667.
Wittwer, S.H. 1974. Maximum production capacity of food crops. Bio. Sci. 24:216.

TABLE 8.1. CONSUMPTION OF FEEDS BY LIVESTOCK IN 1974 (MILLION METRIC TONS OF FEED UNITS[a,b])

Animal type	Concentrates	Forages	Total feed	% Forage
		Metric tons (10^6)		%
Dairy cattle	23.8	41.7	65.5	63
Beef cattle	33.2	171.8	205.0	84
Sheep and goats	.7	6.1	6.8	90
All livestock	157.8	240.3	398.1	60
Hog	43.9	7.0	50.9	16
Poultry[c]	36.6	.7	37.3	3
Other	19.6	13.0	32.6	40
All ruminants	57.7	219.6	277.3	80
% by ruminants	36	91	70	--

[a]A feed unit as used herein is the nutritional equivalent of .45 kg (1 lb) of corn.
[b]Hodgson, 1978.
[c]Hens and pullets, chickens raised, broilers and turkeys.

TABLE 8.2. TRENDS IN MAJOR USES OF LAND IN THE UNITED STATES, SELECTED YEARS FROM 1950 to 1974

Major land use	1940	1950	1959[g]	1969	1974
			ha (10^6)		
Cropland[a]	162	166	159	156	155
Available grassland pasture and range[b]	291	284	283	280	276
Forest and woodland[c]	294	292	295	293	291
Other land[d]	172	179	183	188	196
Special use areas[e]	---	54	59	70	74
Unclassified areas[e]	---	125	124	118	122
Total[f]	919	921	920	917	918

[a] Excludes cropland used only for pasture.
[b] Grassland pasture and other nonforested grazing land plus cropland used for pasture.
[c] Excludes reserved forest land in parks, wildlife refuges and other special land uses.
[d] Includes such special land uses as urban areas, highways, roads, farmsteads, parks and military reservations, and also land having little value for surface use (desert, rock, marshes, tundra, etc.).
[e] Indicates data are not available.
[f] Changes in total land area are attributable to changes in methods and materials used in occasional remeasurements and to increases in the area of artificial reservoirs.
[g] Estimates for 1940-1959 are based primarily on Major Uses of Land and Water in the United States: Summary for 1959 (USDA, Economic Reserach Service) and assume essentially no change in Alaska and Hawaii prior to 1950. The estimates are only approximately comparable.

Source: USDA 1980. SWRCA Program Report and Environmental Impact Statement - Review Draft.

TABLE 8.3 PRODUCTION OF FEED GRAINS, WHEAT, RYE, SOYBEANS, HAY AND SILAGE, 1979[a]

Crop Commodity	ha (10^3)	Metric tons (10^3)
Feed grains	40,967	233,881
Corn	28,726	197,209
Oats	3,979	7,757
Barley	3,022	8,231
Sorghum	5,240	20,684
Wheat	25,334	58,288
Rye	384	624
Soybeans	28,543	61,715
All hay	24,752	132,338
Alfalfa	11,137	79,452
Other	13,615	52,886
Corn for silage	3,238	103,151
Corn for forage	168	---
Sorghum for silage	310	8,175
Sorghum for forage	512	---
All principal crops	136,236	---

[a]USDA, 1980.

Table 8.4. DISAPPEARANCE OF SUPPLIES OF FOOD AND FEED GRAINS, 1971 AND 1977[a]

	Food, seed and industrial %		Feed %		Export %		Tables[b]	
	1971	1977	1971	1977	1971	1977	1973	1979
Feed Grains								
Corn	9	9	76	58	15	30	45	41
Oats	8	11	90	68	2	1	52	50
Barley	30	38	59	40	11	14	65	58
Sorghum	2	1	83	60	15	27	76	67
Wheat	37	33	20	9	43	55	13	5
Rye	28	59[c]	66	41	6	0	23	19

[a] USDA Agricultural Statistics 1973 and 1979, Washington, DC.
[b] Table number in Agricultural Statistics.
[c] 28% used as seed - indicates large usage for grazing.

TABLE 8.5. TOTAL ANIMAL FEED USAGE FROM FORAGES[a] AND CONCENTRATES IN 1972[b] VS 1978[c].

Animal type	% of Diet: Forages 1972	Forages 1978	Concentrates 1972	Concentrates 1978
All livestock and poultry	54	62.5	46	37.5
All dairy cattle	63	61.2	37	38.8
All beef cattle	73	83.0	27	17.0
Beef cattle on feed	--	27.6	--	72.4
Other beef cattle	--	95.8	--	4.2
Sheep and goats	89	91.1	11	8.9
Hogs	14	14.7	86	85.3
Hens, pullets and chickens raised	3	0	97	100.0
Broilers	0	0	100	100.0
Turkeys	5	0	95	100.0
Horses and mules	--	72.2	--	27.8

[a]Referred to as roughage in CAST Rep. No. 82 (1980).
[b]Wedin *et al.* (1975).
[c]CAST Rep. No. 82 (1980).

Table 8.6. AGRICULTURAL RESIDUES AVAILABLE BY REGIONS[a]

Region	Crop residue	Manure	Forestry	Total	% of interregional total
			Metric tons (10^6)		
New England/ Mid-Atlantic	6.2	2.7	8.8	17.7	4.1
Southeast	28.4	4.3	53.0	85.7	19.9
Corn Belt	108.4	7.2	9.1	124.7	29.1
Plains	103.9	7.0	8.2	119.1	27.8
Intermountain/West	30.8	5.3	45.7	81.8	19.1
Total	277.7	26.5	125.8	429.0	
% of total	64.7	6.2	29.1		

[a]Vetter and Boehlje, 1978.

TABLE 8.7. SOME CHARACTERISTICS OF 1980 AGRICULTURE RESULTING FROM CHANGE IN CROPPING PATTERNS AND LAND USE

Increased mechanization
Increased size of equipment
Narrowed germ plasm base
Expanded use of synthetic nitrogen
Increased pesticide use
Reduced number of farm enterprises
Increased labor costs
Decreased use of legumes in cropping systems
Increased demand for export of corn and soybeans

TABLE 8.8. CROP AND FORAGE YIELDS WHEN EITHER TEMPERATURE OR PRECIPITATION OR BOTH ARE SHIFTED ABOVE AND BELOW NORMAL AVERAGES[a]

	Units of Measure	Change in yields from normal production					
		Change in precipitation		Change in temperature		$+2^{\circ}C$,	$+2^{\circ}C$,
		-10%	+10%	$+2^{\circ}C$	$-2^{\circ}C$	$-10\% H_2O$	$+10\% H_2O$
Winter wheat (N. Central U.S.)	kg/ha	-13	-13	-188	+54	-269	-202
Soybeans (OH,IN,IA)	%	-5	+4	-6	-8	-9	-1
Winter oats (GA; maximum forage)	t/ha	-.2	+.2	-.6	+.5	-.7	-.4
Grain sorghum (KA)	%	-8	+6	-3	-9	-9	+2

[a]Adapted from CIAP Monogr. 5, Part 2: Climatic Effects, from Imports of Climatic Change of the Biosphere, Final Report of the Climatic Impact Association Program, Dept. of Transportation, Washington, DC. (1975).

TABLE 8.9. THE NATION'S FRESH GROUND WATER SUPPLIES (USDA, 1980)[a]

Water resources regions	Ground water in storage that can feasibly be withdrawn[b]	Portion of total withdrawals from ground water	Ground water withdrawals: Total ground water withdrawals	Withdrawals in excess of recharge: Amount mined	Withdrawals in excess of recharge: Percentage of ground water mined	Water resource subregions: In region	Water resource subregions: With mining
	10^{12} liters	%	liters/day x 10^6	liters/days x 10^6	%	No.	No.
New England----------------------	NA[c]	12	2,407	0	0	6	0
Mid-Atlantic---------------------	1,326	15	10,085	121	1	6	3
South Atlantic-Gulf--------------	4,245	22	20,652	1,285	6	9	8
Great Lakes----------------------	985	3	4,605	102	2	8	1
Ohio-----------------------------	417	5	6,985	0	0	7	0
Tennessee------------------------	2,009	4	1,027	0	0	2	0
Upper Mississippi----------------	815	19	8,967	0	0	5	0
Lower Mississippi----------------	4,813	33	18,336	1,561	9	3	3
Souris-Red-Rainy-----------------	417	26	326	0	0	1	0
Missouri-------------------------	1,706	27	39,442	9,691	25	11	10
Arkansas-White-Red---------------	1,895	69	33,526	20,682	62	7	7
Texas-Gulf-----------------------	2,464	43	27,371	21,141	77	5	5
Rio Grande-----------------------	4,775	37	8,850	2,490	28	5	4
Upper Colorado-------------------	133	2	478	0	0	3	0
Lower Colorado-------------------	NA	56	18,980	9,153	48	3	3
Great Basin----------------------	379	18	5,397	2,240	42	4	4
Pacific Northwest----------------	682	20	27,849	2,376	9	7	6
California-----------------------	208	48	72,616	8,327	12	7	5
Conterminous U.S.----------------	NA	24	307,900	79,169	26	99	59
Alaska---------------------------	4,245	14	167	0	0	1	0
Hawaii---------------------------	NA	42	2,994	0	0	4	0
Caribbean------------------------	61	28	963	49	5	2	1
Total U.S.------------------	NA	24	312,023	79,218	25	106	60

[a]Geological Survey Professional Paper 813 Series (USDI, 1974-1979). Second National Water Assessment (U.S. Water Resources Council, 1978).
[b]Depth, quality and ease of extraction are not nationally consistent.
[c]NA means data not available.

TABLE 8.10. SOME ESTIMATED FORAGE-CROP LOSSES FROM PESTS[a]

Commodities	Kinds of pests		
	Diseases	Insects	Weeds
Hay			
Alfalfa	24[b]	15	--
Pasture and range	3-9	--	13-20
Seed			
Alfalfa	9	--	--
Other	4-50	15-38	12-18

[a]USDA, 1965.
[b]Values are estimated percentages of the commodities lost by indicated causes.

TABLE 8.11. LOSSES DURING HARVESTING AND STORAGE OF HAY[a]

Causes of loss	Loss in dry matter, %
Respiration	
In good drying conditions	2-8
In poor drying conditions	up to 16
Mowing and conditioning	2-5
Raking	5-25
Baling	
Conventional	3-8
Large-round	1/2-15
Transporting large packages	1-10

[a]N.P. Martin. 1980. Harvesting and storage of quality hay, p. 177. In: Proc. Amer. Forage and Grassland Conf., Louisville, KY.

TABLE 8.12. ESTIMATED EFFECTS ON MILK (DAIRY) AND LIVE WEIGHT GAIN (BEEF) OF VARIOUS LEVELS OF LOSSES IN DRY MATTER AND DIGESTIBILITY OF ALFALFA[a]

Dry matter loss (%)	Digestibility % of dry matter	Milk production, kg[c]		Live weight gain, kg[d]	
		Per day	Per ha	Per day	Per ha
10	60	11.5	6,093	.4	448
15	57	7.8	4,435	.2	268
25	50	1.4	874	0	0

[a]H.J. Hodgson, (Personal communication). 1980.
[b]Assumes an alfalfa crop harvested at first bloom and yielding 8,297 kg/ha of dry matter with 66.6% digestible dry matter.
[c]Based on 650 kg cow weight.
[d]Based on 300 kg growing animal.

9

R. N. Van Arsdall (Co-Chairman), W. B. Sundquist (Co-Chairman), J. R. Black (Rapporteur)

F. H. Baker, B. Bullock, T. Graf, P. J. Luby, A. Paul, J. C. Purcell, V. J. Rhodes, M. E. Templeton, J. K. Wise, and M. B. Wise

PRODUCTION, MARKETING AND DISTRIBUTION

SUMMARY

In the decade just ended, the U.S. animal products economy has been severely shocked by several unanticipated events, emanating from different quarters, over which individual participants have had no control and to which they (as producers, investors or consumers) have had to adjust as best as they could. There were massive grain exports in 1972 and 1973 coupled with a poor U.S. grain harvest in 1974; Federal price controls in 1971-1974; a huge rise in energy costs in 1973 and 1979 with its fairly direct impacts on nitrogen and other inputs; a sudden embargo on grain exports to the USSR beginning in 1980; a sharp rise in borrowing costs in 1979-1980; and, at various times through the decade, interruptions of established ways of business by government because of public concerns over food safety or polluted environments.

The resulting shocks to a modern animal food products system are much greater than they would have been to the more traditional system of the 1960s and before. The greatly increased specialization of production, enlargement of scale and application of modern technology in almost every line of the animal food products chain has rendered the individual enterprise more vulnerable to price and other disruptions.

Put another way, the modern system is strong in one respect -in its great present and future productive potentials - and weak in another, namely, in its great vulnerability to the unwelcome and, often, sudden events that have occurred and probably will recur when least expected.

What can be done? One possibility is to do nothing, but this may entail a large social cost and large individual privation. The other possibility is to do a serious job of contingency planning. This should enable more intelligent choices to be made by individuals and firms in managing their affairs before the fact, and by governments, at different levels, in having at their command a wider choice of well-considered options on how to best cope with unexpected events and ameliorate shocks.

To serve this contingency planning need, major research imperatives have been developed in this paper. The first imperative is to develop the blueprint of the present production-distribution-marketing system (for each of the animal products) that is relevant for each serious contingency, e.g., a further cut-off of Mideast oil supplies. This blueprint would be an aggregative model that relates the different components of the system in an integrated way, and upon which shocks and stresses could be evaluated. It should provide invaluable insight into sensible types of choices and adjustments that might be made and, in turn, forestall action potentially detrimental to industry and society.

A second imperative is to develop a fuller range of information about the characteristics of demand for animal products to be fed into the analytic systems and subsequent decision processes to achieve greater forecasting accuracy and improved operating performance.

A third imperative is to develop models of farm-level production systems at the farm level where science-based technology from several disciplines specify with greater reliability the possibilities for the most efficient production in different situations.

A fourth research imperative is to examine carefully the market coordinating mechanisms through which individual decisions are adjusted to one another so that the whole animal food products system moves forward with greatest efficiency. This research effort would be particularly valuable if price controls were imposed once more as an inflation control device.

Finally, a fifth research imperative is to develop much better insight into how marketing services between farmer and consumer (the full range of assembly, storage, processing, transport and distribution activities) could be provided at a lower cost through innovation and improved systems design. This research would have to be a joint effort between private industry and public institutions.

PRESENT STATE OF THE ART AND SCIENCE

HIGHLIGHT DIMENSIONS OF PRODUCTION

Over the last decade or more livestock production in the U.S. has generally leveled off, while crop production has increased dramatically (figure 9.1). While crop production had increased by about 40% from its 1967 level by 1979, livestock production had increased by only 7%. The increase in livestock production was restricted by an absolute decline in the number of breeding units which largely offset the increase in production per breeding unit (figure 9.2). However, there were big differences in the rates of change between livestock product categories and even in their direction. The rapid increase in crop production, particularly during the 1970s, has generally been attributed to a greatly expanded demand from the commercial export sector for grains and soybeans. There was no similar incentive to expand food animal production.

Production changes between 1960 and 1979 may be summarized as follows (table 9.1). Beef production increased dramatically over the period largely as the result of demand increases, a major increase in the national beef herd up to 1975, and a shift to more feedlot finishing of cattle which increased both the rates of gain and the marketing weights for cattle. The decline in veal production over the same period was mainly the result of: 1) a decline of 45% in the national dairy herd from which most veal calves were produced; and 2) a higher proportion of these potential veal calves were diverted to the feedlot for finishing.

Total pork production changed relatively little although the structure of the swine industry changed dramatically. The amplitude of the "hog cycle" diminished somewhat, but its impact continued to exceed by far any secular changes in total pork production.

Lamb and mutton production declined by more than 60%. This major decline has several causes, both internal and external, important to the sheep subsector. These causes include high labor and land costs, high losses to predators and strong competition from the poultry, beef and pork subsectors and synthetic fibers.

Poultry meat (particularly broilers) grew the most rapidly. This growth continues as a result of increased production efficiency (particularly in feed to meat conversion and labor use) and specialization of production and vertical integration. While the change in total egg production was not great, integration and specialization of the egg subsector contributed to the rapid disappearance of small producers.

Finally, although total milk production changed very little, the secular decline in dairy cows and farms and the secular increases in size of enterprise and degree of specialization continued. Major changes occurred in the mix of dairy products to which the milk was converted for human consumption. Although milk production rose substantially in response to much higher price supports for milk in 1976-1977, there is considerable evidence that total production has now leveled off again.

CONSUMPTION

That total supply equals total demand at an equilibrium price for food animal products is an abstraction that does not in itself reveal the real world. Not all food animal products are consumed in the same period in which they are produced. Moreover, domestic production may be augmented by imports or may be exported. Thus, the identity equations which balance product supply with demand in the short run must include storage and trade. A variety of governmental and private entities may intervene on both the supply and the demand side, particularly on the latter. Despite these and other complexities, the aggregate annual demand for and consumption of domestically produced food animal products is mainly the product of the U.S. population times the average annual per capita consumption rate. Of course, prices for individual food animal products (both absolute and relative to substitute food products) and consumer incomes, product availabilities and consumer tastes and preferences are important determinants of per capita demand.

Table 9.2 shows per capita consumption rates for the same animal product categories and the same time period (1960-1979) for which aggregate production figures are shown in table 9.1. The retail price index of all food increased by 67% during this period. Corresponding increases for individual items were 70% for poultry, 53% for eggs, 134% for dairy products, 177% for meat and 256% for fish, with fish being a major substitute for red meat and poultry. Thus, meat and dairy product prices increased at two or three times the rate for poultry and eggs, but at a much slower rate than for fish. Also, during the 1960-1979 period the consumer price index (CPI) rose 162%. So, using the CPI as a base measure of value, the real price of poultry and eggs at retail dropped substantially, the real price of meat and dairy products changed only slightly and the real price of fish increased dramatically.

Per capita consumption of beef increased substantially (about 24%) from 1960-1979 while the real price of beef remained fairly constant. The 1979 per capita consumption level was, however, down about 20% from its high in 1976. Although per capita consumption of beef (retail cut equivalent) exceeded that for pork only slightly in 1960, it exceeded it by about one-fourth in 1979. Thus, among the red meat categories, beef has increased significantly in per capita consumption. This increase probably reflects a strong consumer preference for beef along with a high producer capacity to provide beef with new feedlot-type production technology. Although consumer preference for veal may have declined during the 1960-1979 period, the reduction in per capita consumption of about 70% from 1960-1979 was probably mainly an "availability" phenomenon.

Per capita pork consumption from 1960-1979 remained nearly constant except for cyclical movements. There was a continuation of the long-term shift of production from lard-type hogs.

Per capita consumption of lamb and mutton declined to 1.3 kg (2.9 lb.) in 1979. This decline indicates that that category is now of only minor importance to most U.S. consumers except as a "variety" or "specialty" food item.

The major expansion in per capita consumption of chicken and turkey meat suggests that per capita poultry meat consumption will soon be exceeded only by beef. Effective quality control, widespread availability and relatively lower real prices have all contributed to the increased per capita consumption of poultry meats as have increased consumer concerns about animal fats generally and about cholesterol in particular.

Per capita egg consumption increased to 277 eggs in 1978 and to 284 in 1979 from the 272 egg level of 1977. This increase represents the first break in a long-term decline in per capita egg consumption and may reflect a rather long-term leveling off in per capita consumption.

Per capita consumption of dairy products also appears to have leveled off in the post-1975 period after a long period of decline. Changes in per capita consumption levels of major dairy product categories for the past decade (1968-1978) are shown in figure 9.3. Clearly the strengthening influence in per capita consumption of dairy products has come from low fat fluid milk, cheese and other low fat products. Per capita consumption of high fat products such as butter, fluid cream and fluid whole milk continues to decline.

The decline in per capita consumption of butter is part of a broader phenomenon relating to consumption of fats and oils. In 1960 per capita consumption was 3.40 kg (7.5 lb) for butter and 3.44 kg (7.5 lb) for lard. They had declined by 40 and 70%, respectively, by 1979. Meanwhile, per capita consumption of fats and oils from mainly vegetable sources increased by 62% from its 1960 level of 15.2 kg (33.4 lb). Thus, fats and oil from animal sources have fallen into disfavor with consumers in recent years and this phenomenon represents a major shift in product demand.

THE PRODUCER AND SUPPLY SEGMENT OF FOOD ANIMAL AGRICULTURE

Several key dimensions of the livestock and poultry production sector of major interest from the standpoint of the structure and economics of the sector are briefly discussed here.

Inventory of Livestock Numbers 1960-1980

Table 9.3 shows the number of food animal related breeding livestock on farms on January 1, for the period 1960-1980 and the percentage changes over this period. Because the short production cycle in poultry makes annual inventories relatively meaningless, they are omitted. The data in table 9.3 show several major changes over the decades of the 1960s and 1970s, although they are not designed to show production cycles in cattle and hogs. During the period, beef cow numbers increased by 40% (70% at the cattle cycle peak in 1975) and dairy cow numbers declined by 45%. Ewe numbers dropped by almost two-thirds, and breeding hog numbers mainly cycled over the 1960-1980 period. Never has the aggregate U.S. livestock industry realized changes of these magnitudes over a period of only 20 years.

Structure of the Production Sector

The dramatic changes in livestock numbers over the past 2 decades were a part of pervasive changes in the structure of production.

<u>Beef</u>. The U.S. beef industry is largely made up of: 1) cattle raising, centering on the production of calves from beef-cow-breeding herds but including some yearlings and other animal classes; and 2) the feeding of cattle for slaughter. Although some beef calves move directly from the beef-cow herd to the feedlot, others move into intermediate or final stages of grazing and/or limited grain feeding. These movements occur under a wide range of programs with respect to duration, diet and management systems. Cull cows and bulls find their way to slaughter -- usually without feeding in drylot.

In 1959, almost 2.7 million U.S. farms reported having cattle and calves. At that time many cattle herds were a mixture of dairy and beef animals. In 1964, about 1.3 million farms reported having beef cows with an average herd size of 25 cows. By 1974 only slightly more than 1 million farms reported having beef cows with an average herd size of 40 cows. Also in 1974, as was true a decade earlier, the modal size (over

40% of all units) of beef-cow enterprises was in the 20-99 cow range. Thus, beef cow-calf production operations remain the broadest based and least specialized of all major food animal enterprises. And, although beef-cow enterprises are the major commercial enterprise on many farms and ranches, they are present on many other farms as a claimant for residual pasture, roughage feeds and/or family labor resources. Beef cow enterprises have disappeared from a number of farms during the past 2 decades, particularly in cash crop farming areas. But, they grew in size and number in other areas, particularly in the South. In general, however, beef-cow enterprises have difficulty in competing for the direct use of highly productive cropland which has strong cash crop or feed grain production alternatives.

Cattle feeding, somewhat in contrast to beef-cow herds, has undergone major structural changes. Prior to 1960 farmer-feeders with less than 1,000 head produced most of the fed beef. Now more than one-half of the fed cattle marketed are from about 420 large commercial feedlots; small farm-feeders now account for less than one-third. Most small farm-feeder operations are managed as family-scale operations, as is the case for most beef-cow production enterprises. Large volume, commercial feedlots, on the other hand, exhibit a wide range of organizational structures, including partnerships and corporations. Many (53% in 1974) of the fed cattle marketed from incorporated feedlots with 2,000 head or more capacity were custom-fed under a variety of contractual arrangements. Some of the large commercial feedlots are vertically integrated into one or more other functions, including cattle raising, meat packing, meat retailing and restaurant businesses.

<u>Hogs</u>. Nearly all hogs were produced in small enterprises prior to 1960. Even in 1964, only slightly more than 7% of total hog sales came from farms selling 1,000 or more per year. By 1974 this percentage had increased to 25, and the number of farms selling 200 or less hogs had dropped to about one-half of the 1964 level. Large volume producers, those marketing 5,000 head or more annually, account for a rapidly increasing share of total production. Rhodes et al. (1979) reported 1,340 such large volume operations marketing 13.7 million hogs in the United States in 1978. The rapid growth in large operations has resulted both from the new entry of large operations and the rapid expansion of existing units.

About 80% of slaughter-hog production now comes from complete hog operations (farrow-to-finish) and 20% from split phase operations (pigs produced on one farm and finished on another). Feeder-pig production tends to center in areas and on farms where feed grains are in limited supply, but where adequate labor is available to operate this more labor intensive enterprise. These feeder pigs are then sold to producers who finish them for slaughter.

<u>Sheep and lambs</u>. In 1959 about 340,000 U.S. farms reported having sheep or lambs. They averaged just under 100 head/farm. By 1974 the per farm average had climbed slightly to about 116 head, but only about 140,000 reporting farms (41% of the 1959 number) remained. Some specialized sheep operations continue on the Western range but sheep are a supplementary enterprise on many other farms. As of 1975, about one-half (66,500) of the farms reporting sheep were in the North Central States where flock size averaged only 28 head per farm. Thus, the major trend in the structure of the sheep industry in the U.S. is toward fewer

enterprises without much increase in enterprise size or degree of specialization.

Dairy. The number of U.S. farms with milk cows declined from 4.6 million in 1939 to about 1.8 million in 1959 to 380,000 in 1978 of which only about one-half could be classified as commercial dairy farms. Average herd size on commercial dairy farms was 53 cows in 1979. Most dairy operations remain as a family- or partnership-scale operation, with most of the forages and at least some of the feed grains produced on the same farm. This situation differs, however, by region. A number of large-scale drylot operations, with herds of 2,000 and up to 10,000 cows have been established in California, Arizona and Florida. Many of these large-scale production units purchase all or most of their feed, both concentrates and roughages, from other producers and concentrate their efforts on producing milk. In contrast to the sheep and beef-cow subsectors, few farms have dairy as a supplementary livestock enterprise on farms where it is present.

Poultry. Among the several food animal species reported here, the poultry industry has undergone the greatest structural changes, in the past 2 decades. In 1959 some 2.2 million farms (about 58% of all farms) had chickens and almost 1.1 million produced eggs for sale. At that time over 86,000 farms reported turkeys with an average of 950 birds per farm.

By 1974 commercialization had swept through the poultry industry. Only 5,167 farms with 20,000 or more birds per farm had more than two-thirds of all hens and pullets of laying age, while 1,763 commercial turkey farms selling 16,000 or more turkeys accounted for almost 92% of total output. Less than 40 vertically integrated firms accounted for 70% of output.

Structurally, today's poultry and egg industries involve an extensive network of linkages which have developed between production units and input-supplying and marketing functions. Coordinating systems cover virtually all commercial broiler production and four-fifths or more of all egg and turkey production. In these systems, much production is under contract to marketing firms, or carried out as only one phase within vertically integrated firms. A highly integrated firm can involve all or most of the following: breeding flocks, hatchery, feed mill, production units, assembly of live birds or eggs, poultry slaughtering or packing plants, further processing units, delivery vehicles and distributing centers.

Production Technology and Efficiency

In contrast to an earlier era when food-animal production drew mainly on land and labor inputs, some subsectors now resemble a value-added industrial-type industry in which producers or agribusiness firms purchase most inputs, use mainly borrowed capital (or in some cases corporate-type equity capital) and manage these resources to turn out a marketable product. This organization is most characteristic of the poultry subsector and some specialized firms in cattle and hog feeding, and in large-scale drylot dairy production. It is least characteristic of the beef-cow and sheep subsectors and family-scale dairying where land and labor inputs still figure heavily in the production process.

Economic pressures for using high levels of production technology and for maximizing production efficiencies are probably greatest in those subsectors of animal agriculture which are operated on a large-scale with mainly purchased inputs. They are least intense where the production enterprise uses mainly residual-type resources with low opportunity costs. Most animal production in all categories now comes from commercial producers. As a group they are highly mechanized and use research derived technology and managerial practices in breeding, feeding, housing, health management, product quality control, input buying, product selling, financial management and the coordination of these several dimensions of the production process.

No single measure or even several measures of technology or production efficiency describe adequately the current status of the U.S. food-animal industry. This is true whether one wishes to compare current overall efficiency with that for some historical period or, as is of more interest here, to assess the potential for future improvements. Clearly, the results of past research and technology have been multifaceted. Some have increased output per unit of land, labor, feed or animal. Others have had the effect of eliminating onerous labor tasks. Still others have had their impact via improving product quality. Most have substituted capital for labor in the production process. And, almost all new technology improved producer profits, at least for the early adopters. Clearly, from the producer's viewpoint the major driving forces for adopting new technology have been the dual ones of: 1) reducing per unit production costs; and 2) increasing total output (and thus generating additional income). The new technology which has been adopted most rapidly and most broadly has permitted both to occur simultaneously --as has been the case in the poultry subsector, in large cattle feedlots and in commercial hog production. Table 9.4 illustrates the type of gains made in output per unit of feed and breeding unit over time. Dramatic gains have also been made in labor productivity.

Economics of Size and Specialization

Producers have been able to obtain higher production efficiencies when they have concentrated their management on one, or generally on at most two, food-animal enterprises per farm. Several factors explain the efficiency gains realized via increased size and specialization. First, most new technology, whether an automated milking parlor, a slotted-floor hog house with an automated manure disposal system, or a cattle feedlot complete with feed grinding and mixing equipment, requires a fairly large enterprise in order to exploit fully its efficiency potential. A second set of reasons for the growth in size of food-animal enterprises has been pecuniary in nature. Producers who purchase inputs in large quantities can often obtain discounts in per unit prices. Similarly, they are often able to bargain for price premiums on large-volume product sales. Also, large-scale, specialized producers are more likely to take advantage of special tax provisions, such as investment credits and fast tax write-offs of investments in machinery, buildings and equipment, thus reducing their real costs and maximizing this type of pecuniary benefit. They may more readily engage in futures trading.

Third, the complexity of modern-day food-animal agriculture requires producers to devote a considerable amount of time and energy to gather information (on nutrition, breeding, waste management, disease control, finance, marketing, etc.) and to evaluate this information prior to integrating it into their production system(s). Once acquired, the cost per unit of production of this information can be reduced by spreading it over a larger number of product units.

There is no simple way to measure the economies of size for all animal enterprises. But two types of evidence can be brought to bear on the topic. First, some studies exist which measure size-cost relationships for selected enterprises. Second, "survivorship" data may indicate the future economic viability of different sizes of enterprise. For example, if enterprises of a certain size have declined in number, it is unlikely that they are of an adequate size to remain competitive. We turn now to a brief summary of size economics for major food-animal categories.

Beef. More than one-half of all fed cattle marketed are now fed in 422 feedlots each marketing more than 30,000 cattle. This size will probable continue to grow in relative importance. But, numerous smaller feedlots will continue in the future where farmer-feeders use homegrown feed and family labor. Some per unit cost economies probably exist up to 40,000-50,000 head. Economies of size beyond 30,000 head capacity are, however, probably mainly the result of vertical linkages which result in high rates of feedlot utilization and/or more effective marketing procedures.

Cattle raising (particularly beef-cow herds) is so broad based and so diverse as to almost defy analysis. Clearly, however, the major growth is occurring in herds with 100 cows or more, and herds with less than 20 cows are rapidly declining in number. Available evidence suggests that per unit costs for buildings and facilities, machinery and equipment and, perhaps, breeding stock investments, decline up to herd sizes of 1,000 cows, but may increase for herds beyond this size.

Hogs. Hog enterprises marketing less than 200 head of hogs are rapidly disappearing. Rapid growth is occurring in the number of enterprises which market 1,000 hogs or more annually. In the production of market hogs (complete farrow-to-finish and feeder pig finishing operations) costs per hundredweight of hogs produced probably decline up to a size of 5,000 head or more, although most of the cost economics appear to occur by a size of 1,600 head marketed. Feeder pig enterprises are diverse, but some cost economics probably continue well beyond an enterprise size of 1,000 head produced annually. In production of both feeder pigs and market hogs, the major size economics in production costs are in labor, management and capital costs. Feed requirements per unit of output change very little as the size of operation is increased beyond some minimal level.

Dairy. Dairy farms with less than 30 cows have declined in number rapidly and the largest absolute gain in numbers (1950-1974) has come in the 50-99 cow size range. However, the percentage of farms with more than 100 cows grew the most rapidly during this period and by 1974 they had 26% of all milk cows. Significant per unit cost economies are probably present up to 60-70 cows or more on family-scale dairy farms with on-farm forage production, and up to a much larger size range (several thousand cows) in the specialized drylot dairies of Arizona,

California and Florida. Because dairy production is still a labor intensive enterprise, the major source of cost economies tends to be in the substitution of capital intensive technology for labor.

Poultry. On a survivorship basis only those production units with more than 20,000 laying hens, 60,000 or more broilers and 60,000 or more turkeys sold annually are increasing both in number and in percentage of total production. Laying flocks of 3,200-20,000 hens are still important, however, as are turkey producers marketing from 16,000-60,000 turkeys. Those economies associated with input acquisition (poults, feed, medication and management assistance) and with product marketing via vertical integration are the dominant factors affecting size economies.

Supply Response

It is extremely difficult to unravel the complex interrelationships which exist between the supply response of food animal producers and their cost and profit levels. It is even difficult to determine the relationships between annual enterprise profits and after-tax returns to farmers and other investors in food-animal agriculture. There is, however, a good deal of evidence to suggest that farmers and other investors are willing to accept low current (operating) returns on some investments (such as land, breeding livestock, etc.) in order to realize long-term capital gains which are taxed at a lower effective rate. Certain other animal enterprises, on the other hand, are very responsive to short-term prices and profits. The brief discussion which follows is intended to provide some insight into the supply response situation faced by many producers. A great deal of research is needed into the situation in each of the commodity areas.

Beef. There is widespread agreement that the size of the national beef-cow herd had grown to a size in 1975 (45.4 million head) which would not generate profits adequate for its sustainment. If the national beef-cow herd is to return to its 1975 or higher level, it will need significant profit inducements from higher beef prices or reduced production costs, or both. As reduced production of feeder cattle from the smaller beef-cow herd is realized, calf prices will rise and the cattle-raising subsector will again enter an expansion phase of the beef cycle. Because of higher production costs, the expansion inducement price level of the future will be much higher than in the past. In the expansion phase of the cattle cycle, supply response from the cattle-feeding subsector is, of course, dependent on the cattle-raising subsector for an expanded supply of feeder cattle.

Hogs. Adjustment patterns and other economic factors affecting supply response for hogs have changed dramatically since the 1950s. The expansion and contraction of production by farmers producing hogs has focused increasingly on two groups of producers. One group, including both existing producers and new entrants, has specialized in hog production, enlarging enterprises by increments of substantial size each time favorable profit conditions permitted. The other group, comprised largely of farmers with marginal hog enterprises, older farmers choosing to reduce their farming activities and farmers who have chosen expansion in other enterprises, maintains hog production while returns are favorable, but ceases hog production permanently when returns become unfavorable.

With investment costs now representing a much higher proportion of total production costs than formerly, some measure of the estimated return on investment cost will probably be a better indicator of future supply response for hogs than the hog-corn ratio which predicted supply response so well in earlier periods. As we suggested earlier for beef, because of increased production costs the next expansion phase of the hog cycle will require a higher triggering price level than in the past. Whereas a hog/corn price ratio of 13:1 was considered breakeven in 1950 (and above which production would be encouraged), the current hog/corn price ratio required to encourage expansion is much higher.

Dairy. Dairy, along with beef cow-calf production, is probably least responsive to short-term cost price (profit) fluctuations. Dairy facilities are expensive to construct, and the enterprise is heavily dependent on large roughage and labor inputs. Thus, it is difficult for producers to adjust production levels greatly in response to short-term profit levels and it is virtually impossible to be in and out of dairy production from one year to the next.

More than for any other subsector, future dairy production levels will be dependent on the price support policies of the federal government. These policies, in turn, will likely center on adjusting support prices upward, but only at a rate which will cover rising production costs. Thus, government policies will likely stabilize future incomes for dairy producers, and in so doing will also stabilize milk supplies. These policies, however, will be conditioned heavily by the effective demand by consumers for dairy products.

Sheep and lambs. Sheep producers have suffered from a broad range of economic problems including much higher land and labor costs and other labor problems, heavy losses from predators and increased consumer demand for beef and poultry meats. Synthetic fibers, cotton and foreign-produced wool have applied economic pressures on U.S. produced wool, the joint product produced along with lamb and mutton. The rate of decline in the sheep-lamb subsector has lessened somewhat in recent years. This reduced rate of decline is expected to continue into the future, but there is little evidence to suggest that it will be reversed as some dissatisfied producers continue to shift out of the sheep enterprise.

Poultry. In direct contrast to the sheep-lamb subsector, it seems likely that the poultry-meat sector faces an expanding future. Feed is the largest and one of the most critical inputs in poultry and egg production, accounting for two-thirds to three-fourths of the cost per dozen eggs or per kilogram of live broiler or turkey. Bird costs (hen depreciation or chick and poult costs) are the second largest cost item. Labor costs and overhead costs (buildings, equipment, etc.) are about equal in importance; the former have been declining in importance and the latter are tending to increase.

Poultry and egg producers can adjust output during the year by changing the number of chicks or poults started, changing the frequency of batches raised, adjusting market weights, or culling or recycling layers. Ultimate limits to increases exist, however, in terms of housing capacity and chick or poult supplies from breeding flocks. Year-to-year production responses are affected by past net returns, but there often are several-year lags before large responses occur.

THE CONSUMER AND DEMAND SIDE OF FOOD ANIMAL AGRICULTURE

Effective demand for U.S. produced animal food products has, in the past, been principally for domestic utilization. Table 9.5 summarizes the aggregate domestic disappearance, exports and imports for major food animal product groups from 1960-1979.

Domestic Demand

Historically, the demand by U.S. consumers for food-animal products has been principally a function of consumer disposable income, prices of individual food-animal products and their close substitutes, and the tastes and eating habits of consumers. But, this situation has grown much more complex in recent years. Consumers now eat more of their meals away from home than formerly and their consumption is conditioned by an additional set of factors such as self-imposed dietary constraints, convenience of food preparation and food safety considerations. Many of these and additional population variables are undergoing change currently and are not well quantified. Their identification and quantification are, in fact, among the topics needing effective research attention. It is probably the case, however, that consumer incomes and product prices (including the prices of close substitute foods) are still the most important factors affecting aggregate demand for food animal products. These are followed in importance by health-related dietary considerations, changes in proportion and types of meals eaten away from home and population changes.

Price and substitution aspects of demand. The quantity of food, including animal products, is not as responsive to price change as it is for many nonfood items. Thus, for food-animal products as a group, a 1% change in price will result in a less than 1% change in the quantity of food animal products purchased and consumed. As a result, demand for food animal products as a group is said to be price inelastic. Table 9.6 shows estimated-price elasticities and cross-price elasticities of demand for major food-animal product categories, fish and "other food." The upper left to lower right diagonal of this matrix of elasticities shows the estimated change in the quantity of a commodity purchased as the result of a 1% change in the price of that commodity. For example, if red meat prices increase by 1%, other things remaining constant, red meat consumption is estimated to decline by .6%.

These elasticities show that both red meats and poultry have greater price elasticities (consumption responses to price changes) than do eggs, dairy products, fish and other foods. Thus, if production costs (and subsequently real product prices) can be reduced, one can probably expect greater market response for red meats and poultry than for other major food-animal products. Although not shown in table 9.6, a price elasticity matrix was estimated with further breakdowns in the classification of meats into beef, pork, chicken and turkey. This matrix was inverted to derive a "price flexibility" matrix for these food-animal products. The resulting price flexibilities indicate the following: 1) a change in pork consumption will have a smaller effect on beef prices than will the converse; 2) broiler prices are highly responsive to changes in pork consumption (much more so than for changes in beef consumption); and 3) beef prices are not very responsive to changes in poultry consumption, but pork prices are. These relationships as well as those identified by the price elasticities and cross elasticities of demand can help to show the expected impacts of future research on product prices, on quantities demanded, and consequently, on producers' profits. But, one must bear in mind that these elasticities may now be outdated. And, as national averages they mask a wide range of variation in demand response based on the age, income, family structure, life styles, etc., of consumers, and on the forms in which the various foods are consumed.

Effect of income changes on demand. The effects of changes in consumer incomes on the demand for food products are not easily quantified. They may also be undergoing significant changes at the current time. Nonetheless, some broad generalizations are probably valid. For example, as consumers' incomes rise, other things remaining constant, consumers increase their purchases of beef, turkey, lamb and mutton significantly. They also increase their proportionate expenditures for fed beef, a preferred product. Household Food Consumption Survey data for 1965 show that individual consumers in the higher income brackets of households ($20,000 or more) consumed about 3.9 kg (8.5 lb) of lamb annually at the time of the survey, while those in the lower income brackets (less than $7,000) consumed only about 1.3 kg (3 lb) per person. On the other hand, as incomes rise, consumers decrease slightly their expenditures for chicken and eggs and hold expenditures for pork about constant. Indications are also that consumers increase their expenditures for most low-fat dairy products and cheese as their incomes rise. Consumer demand for animal fats and oils, however, has undergone a major decline in recent years and is unlikely to recover regardless of consumer incomes. Finally, if consumers should face declining real incomes, they may respond by adjusting downward their purchases of food-animal products, particularly meat. They may exhibit revived interest in purchasing a broad range of food substitutes including those from soybean, cereal grain and synthetic sources.

Most food purchases, of course, are made by individual households faced with a unique set of budget circumstances. These individual households (and other consumer decision units) can be classified in numerous ways. Table 9.7 presents selected population and food expenditure data for different income groups for 1973-1974. Among other relationships, this table illustrates the declining portion of income which is spent by households for food at higher income levels. Thus, not only do food-animal products compete with other food products for the consumer dollar, but, because of the income inelastic demand for food, they compete for a much smaller proportion of consumer income as income levels increase. In sum, the relationships between income and demand for food-animal products are complex. In the aggregate, and for most individual products, demand is income inelastic. Yet, gains in the real incomes of consumers are probably critical if per capita demand for food-animal products is to rise much in the future. With higher consumer incomes, fed beef and turkey appear to have the strongest potential as recipients for increased consumer expenditures. Clearly, it is of critical importance that a continuing data base be developed which will permit the estimation of income elasticities of demand for major food-animal product categories. Without such data, estimated projections of the future demand for animal products is virtually impossible.

Food consumed away from home. The increased incidence of away-from-home consumption of food in recent years is the result of several interacting forces. Higher consumer incomes, the rapid emergence of fast-food eating establishments, the increased number of women employed outside of the home and other dimensions of changing lifestyles are some of the factors involved. Table 9.8 shows that about one dollar in four of the overall expenditures of a sample of U.S. households for food was away from home in spring 1977. As might be expected, both income and family size, and even location of residence, have important effects on the incidence of away-from-home food purchases as does age (table 9.9). Lower incomes and larger families both constrain the incidence of away-from-home food consumption as does the post-55 age range.

Diet and health concerns. Health concerns have led to increased per capita consumption of low fat and so called "natural" foods both in the U.S. and in other developed countries. Although sales of health foods currently account for about only 1% of the national grocery bill, an estimated 350-400 health food manufacturers now distribute their products nationally, and about that many more locally --a phenomenal growth since 1960.

A somewhat different dimension of food health concerns is the nitrosamines issue and a number of other food-animal product health-related issues which have been addressed extensively in the literature (see for example CAST (1980), Food from Animals). Their impact on future demand for food-animal products is still unclear and needs effective research attention.

Other factors affecting demand. A complex set of additional factors affect the domestic demand for food-animal products. These include changing life styles and changing population structure. One recent food growth area has been that of household gardens. The value of fruits and vegetables grown in such gardens was estimated at $14 billion in 1977. The significant concentration of this garden production among low-income groups suggests that it is partially an "income problem induced" phenomenon. Yet, some gardening is clearly related to lifestyle preferences and to the desire for natural foods. Its general impact on the demand for food-animal products is negative.

The impact of changing population structure on food demand is complex. Neither the "very young" nor the "very old" in the population is a heavy consumer of meat and other food-animal products, though young children are important consumers of milk and ice cream. Clearly, the effects of changing population structure on the demand for food-animal products needs creative research attention.

In summary, changing life styles can be expected to have a diverse effect on the demand for food-animal products. Some consumers will increase their demand for convenience foods for home use and will resort to more away-from-home food consumption. Others will increase their demand for natural foods, including those produced commercially and from household gardens. But food-animal products may gain if the demand for food convenience increases both at home and via fast-food outlets.

Export Demand

Exports of animal food products to date are rather minor compared with domestic production and consumption. However, there is considerable potential for expansion of export sales, especially to Japan and Western Europe. These markets are currently constrainted by trade barriers in the importing countries. The greatest potential lies in exporting high-quality grain fed beef and poultry products.

Also of considerable importance to the U.S. animal industry is the potential for exporting breeding stock and semen. This potential may be greatest in third world countries striving to increase their production of animal foods.

MARKETING AND DISTRIBUTION OF FOOD ANIMAL PRODUCTS

Marketing and distribution of food-animal products in the U.S. is accomplished via a broad set of activities and services performed by a combination of private firms and governmental agencies. In order to split out the costs of marketing and distributing farm-produced foods, consumer expenditures for farm foods are broken down into their "farm value" and their "marketing bill" components. The marketing bill is then further divisible into sub-components for transportation, processing, wholesaling, retailing and food services.

The above categorization of marketing and distribution within the "marketing bill" excludes from direct scrutiny, however, that set of activities which results in the change in ownership from producers to next owners (principally processors) of food animals and food-animal products. Thus, we give brief attention next to the "sale by producers" function of marketing before considering the marketing bill components in more detail.

Producer Markets

The institutional mechanisms for transferring product ownership from producers vary greatly for the several major food-animal products. Where complete or partial integration exists between production and marketing, the processor may own the poultry or other animals on the farms. In general, two major concerns exist with respect to producer markets for livestock and livestock products. These are: 1) that they be cost efficient; and 2) that they accurately reflect product value. The latter phenomenon is sometimes expressed as a concern about "fair and accurate" price discovery.

Most chickens, turkeys and eggs move through a vertically coordinated production and marketing system. Although both spot and futures market price quotations exist for eggs, spot markets are based on a fairly thin volume of sales transactions. Increasingly, dairy producers market their milk through a system of producer-owned cooperatives which also take on the processing function. By far the broadest set of public market institutions exists for cattle, hogs and sheep. Thus, it is to the market place for these animals that we now turn in detail.

Concurrently with the operation of a "physical transaction" market is the operation of active futures markets in feeder cattle, market cattle, hogs and pork bellies. With the advent of large-size enterprises feeding cattle and hogs, and with extensive decentralization of packing plants into major production areas, a higher proportion of slaughter-ready animals are now purchased via direct (on farm) buying by packers. Terminal and auction markets still perform an important role in the selling of market livestock, as do a variety of contractual arrangements. A broad set of institutional mechanisms also exists for the marketing of feeder livestock. Producer cooperatives play a significant role in these market transactions as do individual feeders, order buyers and both auction and central markets.

The literature on cost and pricing efficiency of markets for livestock and livestock products is voluminous. Without trying to estimate these efficiencies, we list below briefly some of the continuing concerns expressed about the markets.

Livestock markets. There is widespread concern that producers do not use existing marketing institutions and mechanisms as effectively as they might. This is particularily true for futures markets, but for various spot (cash) markets as well. Extensive use of the "Yellow Sheet", a daily market news report of wholesale beef trades, as a guide for the bid price of live cattle by slaughtering firms may be a "thin and risky" source of information for pricing beef. This latter issue resulted in the establishment of a Meat Pricing Task Force by the Secretary of Agriculture in March 1979. There are other concerns about market-pricing information and methods, particularly in areas where livestock markets are few in number and widely dispersed geographically. Finally, increases in energy costs and/or in the location of animal production (particularly for cattle) may signal further changes in the location and structure of livestock markets. These phenomena appear to warrant research study.

Dairy marketing. The current pricing system for milk puts substantial power in the hands of a small number of large producer cooperatives and governmental agencies. This situation has resulted in questions about the adequacy of consumer protection in the pricing process and whether product pricing is regionally equitable and efficient. Some regional price differences for milk do not appear to be the result of a fully competitive pricing system.

Poultry and egg marketing. Most concerns center on the thin market of public sales transactions for eggs to which most other transactions relate by formula. The concern centers on the accuracy of publicly quoted prices as a measure of supply and demand.

The Marketing Bill

Compared to all farm food, a lower proportion of consumers' expenditures for food-animal products goes to pay the "marketing bill" for these products and more is allocated to "farm value" (table 9.10). But the marketing bill component for food-animal products is still high (33-49% of prices at retail) and, on an absolute basis, rising rapidly. Thus, marketing and distribution costs are critical components of food-animal product prices at retail. Moreover, the economic activities in the marketing and distribution of food-animal products are important in their own right generating a large volume of income and employment.

Figure 9.4 provides some useful perspectives on the major items included in the farm-food marketing bill as of 1978. As an individual component, labor costs at 47% of the total in 1978, represent much the largest item. In looking ahead to 1980, industry analysts have projected an increase of 9-12% in the marketing bill for farm food from the level of 1979. Declining labor productivity and much higher energy costs are heavy contributors to this large increase in the marketing bill.

A somewhat different perspective on the incidence of the marketing bill for food-animal products is presented in table 9.11. These 1978 data show that the marketing bills for meat and poultry products consumed away from home exceed by far those for meat and poultry products consumed at home. Even for dairy products, almost 43% of the total marketing bill was associated with products consumed away from home.

In addition to the regular marketing and distribution services performed by the private sector, federal, state and local governmental agencies provide a broad set of services including the provision of health inspection, maintenance of sanitation standards and operation of a broad set of animal health programs, price support programs, market orders, market news, import controls, acquisition and disposal of surplus products by the Commodity Credit Corporation, PL-480 programs, domestic food aid programs and many others. Although most of these services are not identified directly as charges in the food marketing bill, they are costs and they do affect the economic environment in which the marketing and distribution processes for food-animal products occur.

Transportation. The U.S. animal industry is highly dependent on a complex transportation system both to move massive supplies of production inputs, including feedstuffs, to producers and to move intermediate products (particularly feeder cattle and feeder pigs) and final products to their respective destinations. Since most animals and animal products have a high degree of perishability and bulk, transportation is a major cost item for the food-animal industry. The transportation component for the farm food marketing bill was estimated at $10.9 billion in 1978. Although it is difficult to break transportation costs down by individual food-animal categories, the annual cost of shipping livestock and meat products to and from slaughter plants alone was estimated at $800 million for 1975-1977. Transportation costs for the widely dispersed dairy sector, with its highly perishable raw milk and manufactured dairy products, represent a major component in the market bill for that sector.

Of particular concern is the rapid increase in energy and labor costs which are important in transportation. Also, of critical importance is the need to keep the nation's rail and highway system in effective operating condition. In addition, there is a critical need to maintain a system of efficient pricing, rate structure regulation and performance in both intrastate and interstate transportation. Thus, the outcome of current discussions relative to deregulation in trucking and rail transportation could be of crucial significance to the animal industry.

Processing. Processing was the largest cost component in the farm food marketing bill until it was surpassed by retailing in 1977. The processing component of the farm food marketing bill still remains a major item totalling $37.6 billion in 1978. Moreover, processing, broadly defined, still remains a major functional activity in the marketing of food-animal products, particularly red meat, poultry and dairy products. The meat-packing industry is the largest single component of the food-animal processing sector, with total operating costs (exclusive of the cost of livestock and other raw materials) of $8.6 billion in 1978 of which 52% went for wages and employee benefits (AMI, 1979). An estimated 163,000 persons were employed in meat packing plants and 148,000 in the fluid milk and cheese industries alone in 1979 (USDL).

Among the economic topics receiving attention in the food-manufacturing industry are those relating to degree of concentration within the industry and the relationship of size of firms to per unit costs. In general, the average industry concentration in food and kindred products is high relative to most industries. For example, the share of total food manufacturing industry assets owned by the 50 largest firms was nearly 64% in 1978 compared to 41% in 1950 (Connor, 1980). There have been large declines (of the order of one-half) in the total number of food manufacturing firms in the U.S. since 1947. A recent study has estimated that market power in food processing has resulted in multibillion dollar overcharges to consumers (Parker and Connor, 1979). There remain some 17 national meat packers (those with $250 million or more in annual sales) and more than 40 regional packers (those with product sales of $25 million to $250 million annually). Overall, it does not appear that concentration in food animal product processing is nearly as acute as is the case, for example, in the dry cereal manufacturing sector of food processing. Raw material assembly costs are probably an important factor in keeping some dispersion in food-animal product processing plants.

The technical studies of the National Commission on Food Marketing provided the most comprehensive perspective available on cost structure in the food manufacturing industry as of the mid-1960s. These technical studies presented estimates of cost-scale relationships for cattle and hog slaughter, broiler and turkey processing and processing of fluid milk and major manufactured dairy products. The conclusions reached in these studies were that important scale economies exist in these phases of food marketing, but are achieved by "medium-size" plants relative to the size of the total industry. For example, most in-plant economies were estimated to be achievable by plants handling 1% of the poultry supply, 2% of the turkey supply and a similar or smaller portion of the milk supply. These old data obviously need updating.

There was clearly a great need for consolidation and growth of individual firms in food-animal product processing in the period following World War II. This was particularly true for dairy and poultry processing. Such consolidation continues in dairy processing where the number of butter plants decreased by 34% between 1972 and 1977, while production per plant increased by 50%. Both a high proportion of new, efficient plants in the industry and the extremely high cost of new plant construction suggest a likely moderation in the rate of future change. Some additional trend to decentralization of livestock marketing and meat packing on a selective basis may occur in response to higher energy costs and/or to changes in the location of animal product supplies. The scale, technology and location of dairy and poultry processing facilities, on the other hand, are probably pretty well set for the near future.

There is strong evidence that labor union practices have retarded the rate at which some cost efficient operations (principally those involving centralized meat cutting and boxing) have been adopted in the meat-packing sector. But a major part of the cost increases due to packaging, precooking and other built-in convenience features for food animal products, have resulted from the consumer demand for them. Labor productivity continues to be a concern since, despite the adoption of much labor-saving technology, labor productivity in the manufacture of farm originating food increased by only about 1% per year from 1972-1977.

Wholesaling and retailing. Wholesaling of farm food employed an estimated 655,000 people in 1978 and added an estimated $21.9 billion to the marketing bill. Retailing, the largest single functional component of the farm food marketing bill, employed 1,743,000 persons in food stores alone and added $40.0 billion to the market bill. Both wholesaling and retailing are labor intensive activities with labor costs representing 43 and 48% of total costs, respectively. Labor productivity in food stores is a continuing problem, as it decreased an average of 1.2% per year between 1972 and 1977 and the drop continued through 1978 and into 1979.

Table 9.12 provides a breakdown of the distribution of marketing costs for key food-animal products sold at retail and points up the importance of the wholesaling and retailing functions, particularly the latter, to consumer prices for these products. One needs to keep in mind, moreover, the high proportion of marketing bill costs for meat, poultry and dairy products that are incurred in the away-from-home sector (table 9.11).

Our discussion of the food wholesaling sector will be brief because it is difficult either to separate this functional sector or to generalize about it. For example, most large retail chains are integrated into wholesaling and perform this latter function even though their primary business is food retailing. Some food manufacturing firms are also integrated into the wholesaling business. Food manufacturers, through their sales branches and offices, handled about 20% of the food wholesaling business in 1972. Also, some of the big firms in the rapidly-growing food service industry have integrated into the wholesaling business and even into the contracting for supplies from producers.

In terms of function, food wholesalers in the future can be expected both to broaden their services and to provide more specialized attention to individual firms in the food retailing and food service sectors. Financing, promotion, site selection, inventory controls and general computer services are among the services which likely will be expanded by wholesalers.

There has been a big increase in concentration in the food retailing sector since World War II largely as a result of the growth in supermarket chains. This structural adjustment has continued through the 1970s. In U.S. metropolitan areas the four largest grocery firms' share of the total grocery sales increased from 46 to 52% from 1954 to 1972.

The 1970s also saw a rapid growth in so-called "superstores" in which nonfoods and general merchandise are also sold. At the other size extreme, "convenience stores" have grown rapidly in number as food outlets during the past 20 years. These stores do not, however, account for a major share of total food sales. Finally, although many changes are still occurring in food retailing, some adjustments are now virtually complete. This is true, for example, of the shift from home delivery to bulk store sales of dairy products and eggs.

As in the case of food processing, the emergence of large-scale food retailing firms coupled with increased concentration within the industry has led to charges of abusive monopoly powers. A recent study prepared by the Joint Economic Committee of the Congress, estimated that "monopoly overcharges" for the four largest food retailing firms totaled $662 million in 1974 or 1.6% of sales (Marion et al. 1977). Regardless of the degree of market power, the after-tax profits of firms in food processing and retailing are, at most, a modest proportion of the total final costs to consumers of food-animal products. The paucity of good evaluative information on profits in the marketing and distribution sectors of the food-animal industry does, however, suggest the need for solid economic research on this topic.

Food service. No single portion of the U.S. food industry has changed as drastically or grown as rapidly in recent years as the food service industry (Van Dress, 1979). This away-from-home eating component of food distribution has both a "public" and an "institutional" component of which the former is much the larger (see table 9.11). For those farm foods consumed away from home, the marketing bill portion represents about 81% of total consumer expenditures leaving only about 19% for farm value. Thus, the food service segment is four times as large a market for agribusiness as for farmers. Among the major contributors to the "public" food service industry are restaurants, cafeterias, fast food outlets, lunch rooms, caterers, food contractors, food vendors and grocery and department stores selling meals and snacks. On the institutional side, schools and hospitals are the largest food-service industry markets, although there are a number of smaller ones.

Fast-food places more than doubled their share of sales between 1963 and 1978 going from 15 to 32%. And, during the 20-year period from 1958 to 1978, real sales by fast-food outlets increased more than 700%. The big increase in fast-food sales was associated with the expansion of franchising in the 1970s. In 1972 some 33,000 fast-food outlets dispensed $7 billion worth of food. Only 6 years later, franchise outlet sales had grown to $17 billion. Thus, we are even now in the midst of a major growth and restructuring of food consumption to which the food-animal industry must direct close attention and for which our past research on food demand is ill-equipped to deal.

Clearly, the food-animal industry has a big economic stake in maintaining or expanding the share of their products which move to consumers via the food service industry. Effective economic research needs to be targeted at this important component of food demand in order to assess both the product volume and the mix of products for which it will represent effective demand in the future. This in turn requires improved knowledge relative to the changing life styles, incomes, tastes, etc., of consumers who generate the demand for food service.

SUMMARY PERSPECTIVE

In the process of evaluating the supply and demand for food-animal products and the changes in their equilibrium levels over time, one finds it easy to conclude that some changes are supply driven, others demand driven and still others are "induced" by the numerous intermediaries in the marketing and distribution system and by the regulatory agencies affecting the whole food-animal industry. Moreover, effects generated from these different sources are not mutually independent. Thus, any information system capable of guiding the future of the industry will need to deal with all of these facets in both their technical and their economic dimensions. Our traditional information system is obsolete.

CONSTRAINTS ON THE PRODUCTION, MARKETING AND DISTRIBUTION SYSTEM FOR FOOD-ANIMAL PRODUCTS

The key goals of the production, marketing and distribution system are to serve human needs by furnishing food-animal products which are safe, nutritious, efficiently produced and fairly priced while providing a reasonable return to all factors of production.

Constraints which deter or prohibit the achievement of these goals can be classified into four groups, institutional, technical, resource related and economic. A number of constraints show up as a mixture of all four. Thus, their resolution or modification requires an integrated approach including technical research, economic analysis and informed action by producers, consumers and units of government. The latter includes data services and research support as well as regulatory activities.

Institutional constraints. Institutional constraints are of several general types; those imposed by governmental or other agencies which are usually regulatory in nature, and those self-imposed by consumers, either as individuals or as groups. Even the operators of the food-animal system impose their own institutional constraints on the system. As of this time, government regulations regarding: 1) air and water pollution; 2) feed and food additives; 3) transportation and trade; and 4) taxes on individuals and businesses are important regulatory constraints. Health and diet related constraints are of critical importance in the consumer area.

Technical constraints. Technical constraints are those imposed by available technology on the alternative methods for production, marketing and distribution. Maximum achievable meat, egg or milk production per unit of feed are examples, as are current capabilities for transportation, food processing, animal disease control and numerous others. Removal of these constraints requires both research achievements in the several technical fields and effective economic analysis and dissemination of this new knowledge to decision makers in the food-animal system. Energy requirements, labor and transportation inefficiencies and high capital requirements are among the key technical constraints which are currently critical in the production, marketing and distribution of food-animal products.

Resource constraints. Resource constraints include those related to natural resources (particularily land and water) and to the available pool of human resources. These resources are limited in total. Nevertheless, their availability to and requirements from the food-animal industry can be modified by technical advances and institutional reforms which improve the competitive position of this industry. Economic research is needed to identify the trade-offs involved in adopting new resource-saving technology and/or in modifying resource related institutional constraints.

Economic constraints. Economic constraints are key determinants of the production, marketing and distribution system for food-animal products. And, in a "market oriented" economy they occur on both the supply and demand sides of the market. Economic research has as a major objective the integration, evaluation and projection of the supply of food-animal products and attendant services under alternative technical, institutional and price conditions. This must then be matched against a similar economic appraisal of the effective demand by consumers for available products and services. Over time, production must be profitable or food-animal product supplies will diminish. Consumers will demand reasonable product values or they will diminish their purchases and shift to other products. It is through the operation of this profit-value system that economic constraints are imposed upon the food-animal system.

RESEARCH IMPERATIVES

RESPONSES TO IMPACTS OF MASSIVE SHOCKS AND INSTABILITY

The objective of this research is to reduce the negative impacts of increased instability within the food-animal sector associated with massive shocks such as energy shortfalls, export fluctuations, catastrophic diseases, adverse weather, regulatory shifts and rampant inflation.

The industry is more threatened by major shocks than ever before. It is more concentrated geographically and more dependent on technology and off-farm inputs, and lacks much of the flexibility of the past. Uncertainty of petroleum supplies threatens interruptions in fuel supplies that could cause large losses. World supplies of grain are in delicate balance. Unfavorable weather anywhere in the world or shocks in international trade can lead to a repeat of the price volatility of 1973-1975. Exotic diseases, such as African Swine Fever, would have tremendous impact on the industry.

RESEARCH NEEDS WITH PAYOUT POTENTIAL

A high priority need is to develop and evaluate contingency plans for producers to deal with temporary shortfalls of energy. Such plans would range from installation of fuel storage as insurance against shortfalls to changes in the type of facilities and management practices.

Longer term curtailment of energy supplies may shift regional comparative advantages of production and processing. Tradeoffs between lower energy use in production against increased transportation costs need to be identified and evaluated as well as the implications of higher energy costs on economies of scale in production units.

The implications for animal agriculture of using large amounts of feed grains in gasohol production need to be evaluated. The impact of major changes in the composition of animal feed supplies with heavy usage of distillery by-products is not known.

Considerable concern is often expressed about feeding grain to animals rather than to humans. However, the use of surplus grain supplies as animal feed provides a method of maintaining reserve grain production capacity for human consumption in periods of grain shortages. Research is needed to evaluate the nature and magnitude of the tradeoffs of idle land and large grain reserves vs fluctuations in livestock production during periods of reduced grain production.

Another major problem is to determine the costs and benefits of various methods of dealing with the occurrence of a major disease outbreak. Included in this evaluation would be identification of alternative indemnification programs for producers and agribusiness firms as input to decisions about how to deal with disease outbreak.

The large capital requirements of both production and marketing units makes them susceptible to shocks to the financial markets. Hyper-inflation or major changes in interest rates threaten the viability of firms and cause disturbances in livestock output decisions with long-lasting consequences. Methods of dealing with these types of shocks and risk need to be identified and evaluated.

Evaluation of the impacts of shocks cannot be made without identifying the nature of interrelationships within the aggregate production and marketing system. This involves identification and incorporation of the interrelationship between resource constraints, transportation systems, biological constraints on production and the economic environment. Aggregate systems analyses can provide useful information and guidance for directions of agricultural research as well as answering other policy questions. For example, a partial analysis of the impact of higher energy costs might suggest the development of animal breeding programs to produce animals that can more efficiently utilize forage, when an analysis of the more complete system might indicate that it is more expensive to expand forage production than to expand grain production.

The development of aggregate systems analyses will expand understanding of the food-animal sector as an interacting economic system. Such knowledge will permit both private and public decision makers to base their decisions on improved predictions. Accurate prediction is the basic method of achieving efficient combination of resources so that the most efficient levels of output can be sustained.

Analysis of the aggregate system does not require the development of one grand model to answer all possible questions. The nature of the analysis required will depend on the particular shock being considered. Effective systems analysis will, however, require that the interrelationships between segments of the industry be identified and understood.

Aggregate systems analysis will require the coordinated development of research results from each dimension of the animal agriculture system to assure that the body of knowledge being developed provides a basis for evaluating the impact of various shocks.

PROJECTING FUTURE DEMAND SHIFTS FOR FOOD-ANIMAL PRODUCTS IN A RAPIDLY CHANGING WORLD

Accurate estimates and projections of demand are imperative for effective planning and resource adjustment in animal agriculture during a period of great instability which is likely to characterize the future.

The demand for animal products is likely to change considerably from that of the past, especially from that prior to 1970. Thus, estimates of factors influencing the demand for animal foods formulated from data prior to 1970 are not likely to be of value in the future. Knowledge of consumer markets is outdated and inadequate for assessment of the future structure and parameters of animal-food markets. Lack of adequate data bases necessary to monitor changes in markets is of vital concern. For example, USDA National Food Consumption surveys of 1965 and 1977 are too infrequent. The nature of the highly aggregated disappearance data is inadequate to identify and quantify the forces operating in the market. Animal agriculture is a huge industry and biological in nature. Considerable expense and time are involved in adjusting to rapidly changing demand conditions. Updated and reliable estimates of demand would conserve considerable resources, and contribute to a more efficient industry in both the producing and fabricating-marketing sectors.

Research Needs with High Payout Potential

Domestic demand for animal foods is affected by, among other factors, population size, real disposable income, tastes and preferences, availability and price of substitute nonanimal foods, the age composition, household size and employment characteristics of the population and the propensity for consuming meals away from home. All of these factors are changing and have impacts on demand.

The influence of changing lifestyles, including a trend toward away-from-home dining, on the demand for food-animal products needs additional analysis. A higher degree of uniformity of products may be required for institutional and fast-food markets than for traditional retail markets.

Age distribution and resultant nutritional requirements are changing and will have a pronounced impact on the demand for food-animal products. Recently declining birth rates and increasing life expectancy will continue to influence demand in terms of both population size and age distribution.

Nutritional and health safety concerns (especially health risks alleged to be related to components of animal food products) will influence consumer demand. More knowledge is needed about these important relationships.

Foreign demand for animal products in the past has been relatively small. The possibility of greater volatility in international relations, the necessity for stricter pollution controls regarding livestock production in densely populated countries, and other factors, may lead to levels and variability of foreign demand for animal foods in the future that are considerably different from the past.

The basic need is an adequate and well-managed data base and continuous analyses and reporting.

DEVELOP AND EVALUATE FOOD-ANIMAL PRODUCTION SYSTEMS FOR MEETING HUMAN NEEDS IN THE 21ST CENTURY

The practical application of new understandings of biological, physical and economic systems, and of new products and technologies requires an integrative, multidisciplinary approach. This entails a systems approach to design, evaluation and decision-making at the producer level for meeting human needs from food animals. Developments in the last decade in biomathematics, in systems science, and in nonlinear and robust statistical estimation techniques offer new analytical methods that are complementary with developments in laboratory techniques and in disciplinary knowledge bases. They are an important tool of research and implementation for the scientific community to maximize its impact on the complicated world in which we live.

RESEARCH NEEDS WITH HIGH PAYOUT POTENTIAL

1. The food-animal production systems approach includes:
   a. Development, evaluation and maintenance of the biological, physical and economic data bases necessary for the estimation of systems parameters and for testing systems validity;
   b. Development of a systems analysis thrust embracing integration of biological, physical and economic processes; and
   c. Development of production systems and management options to find those sets most consistent with resource availability, consumer preferences, environmental constraints and attitudes toward risk.
2. This approach provides a focus for identifying informational gaps in system components for:
   a. Conceptual developments within disciplines;
   b. Interdisciplinary linkages; and
   c. Identifying potential problems in the application of new technology and management concepts at the producer level.

IMPROVED PRICING MECHANISMS

In a large economy, with private property, the market is the main means of communication. In fact, price changes are the only means through which individual plans for production, consumption, savings and investment can be made consistent. An economy without a well-functioning market inevitably will have failures to produce the most desired types of products, to fit the time of output to the demand, to use the most efficient means of production, including new technology, and to channel enough savings into profitable investment. The failures will result in higher unit costs to producers, higher prices to consumers, perpetuation of less desired types of output and/or the greater cycling of output than society needs to endure.

Another large social problem arises from forms of monopoly that raise unit costs, prevent desirable economic adjustments, or otherwise reduce the quantity and quality of output, whether they occur in labor or other input markets or in markets for products themselves.

A third major problem of markets is how to adequately reflect the costs of pollution control at the margin of public acceptability in the pricing of the output of farms, feedlots and factories that produce the animal food-supply.

Research Needs with High Payout Potentials

1. Thin and dispersed markets. Vertical and horizontal integration and formula pricing cause widespread concerns about the adequacy of volume entering into transactions and the accuracy of price reporting. There is much misinformation and misunderstanding about this area and a lack of knowledge about the possible remedies. Evaluation is needed of the methods now in use, including the various price reporting services, auctions, price leadership and committee pricing. New techniques such as electronic exchange systems should be investigated. A sustained effort will be needed to build solid knowledge about these phenomena. Part of the problem is conceptual; and part is the lack of empirical knowledge. Whenever there is a new social loss, the possibilities of remedies through cooperatives or other group action should be systematically examined.

Related to this problem is the problem of increasing dispersion of markets with the aggravation of "pocket-market" pricing by the great increase in transport costs. The latter ultimately will have a bearing on the geographical restructuring of feeding and processing animals and animal products, with its consequent effect on the generation of prices.

The need is to investigate the role of price in the larger market information system, of which it is an integral part. Ultimately at issue is measurement of the costs and benefits of suggested improvements in this entire system.

2. Component pricing procedures and techniques. One of the special features of animal products is the way in which different physical, chemical or biological components of a commodity are valued in the market. Thus, when fat content is no longer as highly valued as other nutrients such as protein, the pricing practices should reflect this. A similar thing is true of feeder livestock practices, which have the added problem of losing the identity of the individual animal as it moves into the feedlots and then to slaughter -- making it more difficult to properly compensate feeder livestock producers for better performing animals. A strong research effort is required to arrive at more appropriate pricing techniques and get them adopted.

3. Federal and state marketing orders. At issue are determination of the appropriateness of negotiated and support-pricing by government including class prices, price differentials, location differentials, compensatory payments and classified pricing systems. While among animal products, this issue now pertains only to milk, it may arise as an issue for some other animal products.

4. Enterprise sharing arrangements. Two conflicting forces make the need to examine the nature and adequacy of enterprise-sharing arrangements in animal agriculture a critical need. One is the persistent instability of agricultural commodity markets that inhibits capital investment in modern operations; the other is the increasing capital required for low unit costs.

Market mechanisms provide various ways to meet this problem by subdividing the enterprise among different participants, thus spreading risk. Forward markets, including the highly institutionalized methods of futures trading, now are becoming a major means to meet this need for cattle, hogs and possibly broilers. But there is much that needs to be understood about the promise and the limitations of such devices, including the alternatives that are available.

5. Labor costs. The current trend in writing labor contracts in many sectors of the animal products economy is to raise the cost of labor and this tends to drive activity into other geographical areas. There is a need for a careful measurement of the impacts (present and potential) of the various provisions of such labor contracts especially in sectors that are labor intensive.

6. Pricing of transport services. Unsatisfactory quality and timeliness of transportation services for feed, livestock and livestock products may be a result of the way such services are priced. Because of the importance of transport in the proper functioning of the animal products economy, there is an increasing need to identify the nature, locus and extent of such effects, and the kinds of changes in the formation of transportation prices that would lead to improvements.

REDUCE COSTS IN MARKETING FOOD-ANIMAL PRODUCTS

More than 40% of food-animal costs to consumers occur in the marketing system. It is unlikely that food-animal prices to consumers can be contained unless the marketing system can be made more efficient. Marketing costs can be reduced by: 1) improved organization and management; 2) mechanization and automation of marketing functions; 3) optimal location of fabricating, wholesaling and retailing facilities; 4) efficient use of transportation; 5) economies of size in fabricating, wholesaling and retailing facilities; and 6) optimal storage strategies to meet consumer demand over time and space.

Research Needs with High Payout Potential

1. Develop organizational and management strategies to make more effective use of labor, utilities, transportation and storage.
2. Develop information on interregional competitive relationships for various animal products in both production and fabrication.
3. Develop and evaluate alternative mechanization schemes and electronic aids to more effectively utilize transportation; i.e., maximize backhauls and capacity loads to coordinate transportation among commodities.
4. Develop information on size and scale economies in fabricating, storing, transporting and retailing.
5. Examine the option of electronic purchasing and the delivery of products to households vs traditional shopping in light of escalating energy costs and impending declining use of private automobile transportation.

6. Develop information on optimal shipment patterns given the location of production, fabrication, storage and final consumer markets.
7. Evaluate alternate trans-shipment systems for feed grains and soybeans, i.e., wind or direct solar powered pipelines, barge and reactivation and expansion of railroad transportation.
8. Examine impact of government regulations on marketing costs.
9. Examine the availability and cost of credit on marketing costs.

REFERENCES

AMI. 1979. Annual Financial Review of the Meat Packing Industry for 1978. Amer. Meat Institute, Washington, DC.

Connor, J. M. 1979. Competition and the Role of the Largest Firms in the U.S. Food and Tobacco Industries. Regional Research Project NC-117, Working Paper Series WP. 29, Univ. of Wisconsin.

CAST. 1980. Food from Animals, Quantity, Quality, and Safety. Council for Agricultural Science and Technology Rep. No. 82, Ames, IA.

CAST. 1979. Impact of Government Regulations on the Beef Industry. Council for Agricultural Science and Technology Rep. No. 79, Ames, IA.

Duewer, L. A. and T. L. Crawford. 1977. Alternative Retail Beef-Handling Systems. ERS-661, USDA, ESCS.

Farrell, K. R. 1977. Market Performance in the Food Sector, ERS-653, USDA, ESCS.

Freebairn, J. W. and G. C. Rausser. 1975. Effects of changes in the level of U.S. beef imports. Amer. J. Agr. Econ. 57(4) 676-688.

Frye, R. E. 1978. The role of large firms and large stores in grocery retailing in the United States. Unpublished paper, USDA, ESCS.

Gee, C. K. and R. S. Magleby. 1977b. Sheep and Lamp Losses to Predators and other causes in the Western United States. AER-369, USDA, ESCS.

Gee, C. K., R. S. Magleby, D. B. Nielsen and D. M. Stevens. 1977. Factors in the Decline of the Western Sheep Industry, AER-377, USDA, ESCS.

Gee, C. K. and R. Van Arsdall. 1978. Structural Characteristics and Costs of Producing Sheep in the North Central States, 1975. ESCS-19, USDA, ESCS.

Gee, C. K., R. Van Arsdall and R. A. Gustafson. 1979. U.S. Fed-beef Production Costs, 1976-1977, and Industry Structure, AER-424, USDA, ESCS.

George, P.S. and G. A. King. 1971. Consumer Demand for Food Commodities in the U.S. with Projections for 1980. Univ. of California, Giannini Foundation Monogr. 26.

Graf, T. F. 1979. Major policy issues facing the dairy industry. Staff Paper, Dept. of Agricultural Economics, Univ. of Wisconsin.

Grinnell, G. and T. Crawford. 1977. An analysis of overhead expenses of food retailers at headquarters, warehouse and store levels. Staff Paper, Dept. of Agricultural Economics, Univ. of Wisconsin.

Grinnell, G. E., R. C. Parker and L. A. Rens. 1979. Grocery Retailing Concentration in Metropolitan Areas, Economic Census Years, 1954-1972. USDA, ESCS and Federal Trade Commission, Bureau of Economics.

Hoffman, L. A., P. P. Boles and T. Q. Hutchinson. 1975. Livestock Trucking Services: Quality, Adequacy and Shipment patterns. AER-312, USDA.

Jacobs, V. E. 1974. Needed: a systems outlook in forage - animal research. In: Crop Science of American Special Rep. #6, p. 33.

Jacobson, R. E. 1979. These are major dairy issues in 1980. Hoards Dairyman, Nov. 25.

Marion, B.W. et al. 1977. The Profits and Price Performance of Leading Food Chains. Joint Economic Committee, U.S. Congress, Government Printing Office, Washington, DC.

McCoy, J.H. 1972. Livestock and Meat Marketing. The Avi Publishing Co., Inc. Westport, CT.

Mueller, A.G. 1978. Hog/corn ratio: No longer your best pork profit guide. Farm Manage. Monthly.

National Cattlemen's Association. 1979. Beef cattle research needs and priorities. Beef Business Bull., Nov. 2.

Parker, R.C. and J.M. Connor. 1979. Estimate of consumer loss due to monopoly in U.S. food manufacturing industries. Amer. J. Agr. Econ. 61(4), 626-639.

Rhodes, V.J., C. Stemme and G. Grimes. 1979. Large and Medium Volume Hog Producers. Univ. of Missouri Agr. Exp. Sta. SR-223.

Salathe, L.E. and W.T. Boehm. 1979. Food Prices in Perspective. Agr. Information Bull. No. 427, USDA, ESCS.

Schertz, L.P. et al. 1979. Another Revolution in U.S. Farming? Agric. Econ. No. 441, Dec., USDA, Washington, DC.

Stout, T. (Ed.) 1970. Long-Run Adjustments in the Livestock and Meat Industry: Implications and Alternatives. Ohio Agr. Research and Development Center Research Bull. 1037.

University of California. A Hungry World: The Challenge to Agriculture. 1974. Univ. of California, Division of Agricultural Sciences.

USDA. 1978. Beef Pricing Report. USDA, AMS.

USDA. 1979. Agricultural Outlook. USDA, ESCS.

USDA. 1978. Food Consumption, Prices and Expenditures. AER No. 183, 1977 Suppl., USDA, ESCS.

USDA. 1979. 1979 Handbook of Agricultural Outlook. USDA, ESCS.

USDA. Livestock and Meat Situation. Various Issues, USDA, ESCS.

USDA. National Food Review. Various issues and especially those for June, 1978 and Summer, 1979. USDA, ESCS.

USDL. 1972-74 Consumer Expenditures Survey. 1977. U.S. Department of Labor, Bureau of Labor Statistics.

USGP. Series of Cost of Production Reports made by USDA ESCS to the Committee on Agriculture, Nutrition and Forestry of the U.S. Senate. These reports printed by the U.S. Government Printing Office cover costs of producing milk, hogs, feeder cattle and fed cattle in the U.S., 1976-1979, selected size and regional breakdowns. U.S. Government Printing Office.

Usman, M. and C.K. Gee. 1978. Prices and Demand for Lamb in the United States. Colorado State Univ., Agr. Exp. Sta. Tech. Bull. 132.

Van Arsdall, R. 1978. Structural Characteristics of the U.S. Hog Production Industry. AER-415, USDA, ESCS.

Van Arsdall, R., R. Gustafson and H. Jones. 1978. The future for livestock, poultry production. Feedstuffs, June 13.

Van Dress, M. 1979. An overview of the food source industry. National Food Review. Summer Issue. USDA, ESCS.

"...realistic research policies should be directed toward increasing the opportunities for livestock industries to make a reasonable return on investment. This is a business and unless we understand this, nobody wins."

Carol Foreman

"...our more successful agricultural production policies... have usually been based on sound economic and policy research."

Carol Foreman

TABLE 9.1. PRODUCTION OF MAJOR FOOD ANIMAL PRODUCTS IN THE U.S. (1960-1979)[a]

Product category	Unit of measurement	1960	1970	1975	1979	Percentage change 1960-1979
Beef	Million kg[b]	6,706	9,857	10,898	9,748	+45.4
Veal	Million kg[b]	504	267.3	397	197	-60.9
Pork	Million kg[b]	6,320	6,681	5,354	7,023	+11.1
Lamb and mutton	Million kg[b]	349	250	186	133	-61.8
Chicken	Million kg[c]	2,338	3,848	4,015	4,964	+112.3
Turkey	Million kg[c]	530	787	820	1,066	+101.1
Eggs	Million doz	5,339	5,710	5,365	5,769	+8.1
Milk	Billion kg	56.0	53.2	52.5	56.2	+.4

[a]Food Consumption, Prices and Expenditures, ESCS, USDA.
[b]Carcass weight equivalent excluding edible offals.
[c]Ready to cook basis.

TABLE 9.2. PER CAPITA CONSUMPTION OF MAJOR FOOD-ANIMAL PRODUCTS, 1960-1979[a]

Product category	Unit of measurement	1960	1965	1970	1975	1979	Percentage change 1960-1979
Beef	kg[b]	29.3	33.5	38.2	40.4	36.2	+23.8
Veal	kg[c]	2.4	2.0	1.1	1.6	.7	-69.2
Pork	kg[c]	27.4	24.9	28.2	23.3	29.6	+8.0
Lamb and mutton	kg[c]	2.0	1.5	1.3	.8	.6	-69.8
Chicken	kg[d]	12.6	15.2	18.4	18.5	23.4	+85.3
Turkey	kg[d]	2.8	3.4	3.6	3.9	4.6	+62.9
Eggs	Number	335	314	311	279	284	-15.2
Dairy products	kg Milk equivalent	297.0	281.7	255.3	248.0	255.0	-14.2

[a]Food Consumption, Prices and Expenditures, ESCS, USDA.
[b]Computed from rounded data.
[c]Retail cut equivalent.
[d]Ready to cook basis.

TABLE 9.3. BREEDING LIVESTOCK ON FARMS AND PERCENTAGE CHANGE 1960-1980[a]

Year	Beef cows	Dairy cows	Ewes	Hogs kept for Breeding[b]
		Million head		
1960	26.3	19.5	22.4	NA
1965	33.4	15.4	17.5	8.2
1970	36.7	12.1	13.9	9.2
1975	45.7	11.2	10.1	7.4
1979	37.0	10.8	8.2	9.6
1980	37.0	10.8	8.4	9.6
Percentage change 1960-1980	+40.7	-44.6	-62.5	+17.0[c]

[a]ESCS, USDA.
[b]Jan. 1 inventory.
[c]Change is given for 1965-1980.

TABLE 9.4. IMPROVEMENTS IN EFFICIENCY OF PRODUCING FOODS OF ANIMAL ORIGIN[a]

Species and measure of productivity	Value in indicated year		
	1925	1950	1975
Beef cattle			
Live weight marketed per breeding female, kg	100	144	219
Sheep			
Live weight marketed per breeding female, kg	27	41	59
Dairy cattle			
Milk marketed per breeding female, kg	1,904	2,445	4,773
Swine			
Live weight marketed per breeding female, kg	727	1,105	1,295
Broiler chickens			
Age to market weight (weeks)	15.0	12.0	7.5
Feed to gain ratio	4.0	3.3	2.1
Turkeys			
Age to market weight (weeks)	34	24	19
Feed to gain ratio	5.5	4.5	3.1
Laying hens			
Eggs per hen per year (number)	112	174	232
Feed per dozen eggs, kg	3.6	2.6	1.9

[a]CAST, Foods from Animals: Quantity, Quality and Safety. Rep. No. 82, March, 1980.

TABLE 9.5. DOMESTIC DISAPPEARANCE, EXPORTS AND IMPORTS OF MAJOR FOOD-ANIMAL PRODUCTS, 1960-1979[a]

Product	Unit in millions	1960	1970	1975	1979	Item as percentage of domestic disappearance in 1979
Beef and Veal:						
Domestic disappearance	kg[b]	7,549	10,909	12,074	10,974	100.0
Exports and shipments	kg[b]	26	48	56	101	.9
Imports	kg[b]	352	825	810	1,105	10.0
Pork:						
Domestic disappearance	kg[b]	6,390	6,760	5,436	7,028	100.0
Exports and shipments	kg[b]	75	88	144	204	2.9
Imports	kg[b]	101	223	200	227	3.2
Lamb and mutton:						
Domestic disappearance	kg[b]	389	301	196	152	100.0
Exports and shipments	kg[b]	1	3	4	1	.6
Imports	kg[b]	40	55	12	20	13.2
Chicken:						
Domestic disappearance	kg[c]	2,282	3,741	4,081	5,129	100.0
Exports and shipments	kg[c]	70	83	124	271	5.3
Imports	kg[c]	0	0	0	0	0
Turkey:						
Domestic disappearance	kg[c]	514	755	832	1,010	100.0
Exports and shipments	kg[c]	11	20	24	26	2.6
Imports	kg[c]	0	0	0	0	0
All dairy products:[c]						
Domestic disappearance	kg[d]	55,201	53,333	53,636	56,862	100.0
Exports and shipments	kg[d]	468	450	475	464	.8
Imports	kg[d]	275	852	883	1,045	1.8
Eggs:						
Domestic disappearance	doz	5,312	5,688	5,323	5,676	100.0
Exports and shipments	doz	44	45	62	104	1.8
Imports	doz	3	28	6	10	.2

[a]Food Consumption, Prices and Expenditures, Agr. Econ. Rep. No. 138, ESCS, USDA.
[b]Carcass weight equivalent exclusive of edible offals.
[c]Ready to cook basis.
[d]Milk equivalent.

TABLE 9.6. PRICE AND SUBSTITUTE PRICE ELASTICITIES OF DEMAND, COMPOSITE GROUPS[a,b]

Corresponding percentage change in consumption of	Given a 1% change in the price of					
	Red Meat	Poultry	Fish	Eggs	Dairy	Other Food
Red meat	-.6	.1	.0	.0	.0	.0
Poultry	.6	-.7	-.1	.0	-.4	.0
Fish	.4	-.2	-.4	.0	-.6	.2
Eggs	.0	.0	.0	-.1	.2	-.2
Dairy	.0	-.1	-.1	.0	-.4	.4
Other food	.0	.0	.0	.0	.1	-.2

[a]Based on data for 1953-75.
[b]Data are rounded.

TABLE 9.7. RELATIONSHIP BETWEEN INCOME AND EXPENDITURES FOR FOOD, 1973-1974[a,b]

Income class	Total population	Total income	Total food expenditures	Food as % of income
Dollars		Percentage		
Under 5,000	18.2	6.5	15.4	38.9
5,000-8,000	14.1	9.3	13.1	23.0
8,000-12,000	21.1	17.8	20.3	18.7
12,000-15,000	14.5	14.6	14.1	15.8
15,000-20,000	16.1	19.9	17.3	14.3
Over 20,000	16.0	31.9	19.8	10.2
Percentage of the total	100	100	100	XX

[a]USDA. 1978. National Food Review. ESCS. June

[b]Data from 1973-1974 Consumer Expenditure Survey, Bureau of Labor Statistics.

TABLE 9.8. PER CAPITA VALUE OF FOOD USED IN A WEEK BY HOUSEHOLDS, SPRING 1977[a,b]

Item	Total	At home	Away from home
	Dollars		
All households	19.91	15.17	4.74
Region:			
Northeast	22.56	16.77	5.79
North Central	19.19	14.61	4.58
South	18.40	14.46	3.94
West	19.99	15.08	4.91
Urbanization:			
Central city	20.69	15.75	4.94
Suburban	20.91	15.54	5.38
Nonmetropolitan	18.17	14.32	3.84
Before tax income (1976):			
Under $5,000	17.51	14.99	2.52
$5,000-$9,999	17.26	14.20	3.06
$10,000-$14,999	18.50	14.15	4.35
$15,000-$19,999	19.99	14.99	4.99
$20,000 and over	23.19	16.36	6.83
People living in household:			
One	26.34	20.81	5.53
Two	24.28	18.36	5.93
Three	20.80	15.41	5.39
Four	18.88	14.22	4.66
Five	18.07	13.80	4.27
Six or more	15.52	12.36	3.17

[a] USDA. 1979. National Food Review. ESCS. Summer.
[b] Data from USDA 1977-1978 Nationwide Consumption Survey.

TABLE 9.9. RELATIONSHIP BETWEEN AGE, INCOME AND EXPENDITURES FOR FOOD[a]

Age of head of household	Per capita median income,	Percentage of income spent on		
		Food at home	Food away from home	Total food
		Dollars	%	%
Under 25	2970	11.2	6.3	17.5
25-34	3210	11.5	4.9	16.4
35-44	2850	13.9	5.0	18.9
45-54	3600	12.6	4.9	17.5
55-64	4080	12.4	4.1	16.5
Over 65	2950	17.4	4.1	21.5
All	3260	13.0	4.8	17.7

[a]USDL. 1972-1974 Consumer Expenditure Survey. Bureau of Labor Statistics.

TABLE 9.10. FARM AND MARKETING BILL SHARES OF RETAIL FOOD PRICES FOR SELECTED FARM FOODS, 1978[a]

Farm food	Farm share	Marketing bill share
	Percentage	
Eggs	67	33
Meat products	58	42
Poultry	58	42
Dairy products	51	49
Average of market basket of farm foods	39	61
Fresh fruits and vegetables	19	81
Baking and cereal products	14	86

[a]USDA. 1979 Agricultural Chartbook. ESCS.

TABLE 9.11. CONSUMER EXPENDITURES, MARKETING BILL AND FARM VALUE BY WHERE CONSUMED FOR SELECTED COMMODITIES IN 1978

Item	Meat	Poultry	Dairy products	Total of all farm foods
		Millions of dollars		
Farm value	27,794	7,269	12, 141	68,346
At home	21,231	6,571	10,162	55,847
Away from home	6,563	698	1,979	12,499
Public eating places	5,499	572	1,410	9,864
Institutions	1,064	126	570	2,636
Marketing bill	35,952	7,882	17,138	144,076
At home	14,435	3,659	9,803	90,585
Away from home	21,517	4,223	7,335	53,491
Public eating places	19,150	3,795	5,653	44,260
Institutions	2,367	428	1,681	9,229
Consumer expenditures	63,746	15,152	29,279	212,425
At home	35,666	10,231	19,965	146,435
Away from home	28,080	4,921	9,314	65,990
Public eating places	24,649	4,367	7,063	54,123
Institutions	3,431	533	2,251	11,866

[a] USDA. Agricultural Outlook, Nov. 1979. ESCS.

TABLE 9.12. DISTRIBUTION OF RETAIL PRICE BY MARKETING FUNCTION, 6 FOOD-ANIMAL PRODUCTS, 1979[a]

	Marketing function					
Food item	Assembly and procurement	Processing	Intercity transport[b]	Wholesaling	Retailing[c]	Retail price
	Price in cents					
Beef, choice, kg	6.4	14.6	9.7	36.4	137.6	498.9
Pork, kg	5.5	68.1	9.3	25.1	77.4	317.7
Broilers, kg	2.9	21.2	3.5	9.3	35.9	149.9
Eggs, grade A or AA large, doz	1.1	11.6	1.8	3.8	13.7	84.0
Milk, sold in stores, liters	1.8	7.9	--[d]	7.5	7.2	51.9
Butter, kg	8.2	31.0	6.4	14.8	48.5	362.2

[a]USDA Agr. Econ. Rep. 449. 1980. Development in Marketing Spreads in Food Products in 1979 March, 1979.
[b]Includes inter-and intra-city transport for beef and pork.
[c]In-store costs only.
[d]Included in wholesaling.

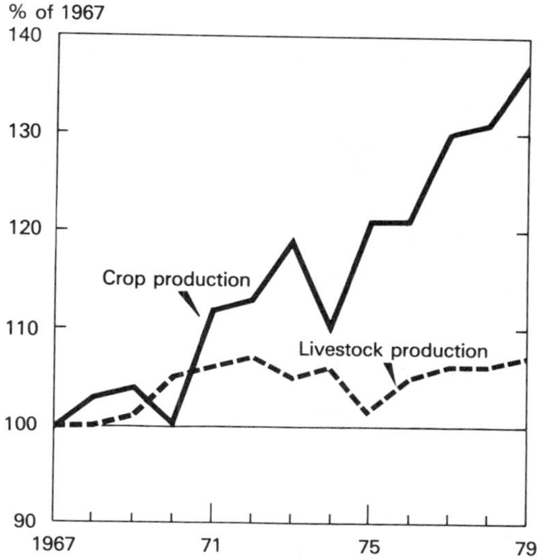

Figure 9.1. Crop and livestock production as a percentage of 1967. (USDA. 1979. Handbook of Agricultural Charts. ESCS.)

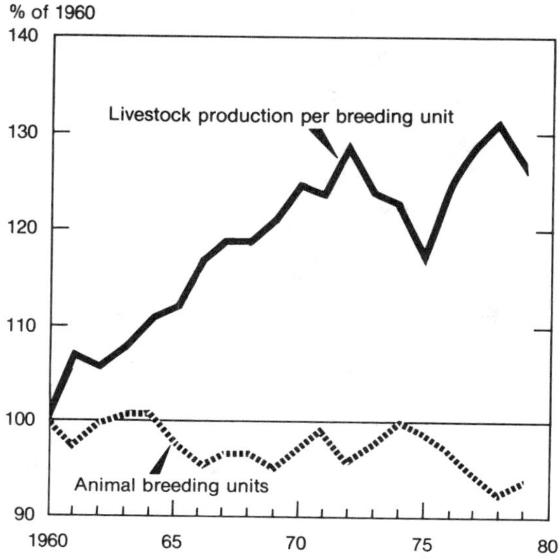

Figure 9.2. Livestock production per breeding unit as a percentage of 1960. (USDA. 1979. Handbook of Agricultural Charts. ESCS.)

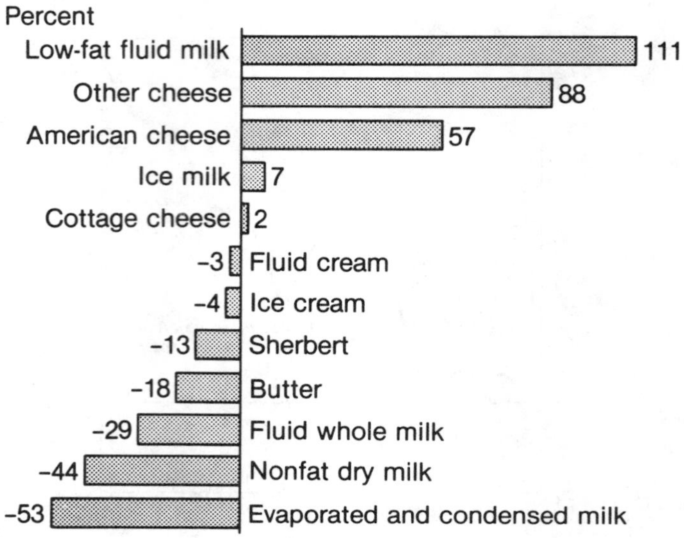

Figure 9.3. Changes in per capita dairy product sales, 1960 to 1978. (USDA. 1979. Handbook of Agricultural Charts. ESCS.)

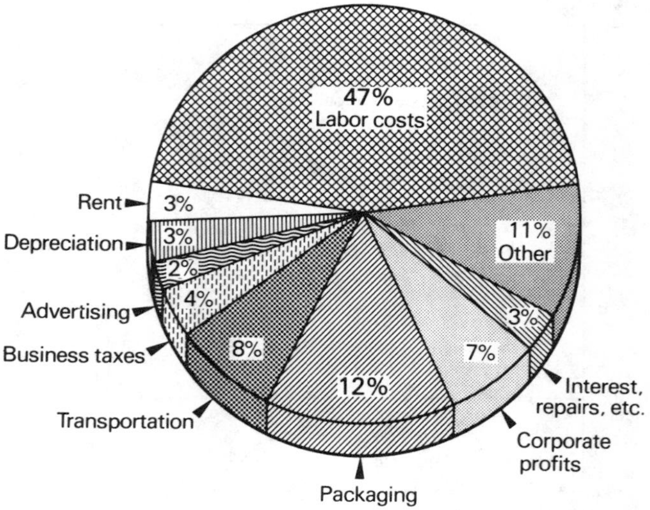

Figure 9.4. Major items included in the farm-food marketing bill. Transportation is intercity rail and truck. Corporate profits are before taxes. Other includes utilities, fuel, promotion, local hired transportation, insurance, etc. (USDA. 1979. Handbook of Agricultural Charts. ESCS.)

10

L. W. Smith (Co-Chairman), G. W. Thomas (Co-Chairman), A. B. Carr (Rapporteur)

D. Badger, J. F. Bartholic, R. D. Dunlop, J. P. Fontenot, R. Harris, H. F. Heady, W. Krejci, J. R. Miner, J. G. Morris, J. E. Newman, L. Rittenhouse, L. B. Safley, J. N. Walker, and R. K. White

RESOURCES AND ENVIRONMENT

SUMMARY

The key to sustained animal agriculture is proper use of resources: air, water, land and energy. Animals are kept for food production under a broad diversity of systems. The extremes vary from ruminants grazing rangeland to highly integrated confinement systems for ruminants, swine and poultry. Over the next several decades, energy will likely be the most critical resource in shaping animal production. However, as we enter the 21st century, competition for water will become an increasingly important factor.

Animal production systems must be designed to adjust to climatic variation and to minimize adverse, environmental impacts. Problems relating to environmental quality that deserve the most attention are air and water contaminants, vegetative change and erosion by wind and water. Man and livestock must live in harmony with other factors of the environment subject to a limited supply of natural resources. Following are the major research imperatives relating to resources and the environment as we strive to meet human needs through animal agriculture in the 21st century.

Increase Research on Conservation in Use of Natural Resources

A. Evaluate the impact of and develop response to a decreasing supply of land, water and energy on animal agriculture.

B. Develop techniques for the production of animals requiring minimal cultural energy (fossil fuel) inputs by the optimal use of range forage, forage crops, crop residues and concentrates.

C. Assess the impact of climatic variables on animal production with emphasis on minimizing the impact of adverse climatic fluctuations such as drought.

Develop Innovative Techniques for the Improvement of the Environment

A. Develop alternative strategies for the utilization of manure and other organic wastes as a resource in animal production systems and to minimize adverse impacts on the environment.
B. Develop conservation technologies for animal agriculture which improve the production of vegetation to prevent wind and water erosion.
C. Develop innovative building designs to reduce fossil fuel inputs and to control temperature, relative humidity and toxic gas concentrations to promote maximum performance of animals confined in buildings.

Design Research to Define Interrelationships Between Resources and the Environment

A. Develop standard techniques for evaluation, monitoring and assessment of the basic elements of the environment associated with range and pasture production.
B. Utilize systems analysis techniques to develop dynamic models for the evaluation and optimization of animal production.
C. Conduct long-term research on range and pasture ecosystems to elucidate the complex interrelationships between soils, vegetation, animals and variable climatic conditions.

INTRODUCTION

The production of animal products for food in the 21st century is dependent upon an adequate resource base - particularly the availability of land, water and energy. In addition, production systems must be designed to adjust to climatic variations and to minimize adverse environmental impacts.

For the purpose of this discussion the various systems of animal production are broadly divided into 1) "intensive" or confinement methods, such as those used in dairy, poultry, swine and feedlot ruminant production and 2) "extensive" systems such as range and pasture production. It is also recognized that, in the case of ruminant animals, part of their life may be spent on range- or pasture-land and part under feedlot or more intensive management.

NATURAL RESOURCES

THE LAND BASE

The amount of land required to support animal agriculture varies with the class of livestock, system of production, climate, soil conditions and the design of physical facilities. Land requirements range from a fraction of a hectare per animal under confinement to more than 50 hectares (125 acres) per animal on some rangelands.

In the more intensive systems, most of the feeds originate on cultivated lands and are transported to the confined animals. Types of feed and associated resource requirements for these systems are more adequately discussed by the panel on Feed Production. However, it should be emphasized that there are many interrelationships between cropland, improved pasture and rangelands for many animal production systems. Therefore, a thorough analysis of the land-base must take these requirements into consideration.

The total land area in the United States adapted to livestock grazing is more than 405 million hectares (1 billion acres). This vast resource can only be converted effectively to human food through the unique ability of the ruminant animal to harvest grass, browse and forbs. Most of this land is not suitable for sustained crop production but is adapted to range and pasture grazing systems.

Recent reports provide estimates of the land-base for grazing. An Interagency Work Group for USDA in July, 1979, stated that the range resource covers 630 million hectares (1.56 billion acres) in the 50 states. A second report (RPA, 1980) more accurately evaluates the actual "grazed" area, and shows the relationships among range, pasture, cropland and aftermath. Data from this RPA study show that rangelands contributed 1,067 million animal unit months (AUMs) of grazing each year during the study period, 1974-1976, while pasture, cropland pasture and aftermath contributed 701 million AUMs during the same period. This analysis indicates that, by the year 2030, the projected demand for grazing on cropland used for grazing and pasture will exceed 1 billion AUMs. The demand for roughage from rangelands is projected to increase to 1,966 million AUMs by the year 2030. In spite of this sizeable projected increase in demand there is a slight downward trend in grazing on rangelands. Also, there are changes in the roughage used among the various classes of livestock. For example, dairy cattle production shifted toward more concentrates, actual numbers of sheep and goats declined and more cattle moved into feedlots.

Ownership of grazed land is also an important consideration in the use of the land-base. In the contiguous U.S. in 1976 (figure 10.1), the Federal Government owned 154 million hectares (386 million acres) of rangeland, while 330.2 million hectares (816 million acres) remained in private ownership. Federal lands, for the most part, must be planned for multiple use values, placing livestock in competition with uses such as for timber, wildlife and recreation. The use of private land, on the other hand, is under the control of the land owner.

An examination of the trends in land-use leads to the following concerns as related to livestock production:

1) There is a continuing trend to transfer private land from agricultural uses (including animal agriculture) to industry, housing, urban development and to other uses. This rate of transfer is now in excess of 404.7 thousand hectares (1 million acres) per year. The implication of this trend needs to be evaluated. At the same time more pressure will be placed on existing land resource to increase the production of roughages and other feeds for animals. Thus, research designed to develop the full potential of the remaining agricultural lands continues as a high priority need. Competition for land in the

United States also affects confined animal agriculture because of zoning and environmental regulation. Municipalities and industry are increasing application of waste materials to land. Such activity will require increased levels of agronomic management to insure that land productivity is not permanently impaired. This land must be carefully monitored to manage salts, heavy metals, organic and pathogenic bacteria. Likewise, sound judgment should be exercised in manure handling systems to control pollution.

2) On over 404.7 million hectares (1 billion acres) of federal lands with potential for livestock grazing under the multiple-use concept, there is a trend to reduce livestock numbers to accomodate more wilderness, wildlife and other uses. A thorough analysis of the economic importance of livestock grazing on public lands and the resulting ecological changes is published in a CAST (1974) report. One conclusion in this report was that, on vast areas of public lands, grazing is compatible with other uses and in many cases provides measurable benefits to game animals, water yield, fire abatement, nutrient cycling and human enjoyment. Certainly, more research is needed on the issue of multiple use, and the implications go far beyond the individual livestock operator to the welfare of people in the local communities and even to the farmer who produces feed supplements for range animals.

CLIMATE

Climate is very important to animal-production systems because it affects the animal directly and influences the amount of forage and feed production. Climate is often defined as average weather. But this definition is too restrictive for resource planning. A more complete consideration of climate as a resource begins with these average values of weather events for a given time and place, and includes the probability distributions of the weather occurrences.

Climatic variations have many time and place dimensions. Elements vary on a daily, seasonal, annual or longer basis. In space, climatic elements vary in the immediate area surrounding the animal, over the farm or range, region, continent and globe. A climate change caused by the continued rise in atmospheric CO_2 can produce changes in climate risks and, thus, induce changes in animal stress for any time or place. Therefore, the assessment of possible range of climatic risks is essential to evaluation of climatic stress on animal agriculture for the coming decades.

Most climatic stress on animals and plants is related to unexpected variations in the elements of climate - temperature, humidity, precipitation, solar-terrestrial radiation balance and wind. Normal diurnal and annual terrestrial energy balance exchanges produce expected daily and annual climatic variations. But unexpected or changing climatic variations on any time and space scale can produce increased stresses on animal agriculture.

To optimize the value of climate as a resource will require risk analysis that takes into account climatic variability and its impact on livestock systems. Failure to produce and use such analyses leads to mismanagement, for example overstocking, which results in lower sustained productivity and increased environmental problems.

Many factors contribute to inefficient use of pasture and range forage. Climatic fluctuations are among the foremost in their effects on forage quantity and quality. Additional investigations and further testing is needed on forage composition and the mix of grazing animals or classes before the most efficient range-pasture system can be determined for many areas of the U.S. There is a special need to integrate biological and socio-economic research efforts to ascertain short- and long-term consequences of climatic extremes. Pasture and range management strategies are needed that can be responsive to fluctuations in forage resources and to the maintenance of a stable ecosystem.

A major concern of climatic variation is the impact of drought. There are several definitions of droughts. These definitions usually relate to their effects rather than the causes. An agricultural drought exists when soil moisture or water supplies are inadequate for a particular production system. Droughts vary in their frequency, intensity and length. Strong evidence indicates droughts are cyclic in many continental areas and agricultural regions of the world. This is particularly true for the Great Plains of North America. Large-scale droughts are less frequent and more cyclic than small-scale droughts that are both frequent and more random in occurrence.

Before climatic risk of drought can be successfully estimated, drought must be defined in terms of its impact on a particular agricultural production system. Reliable estimates of seasonal water balance deficiencies are needed before climatic drought frequency can be estimated. Drought indexes should be developed for the major agricultural and ecological systems of an area or region. Drought index development will require both a meteorological and biological response data base. These data bases must be developed before drought risks can be computed and subsequent management strategies developed.

Rapid and extreme changes in weather impact animal agriculture production systems in many ways. Rapid temperature changes increase animal stress. This is particularly true for animals in feedlots that are not properly designed for limiting animal stress during extreme weather occurrences. Each year millions of kilograms of animal products are not achieved due to adverse effects of weather. Control over these unusual weather events must be developed in the coming decades. These methods of control must be less dependent on fossil fuels.

Research and development efforts must concentrate on the most efficient use of topography in the location of animal-production systems. Facilities should be designed and located to take greater advantage of solar radiation, winds and other specific natural features.

In the past, weather modification in the form of cloud seeding has been attempted primarily in climatic areas with low probability for success. Furthermore, many of these cloud seeding operations have occurred under prevailing drought conditions which ensure a low-frequency of seedable cloud conditions. Experience in cloud-seeding suggests that more basic research is needed on the question of weather modification. Improved understanding of cloud physics is needed before further field experimentation.

WATER

The amount and quality of water is critical to all aspects of the livestock industry. While the direct consumption of water to satisfy the thirst of the animal is relatively small, large quantities of water are used in getting animal products to the consumer.

The largest demand for water, however, is for production of forage and feeds. Some estimates show that about a metric ton of water is needed to produce a kilogram (2.2 lb) of grain. On part of the Western Range, over 100 metric tons of water are associated with the production of a kilogram (2.2 lb) of beef. Much of the water involved in this process on range and pasture lands evaporates from the soil surface or is transpired through brush and undesirable weeds. Therefore, improved water management and better water conservation practices are essential to the improvement in quality vegetation. Herein lies an opportunity for research.

Irrigated cropland in the United States has increased dramatically in the past 2 to 3 decades. Much of the alfalfa production and part of the production of other feeds and forages is on irrigated land. Some of these irrigated areas are drawing water from geological aquifers (such as the Ogalalla Formation in Texas, New Mexico, Kansas, Oklahoma, Colorado and Nebraska) which are being rapidly depleted. If water withdrawals from these nonrecharged aquifers continues at the present rate, recoverable water supplies will be exhausted early in the 21st century. With the increased cost of energy, some scientists predict that economic costs will prohibit irrigation from deep wells for a considerable time before the actual supplies are exhausted. Forage production from irrigated lands will likely diminish before more intensive crop production is reduced.

A major concern of the panel on resources and environment is the fact that water is being transferred from the agricultural sector to other uses. Business, industry, municipalities and new energy developments can pay more for present water supplies than animal agriculture. These water transfers, combined with serious problems of depletion and increased costs, may lead to regional shifts in feed production to the higher rainfall zones.

ENERGY

The activities which encompass on-farm production of food and fiber are basically all associated with different forms of energy. All forms of forage production capture solar energy. Animals are utilized in the process to modify plant materials, which cannot be utilized directly by man, into an acceptable form or to enhance the appeal of material to the consumer. Food production requires energy expenditures in the form of human and animal labor, materials such as fertilizer, chemicals and pesticides, fuel for tilling the land with machines, structures to protect crops or livestock, and also for activities such as drying or processing the harvested crops, supplying irrigation water during rainfall deficient periods, and processing and transporting products to the consumer.

Generally these activities require an energy subsidy from fossil fuels. The energy subsidy input is often referred to as "cultural energy". Due to the increase in energy costs, the economic significance of this input is increasingly important. On-farm activities use about 3.0% of total U.S. energy usage, but its cost is significant in the total cost of agricultural production. Modern agriculture with its high productivity is, nevertheless, highly dependent upon energy, and curtailment or restriction in supply will directly result in a decrease in food production.

The biggest energy use on farms and ranches is that associated with agricultural chemicals (CAST, 1977). Fertilizer production uses the most energy. After chemicals, the largest energy uses are associated with machinery operation, transportation, irrigation and livestock production (not including feed production). Of all the different forms of fossil energy, oil and natural gas are the most important.

The energy use in production agriculture is only one element of the total energy involved in the food chain. When the total food chain is considered, U.S. energy use by agriculture rises to 16.5%. The energy uses in food processing, distribution and home preparations are each larger than that involved in agricultural production. However, there are vast opportunities for energy savings and possible alternatives to fossil fuels at every step in the total food system.

ANIMAL ENERGY EFFICIENCY

Two measures of energy efficiency for animal agriculture can be applied. One is the direct energy efficiency, i.e., the energy subsidy (cultural energy) required per unit of agricultural production. The other is the biological efficiency, which can be measured in terms of the units of human digestible food produced per unit of plant energy consumed by the animal. A number of analyses have also been made where animal energy efficiency was evaluated by determining the energy value of the food produced relative to the cultural energy input. Such a value, although useful in comparing the relative energy efficiency of various production practices for producing comparable food products, is of little value when comparing the values of differing foods. Meats, for instance, play an entirely different role in the dietary needs of people than do grains or fruits. Since farming is conducted to provide the most basic human need of food and not energy alone, the critieria for evaluation should encompass other parameters which measure total food value including protein, vitamins, etc. The first obligation of agriculture is to meet the food needs of people. With an ever-expanding population, ways must be found to expand and intensify agricultural production. Within these constraints the most energy efficient practices can and should be selected.

Different measures of agricultural efficiency, either direct energy efficiency, or biological efficiency, cannot be used in themselves to make decisions relative to alternate production practices. It has been argued from a number of studies that subsistence forms of agriculture, where virtually no inputs of cultural energy are made, are highly efficient. Although this may be true, the productivity of such agriculture per unit of land area is comparatively low.

Ruminant animals have the capability of converting roughages to human food. Ruminants include cattle, sheep, goats, deer and other species. Their digestive tract has the capability for microbiological fermentation of plant materials including roughages, changing them to usable nutrients. This unique capability allows them to convert the cellulosic plant material to milk, meat and fiber that would otherwise be unavailable for human sustenance. Even with ruminants, some cereal grains are often needed to supplement the nutrient value of roughages and to fully meet the dietary requirements of the animals or to improve the quality of the final product.

The proportions of dietary energy provided by certain feedstuffs are shown in table 10.1. These data show that the percentage of dietary energy from grains in the diet of beef cattle is only 21.8% of the total. Almost 75% of their dietary energy is currently obtained from roughages. In contrast to beef cattle, the utilization of grains by poultry and pigs is much higher and roughages play only a minor role in their diets.

Pigs and poultry are considerably more efficient than ruminants (cattle, sheep and goats) in converting dietary energy (digestible energy) into animal gain (table 10.2). However, pigs and poultry are grain dependent, and do not make effective use of roughages.

Table 10.2 also shows that animals such as rabbits are more efficient by some measures than ruminants primarily because they have multiple births and are dependent upon mother's milk for only a short period of time (Cook, 1972). By the same measure, small ruminants such as sheep and goats, if properly managed, are more efficient than cattle in producing meat per kilogram of metabolizable energy from range and roughages. This efficiency is due to potential for twinning and to the fact their progeny attain market size in a shorter period of time than cattle.

Biological efficiency of rangelands and unimproved pastures is dependent upon producing usable forage that furnishes adequate nutrients for each of the various physiological functions during the animal's annual reproductive cycle. It also requires the use of animal species that are effective in obtaining and converting nutrients from plants to animal products for human use. Animals that produce an acceptable product directly from range or pasture, with minimal input of concentrate feeds, have an advantage.

INTEGRATED USE OF RANGE AND IMPROVED PASTURES AND CROP FORAGES

Plants furnish food for all forms of life on earth by converting solar energy into organic material. Unimproved pastures furnish herbivores a diet with minimal input of cultural energy compared to confinement feeding. However, native grasses and shrubs on rangelands alone are generally unsuited to producing a choice grade of meat. Current studies demonstrate that animals grazing forage from permanent pastures and annual forage crops can produce acceptable meat at considerable reductions in monetary cost and fossil fuel use. Finishing beef in a feedlot requires 5.2 kcal subsidized energy per kilocalorie of gross energy in the product. Beef finished to a lower carcass grade, but of acceptable quality on a range/forage regimen, requires an energy subsidy of 3.5 kcal; a 33% reduction in subsidized energy (Cook, 1979).

Since red meat production from range lands and pastures is more energy efficient than confinement feeding, it is possible that forages will become more important in future beef production.

ENVIRONMENTAL QUALITY: INTENSIVE SYSTEMS

The trend in intensive animal production systems is toward concentration in larger facilities in spite of the necessity for long-distance transportation of feed and products. The mechanization and automation of animal handling, feeding and manure handling systems, including the utilization of manure for its nutritive value in crop production, in feeding or for energy production has direct implications on environmental quality. Different manure management alternatives for beef, dairy, poultry, sheep and swine are described and evaluated in an EPA manual (White and Forster, 1978).

BEEF

Changes in both the number and sizes of beef cattle feedlot operations have been dramatic. The increase in commercial confined feedlot operations was stimulated by the availability of relatively inexpensive feed grains, proximity to an adequate supply of feeder cattle, favorable climatic conditions and a strong demand for beef. The advantages of confinement include less space per animal, less labor and economies of scale.

Two main types of cattle-feeding operations exist - the farmer-feeder with a feedlot of less than 1,000-head capacity and the commercial feedlot with greater than 1,000 head in confinement. Neither the number of cattle nor the capacity of the feedlot is the most important criterion to indicate the environmental problems of cattle feedlots. An important factor is whether land is available for application of manure. Generally the farmer-feeders have adequate land available to integrate manure application on land with crop production. Farmer-feeder operations generally are in the Midwest or Corn Belt. Most large commercial feedlots are in the High Plains areas and the Southwestern United States.

Since the early 1960s, more efficient diets and feeding programs have helped sustain the profitability of cattle feedlots. Cattle in feedlots are started on high roughage diets and shifted gradually to high concentrate diets. Manure is removed from a beef cattle feedlot pen whenever new animals are started. Manure characteristics change due to drying, microbial action, wetting by precipitation, mixing with the underlying soil and compaction by animal movement. The manure is scraped as a "dry" material and applied to available cropland or in some cases, used as a source of energy. The runoff from the feedlots is controlled by containment in lagoons to prevent pollution of surface waters.

DAIRY

Most pastures for dairy cattle are in humid areas with a long growing season. Stocking rates of pastures depend on plant growth and management practices but are usually between .5 and 2 ha (1.2-5 acres) per cow. Grazing of dairy cattle is declining because more uniform nutrient quality is possible from harvested feed resulting in higher milk production.

Much of the milk produced comes from farms with stanchion barns and a milk room. This method of production is common in the northeastern and north central portions of the country. Milk cows and replacement heifers are kept in barns and restrained to a fixed location for feeding and milking. The barns generally are insulated and mechanically ventilated. Bedding is used in the stall area to absorb moisture from manure and urine. The total semisolid manure is collected and removed mechanically.

Free-stall barns with milking centers are increasingly used in the larger production systems. Milk cows and replacement heifers are kept in barns, but are allowed free movement between resting stalls and feeding areas. The barns generally are not insulated and ventilated mechanically. Most of these systems use bedding in the resting stalls. Manure is mechanically scraped from the barns. Waste (manure and urine) from free-stall dairy barns normally is more liquid in consistency than from stanchion barns.

Other production systems are also used. In northern regions, solid floor barns with separate liquid manure storage or slotted floor barns with below floor storage are used. In southern regions, flushing systems are used with collection of the diluted wastes and periodic irrigation of fluid wastes onto the land. In dry regions, dairy cows may be kept in open lots and their manure handled the same as for beef cattle in open lots.

In areas where dairy cattle manure cannot be spread on the land during the winter, stanchion barn manure and bedding can be stacked in storage areas until it is hauled to available fields in the spring and summer. Manure and bedding will drain while stacked and seepage liquid has a high pollution potential, and control is necessary to avoid contamination of streams and creeks.

POULTRY

The poultry industry is an example of an intensive animal production system that producers have utilized to remain competitive. All commercial egg production is in confinement with birds spending their egg-laying life in cages. The density per cage ranges from three to five birds with feed and water brought to the birds and the manure falls into pits. Present poultry management permits the concentration of egg-laying birds in buildings containing as many as several hundred thousand birds on a small area. These production operations usually do not grow their grain for feed and may not have enough land for proper land application of manure.

Essentially all broilers produced in the United States are grown on litter that absorbs moisture. The litter may be sawdust, peanut hulls, wood shavings or other suitable material. About 3.68 million m^3 (130 billion ft^3) of used broiler litter must be removed annually. The frequency of removal of litter depends on the number of growing cycles that can be completed without disease problems on one batch of litter. Practices vary among producers, but broiler houses are cleaned one to three times per year. The wastes are normally spread on land and used for fertilization of cropland and pasture.

Turkey production varies seasonally with most production occurring during the summer and fall. Waste accumulation in both confinement houses and in open feeding areas must be considered. Young turkeys are kept in houses on litter for the first 8 weeks and then placed in open range areas at densities of from 170-1,040 birds for the remainder (12-16 weeks) of their growing cycle. Turkey litter from houses is usually removed twice a year and handled in the same manner as broiler litter. Control of runoff from turkey range areas is important to water quality.

The production of ducks for slaughter is a large agribusiness. Duck production facilities are of varying size, from 70,000 to over 500,000 ducks per facility. Ducks primarily are grown in confinement buildings with possible access to outside duck runs. Water can be used to flush the duck manure from the buildings, although many production facilities use water only for drinking and remove the manure and bedding as a "dry" solid waste. Treatment of wastewater is required.

SHEEP

Sheep production facilities may be classified as follows: 1) pasturing of ewe and lamb flocks; and 2) confinement of ewe and lamb flocks to lots and feeding of lambs in lots or buildings. Lot density of ewes or lambs is normally about 3 m^2 (9.8 ft^2) per animal. The major environmental problem is the control of runoff from feedlots. Normally sheep manure is handled in the solid form, because the manure has a higher percentage of total solids than that from other animals.

SWINE

The production of swine varies from raising hogs on pasture to growing several thousands in modern, automated confinement facilities. About 80% of slaughter hog production comes from farrow-to-finish hog operations and 20% from split operations (pigs farrowed on one farm and finished on another). Most hog facilities utilize liquid manure handling systems with application of the slurry to cropland for the recovery of the fertilizer nutrients.

The major environmental problems with swine production facilities are: 1) adequate land for spreading the manure; 2) runoff from open lots and from manure covered cropland; and 3) odor nuisances.

ADDITIONAL CONSTRAINTS

Water quality

The passage of PL 92-500 in 1972 launched a new era of water quality regulations that has had a profound impact on animal production. The most immediate impact has been to require that no manure or manure solutions from confined livestock production units be allowed to enter public waters. For unroofed systems this has meant retention of not only manure but also runoff from precipitation and snowmelt as well. This requirement is generally met by collection and storage of runoff in ponds capable of retaining all runoff except for that which occurs less frequently than once in 25 years. The requirement results in additional land, equipment and management inputs. Information exists to allow accomplishment of this level of pollution control; however, it has definite regional and economic impacts. These impacts will continue as a constraint on animal agriculture.

A further constraint is the imposition of nonpoint pollution control requirements on animal agriculture. Nonpoint pollution control involves the control of contaminated water from small livestock operations and runoff from cropland or pastureland upon which manure has been applied or dropped. Control is most commonly achieved by the adoption of manure and land management practices responsive to individual farm and regional characteristics. Approved practices are involved that are successful in the control of this pollution, but they typically result in increased investment costs to the producer as well as the adoption of management practices which limit the options otherwise available for manure disposal, pasture management or livestock feeding. Most notably this program restricts manure spreading on frozen, snow covered or saturated soils, and restricts access to streams and lakes by grazing or penned animals.

Nuisance problem

As the U.S. population has spread from metropolitan centers to rural and suburban environments the need to control odors, dust, insects, noise and unsightliness related to animal production has become more critical. Although there are no federal and few state or regional standards by which to judge the acceptability of individual enterprises on these matters, the constraints are no less real. These infringements are generally classified as nuisance pollutants and are controlled by a variety of mechanisms. Zoning, as land-use regulation, is commonly used to guide the location, size or design of animal enterprises. When neighboring individuals or groups of people feel use of their property is being unreasonably restricted, they have recourse to legal action under the common law of nuisances to seek damages and/or an injunction to modify or remove the nuisance producing enterprise.

Nuisance problems are most common near new, large or recently expanded animal enterprises and those located near housing areas, shopping facilities or other points of high sensitivity. The net effect of this growing sensitivity of neighbors to nuisance conditions is to restrict the sites upon which animal production is acceptable and to impose more severe requirements on the design and management of animal facilities.

Facility considerations

Intensive animal agriculture involves a complex set of interactions between genetics, feedstuffs, management and the environment. The animal environment includes conditions such as temperature, humidity, air quality, flooring type and other factors controlled by the facility design or the manager. Significant progress has been made in the past 20 years in improving the performance of animals by improved facility design. Additional studies need to be made in optimizing the overall system by integrating the various production and environmental factors influencing animal performance. Due to the complexity of the process and the extent to which interaction is known to occur, a computerized dynamic modeling technique is a logical approach to optimize livestock production systems.

The thermal and gaseous environments are important aspects of animal growth, production and reproduction. Advances in thermal and gaseous control of animal facilities can be expected in the 21st century. Understanding the interactions between the thermal and gaseous environment and the animal in terms of energy balance is essential to improving future facility designs. Creative material selection and design to maintain a year-round environment to allow optimal animal performance is the objective. Future research should be on facility designs that are functionally independent of fossil-fuel energy.

Manure use

Manure needs to be considered as a resource rather than a waste in animal production. Van Dyne and Gilbertson (1978) estimated that about 111 million metric tons of dry manure were voided by farm animals in 1974. The quantities produced and collectible are shown in table 10.3. Range cattle contribute 47% of the total, but less than 4% of this kind of waste is recoverable as a resource. Others estimated there were 300 million metric tons of dry matter in animal waste of which 50% was collectible. Although the estimates differ, the amount is large. Manure, if properly used as a resource could have a substantial inpact on fuel supply and feed protein supply, and contribute toward making livestock production systems energy self-sufficient.

Manures were traditionally applied to the land before 1950. With the advent of confinement feeding, cheap nitrogen fertilizers and before environmental regulations, it was uneconomical to use manure as a source of plant nutrients. More recently, Wilkinson (1979) concluded that replacement fertilizer value can exceed management cost of manure, thus changing manure from a waste to a resource. Wilkinson (1979) also summarized the plant nutrient content of manures by type and comparative crop response relative to commercial fertilizers. Pratt (1979) reviewed and discussed the management restrictions in soil application of manure.

He concluded because of the many factors involved, meaningful recommendations can only be developed for the manures, the crops, the soils and the climate and other pertinent factors, such as water quality objectives and problems, in a local area or region.

Some manures processed in suitable fashions are more valuable as feed ingredients than for other known uses. A documentation of the current state of the art and science of feeding animal waste is contained in a CAST (1978) report. In addition, animal performance and quality of animal products, safety considerations, processing methods, environmental considerations and additional information needed is discussed in another CAST (1978) report. Smith and Wheeler (1979) recently reviewed the nutritional and economic value of animal manure products as feed ingredients. In production trials, animals have performed satisfactorily on diets containing processed manure.

Animal manures, like other organic wastes, can be subjected to anaerobic fermentation to produce biogas which is primarily a mixture of methane and carbon dioxide (Smith et al., 1979). The gas from a properly operating digester will contain 55-65% methane. Thus, it will have a heating value of approximately 5 million cal/m^3 (600 BTU/ft^3). On this basis, approximately 5.6 m^3 (200 ft^3) of biogas has heating value equivalent to 3.8 liters (1 gallon) of gasoline; .4 m^3 (15 ft^3) of biogas is equivalent to 1 kwh of electricity.

The economics of anaerobic digestion are highly size dependent with economies of scale being of major importance. For example, Hashimoto et al. (1979) reported that the capital cost dropped from $371/head of beef cattle for a 1,000 head lot to $76/head for a 50,000 head feedlot.

There are two additional constraints to overcome relative to methane production. One is to find a continuing use for the methane and the other is to have a use for the digestor residue.

A research constraint on use of digestor residue is the lack of separation technology to isolate the protein from the nonnutritive ligno-cellulosic residue. The isolated protein may be a valuable feed.

Biomass, manure being an example, can be converted into energy forms by thermochemical processes that degrade the organic material into simpler compounds at medium or high temperatures, in the presence or absence of air, with or without catalysts. Examples of thermochemical manure conversion processes were reported by Kreis (1979) and Huffman (1978).

Air-dry feedlot manure has an intrinsic heat value similar to lignite and could be burned with coal in a furnace for production of steam or electricity. A current projection for electrical energy yield from a coal/manure-fired utility sized power plant is roughly 756 million calories (3 million BTU) per ton of dry manure (Sweeten, 1980). Sulfur emissions control, mechanical handling, ash-fusion properties and transportation costs are unsolved problems.

Thermochemical processes have the potential to produce as much or more net energy from dirt-surfaced cattle feedlot manure as anaerobic digestion for methane production. Developmental work is needed, and capital costs are high. Only those locations with approximately 200,000 cattle on feed within a 40-80 km (25-50 mile) radius would be candidates for these technologies.

ENVIRONMENTAL QUALITY: EXTENSIVE SYSTEMS

Extensive livestock systems are those in which animals (primarily cattle, sheep and goats) graze pasture and rangeland. Pasture is sometimes the source of environmental problems because of runoff containing fertilizer nutrients and herbicide materials, infestations of noxious plants and burning of debris. Rangeland environmental issues center on erosion by water and wind, application of herbicides, undesirable changes in the vegetation, and damage to the habitats of various wildlife species.

Before examining the current research priorities of grazing animal management, a perspective into present rangeland conditions needs to be stated. The era of greatest damage to arid and semiarid rangelands occurred between 1890 and 1934. By 1934 government agencies were beginning research and management of rangeland grazing. Animal grazing is under better management than ever before. Many large scale examples of improvement in vegetation and production on pastures and rangelands exist as a result of applied science.

Range research, until some 15 years ago, almost exclusively emphasized production of livestock (Lloyd et al., 1970). While still predominant, this objective is no longer primary in some types of rangeland research. Pasture-lands often are combined with recreational facilities, such as fishing in livestock water ponds and hunting of small game. Publicly-owned rangelands have received tremendous pressure by people wanting to fish, hunt or hike. Livestock are considered by some groups as detrimental to those recreational pursuits. The problems in the natural resource area have changed because of the switch from a single focus on livestock production to a multiple-use focus in which the production of livestock must be integrated with other uses. The need for complex research that integrates the impacts of multiple-use of natural resources and the resultant environmental concerns has never been greater. Problems occur in soil, water, air, vegetation and animal populations in the natural resource system.

Livestock producers are affected by several environmental constraints including limitations on the use of pesticides and feed additives, restrictions on tillage practices, erosion of soil, controls on irrigation and water pollution. Environmental constraints and higher energy costs will impact on pasture and range improvements and vegetative manipulation. It is evident that tighter regulations can be expected to increase the real cost of red meat to the consumer.

SOIL

Sediment from erosion of pasture and rangeland is a permanent loss of soil resource as well as a potential pollutant. Eroded materials are the largest pollutant in terms of volume; they carry plant nutrients, pesticides, pathogens and other pollutants. It was estimated (USDA, 1972) that 1.38 tons of sediment were lost annually from each acre of the nation's forest and rangelands.

Accelerated and, therefore, preventable erosion is primarily due to reduced vegetative cover on the soil, which is often a result of overgrazing, poor animal distribution or other faulty grazing practices. It may also result from vehicular use and other mechanical disturbances of the soil. Research should be designed to establish guidelines that maintain animal productivity without causing damage to the vegetative cover and exposing the soil to erosive forces. Especially important are the stream-side sites on rangeland where livestock congregate and tend to do the most damage.

WATER QUALITY AND SAFETY

Water pollution of nonpoint sources is minimal under extensive production systems. The pollutants vary depending on the area and extensiveness of the system. Some contamination of the water will occur due to direct excretion into streams or lakes. However, this will be minimal even in humid areas with high livestock density.

Fertilizing with manures may result in contamination of water with plant nutrients, microorganisms and chemicals. Even though manures contain pathogenic organisms, it appears that these organisms are usually destroyed in the soil and do not present a serious hazard to water. However, application of large amounts of manure on pasture may result in some nutrients reaching streams.

Herbicides and insecticides are frequently used on grazing land to control undesirable plants and insects. These may find their way into streams and lakes if rainfall occurs soon after application. This problem is seldom serious because most of the residues will be leached into the soil where metabolism and destruction will take place.

A research need is to determine the extent and persistence in the food chain of contaminants from nonpoint sources in different physiographic areas with different amounts of rainfall and different stocking rates. The research would include obtaining data on concentrations of chemical and biological contaminants in soil, water and products from grazing animals.

AIR

Dust and smoke are the two principal air pollutants causing environmental problems in the use of pasture and rangelands. Dust concentrations in air are inversely related to the vegetative cover of soil. Where soil is bare, whether caused by cultivation, mechanical disturbance, drought, overgrazing or other factors, winds are likely to erode soil particles. The best solution is to maintain unbroken soil cover, but that is not always practicable. Research is needed on risk associated with wind erosion and techniques for control.

Because of state and federal statutes that regulate and inhibit the use of herbicides to reduce undesirable plant species on rangeland, managed burning has taken on new importance. However, fire to remove brush and undesirable dry materials on range and pasture lands is more frequently being regulated by law because of smoke pollution.

Depositions on range and pasture lands from the burning of fossil fuels and from industrial processes is of increasing concern to animal agriculture. For example, it has been shown that lead concentrations in forage increase as a function of decreasing distance from major highways. Acid rains are a current and widely publicized phenomenon. The extent to which air pollutants from cities, industry and major highways are influencing animal production should be researched.

VEGETATIVE MANIPULATION

There is widespread skepticism about large-scale, intensive range-management schemes, which involve manipulation of vegetation, because of their long-term effects on diversity of plants and animals.

Increasing public pressures and legislative mandates require multiple use management of rangelands, maintenance of species diversity and insurance of nondegradation of soil (U.S. Congress, P.L. 95-514, 1978). Early research developed information on vegetative response to livestock grazing and range improvement practices. Research also has provided knowledge for rehabilitation and management techniques using fire and mechanical and chemical control of vegetation. But little is known about the effects of range practices upon the mix of domestic and wild species, including man, that rangeland can accomodate.

Additional information is needed to integrate various range management practices in ways which enhance a variety of resource uses and minimize adverse environmental effects. Currently the methodology is inadequate for predicting and measuring the impacts of multiple resource management and the resulting resource trade-offs.

ENDANGERED SPECIES

The Endangered Species Act of 1973 protects endangered and threatened animals and plants. By July 1, 1979, 19 of 1,783 proposed plant species were officially classified as threatened or endangered and are under the protection of federal law. Twelve endangered animal species which occur on forest and rangelands have also been listed (U.S. Congress, P.L. 93-294, 1973).

Federal agencies have a mandate to ensure that actions they authorize, fund or carry out will maintain and improve habitat for these species. Some states protect even more species. Protection of endangered species may pose problems regarding management of rangelands. At this time, however, conflicts are minor between well-managed grazing and endangered or threatened plants and animals. None of the listed species has been so categorized because of livestock grazing. However, should any species be jeopardized by grazing, adequate steps must be taken to protect them. Potential impact will depend upon the habitat requirements of the classified species. Further research is needed to determine the habitat requirements and interaction of grazing animals and endangered plant species.

STREAMSIDE HABITATS

Streams in rangeland areas are spawning and rearing grounds for cold water and anadromous fish resources which support significant commercial and sport fishing industries as well as recreation for millions of people. Maintenance or improvement of streamside habitat is essential to the continued productivity of these fish-producing areas. Range management practices can affect sedimentation, stream-bed gravel conditions, water level and temperature and a wide variety of water quality dimensions. Adequate streamside vegetation can modify adverse stream conditions, and grazing management strategies can be developed to maintain and improve vegetative cover. However, quantification of the interaction of livestock and wildlife grazing and damage to streamside habitat is not available for many range ecosystems.

BIOLOGICAL SYSTEMS

An understanding of the biological systems and how they function is essential to the management and use of pasture and rangelands by grazing livestock. Relatively little information is available to managers and livestock producers on the structure and functioning of many pasture and range ecosystems. Knowledge is incomplete on many complex concerns such as energy flows, water availability and nutrient cycling through these ecosystems. Interactions among plants, animals and their relationship to the physical environment are understood in only a few areas.

Much can be done with existing technology to increase production of forage and red meat for people. However, continuing research on such things as renovating deteriorated range, biological control of pests and noxious plants, use of prescribed burning and the development of genetically superior strains of grasses, forage and shrubs is needed to accelerate the rate of improvement and increase the efficiency of management on all rangeland.

RESEARCH IMPERATIVES

I. Increase Research on Conservation and Use of Natural Resources

A. Evaluate the Impact on Animal Agriculture and Develop a Response to a Decreasing Supply of Land, Water and Energy.

There is a continuing trend to transfer private land from agricultural uses to urban and industrial development. At the same time livestock production on public lands is coming under additional pressure because of other multiple use considerations such as recreation, wilderness and wildlife. Water is also being transferred out of forage and feed production to other uses. The cost and availability of energy is placing additional constraints on the animal industry. These impacts should be evaluated, future changes anticipated and alternative solutions developed through systematic research.

As more pressure is placed on the limited resource base, new and imaginative techniques must be developed to increase animal production per unit of land, water and energy. This research should be focused on higher yields of feeds and forages, better feed conversion and improved management systems as they relate to soil nutrients, water and energy supplies.

B. Develop Techniques for the Production of Animals Requiring Minimal Cultural Energy (Fossil Fuel) Inputs by the Optimal Use of Range Forage, Forage Crops, Crop Residues and Concentrates.

Whereas American agriculture in its entirety is now heavily dependent upon fossil fuels, animal production has an opportunity to move rapidly toward fossil fuel independence. Aspects of this research are:

1. To identify the energy inputs into animal production systems, including the production of feedstuffs.
2. To improve the biological and cultural energy efficiencies of various animal production systems, including those involving nonconventional animals and feedstuffs.
3. To develop long term management strategies for sustained yield of forage and animal products from range and agricultural land compatible with multiple use of the resources and minimal energy inputs.
4. To design systems for optimum recycling and use of organic and other waste in the production process.

C. Assess the Impact of Climatic Variables on Animal Production with Emphasis on Minimizing the Impact of Adverse Climatic Fluctuations such as Drought.

A major aspect of this research is to predict the degree of risk and to develop systems such as real time data bases to minimize the impact of adverse climatic variables - particularly drought.

II. Develop Innovative Techniques for the Improvement of the Environment

A. Develop Alternative Strategies to Utilize Manure and Other Organic Wastes as a Resource in Animal Production Systems and to Minimize Adverse Impacts on the Environment.

Some aspects of this research are:

1. Adapt and develop thermochemical technology to convert animal waste and other available ligno-cellulosic materials into energy or useful chemical products.
2. Determine the feeding value of the residue as affected by various processing techniques.
3. Develop animal production systems to minimize pollution in air and water.

B. Develop Conservation Technologies for Animal Agriculture which Improve the Production of Vegetation and Prevent Wind and Water Erosion.

Improved techniques for vegetation management on range and pasture are needed which consider both animal production and conservation. This research should be directed toward the design of better grazing systems, brush and noxious weed control, reseeding of improved plant varieties, proper fertilization and better techniques for vegetative manipulation with due consideration of soil and water conservation.

C. Develop Innovative Building Designs to Reduce Fossil Fuel Inputs with Control of Temperature, Relative Humidity and Toxic Gas Concentrations to Promote Maximum Performance of Animals Confined in Buildings.

Aspects of this research include development of facilities for all food- and fiber-producing animals.

III. Design Research to Define Interrelationships between Resources and the Environment

A. Develop Standard Techniques for Evaluation, Monitoring and Assessment of the Basic Elements of the Environment Associated with Range and Pasture Production.

At present, resources and the environment are measured and evaluated in different ways by different agencies and groups. Standardization of data systems would improve planning for optimum use of our limited resources.

B. Utilize Systems Analysis Techniques to Develop Dynamic Models for the Evaluation and Optimization of Animal Production.

The objective is to combine forage and animal production practices in such a way as to utilize resources efficiently with minimum adverse impacts on the environment. Dynamic system models are needed which describe the interactions among all component parts of the production system in such a way that performance can be predicted, evaluated and optimized.

C. Conduct Long-term Research on Range and Pasture Ecosystems to Soils, Vegetation, Animals and Variable Climatic Conditions.

This research should be designed to continue through several climatic variations in order to evaluate the effect of drought and other environmental parameters.

REFERENCES

CAST. 1974. Livestock grazing on federal lands in the 11 western states. Reprinted in J. Range Manage. 27(3):174.

CAST. 1977. Energy conservation for agriculture. Special Rep. No. 5.

CAST. 1978. Feeding animal waste. Council Agr. Sci. Tech. Rep. No. 75.

Cook, C.W. 1972. Energy budget for rabbits compared to cattle and sheep. Colo. State Univ., Range Sci. Dep., Sci., Ser. No. 18.

Cook, C.W. 1977. Use of rangelands for future meat production. J. Anim. Sci. 45:1476.

Cook, C.W. 1979. Meat production potential on rangelands. J. Soil Water Conserv. 34:168.

Cuthbertson, D. 1970. Role of the ruminants in world food supply. World Rev. Nutr. Diet. 24:414.

Gilbertson, C.B., F.A. Norstadt, A.C. Mathers, R.F. Holt, A.P. Barnett, T.M. McCalla, C.A. Onstad and R.A. Young. 1979. Animal waste utilization on cropland and pastureland. A manual for evaluating agronomic and environmental effect. USDA Utilization Res. Rep. No. 6-EPA-600/2-79-089.

Hashimoto, A.G., V.R. Chen and R.L. Prior. 1979. Methane and protein production from animal feedlot wastes. J. Soil Water Conserv. 34:16.

Huffman, W.J. 1978. Alternate manure recycling system for energy recovery. In Proceedings, Great Plains Extension Seminar on Methane Production, Texas A&M University, College Station TX.

Kreis, R.D. 1979. Recovery of by-products from animal wastes - a literature review. EPA 600/279-142, U.S. Environ. Prot. Agency, Ada, OK.

Lloyd, R.D. et al. 1970. Rangeland ecosystem research: The challenge of change. USDA Agr. Info. Bull. No. 346.

Pimentel, D. et al. 1980. The potential for grass-fed livestock: Resource constraints. Science 207:843.

Pratt, P.F. 1979. Management restrictions in soil application of manure. J. Anim. Sci. 48:134.

RPA. 1980. An assessment of the forest and rangeland situation in the United States. USDA, Washington, DC.

Smith, L.W. and W.E. Wheeler. 1979. Nutritional and economic value of animal excreta. J. Anim. Sci. 48:144.

Smith, R.J., M.E. Hein and T.H. Greiner. 1979. Experimental methane production from animal excreta in pilot-scale and farm-size units. J. Anim. Sci. 48:202.

Stewart, B.A., D.A. Woolhise, W.H. Wischmeier, J.H. Caro and M.H. Frere. 1976. Control of water pollution from cropland. USDA-EPA Rep. ARS-H-5-2.

Sweeten, J.M. 1980. Energy recovery from feedlot manure: An assessment of alternate technologies. Special Rep. Texas Agr. Ext. Serv. Texas Feed Assoc., Amarillo TX.

Thomas, G.W. 1977. Environmental sensitivity and production potential of semi-arid rangelands. ICASALS Publ. 77-2:43.

U.S. Congress. 1973. The endangered species act of 1973. P.L. 93-294.

U.S. Congress. 1978. Public rangelands improvement act. P.L. 95-514.

USDA. 1972. The Nations Range Resources.

USDA. 1979. Secretary's memorandum No. 1999. October.

Van Dyne, D.C. and C.B. Gilbertson. 1978. Estimating U.S. livestock and poultry manure and nutrient production. U.S. Dept. Agr. ESCS-12.

White, R.K. and D.L. Forster. 1978. A manual on evaluation and economic analysis of livestock waste management systems. EPA-600/2-78-102.

White, R.K. 1979. Research needs assessment - livestock manure management in the United States. EPA-600/2-79-179.

Wilkinson, S.R. 1979. Plant nutrient and economic value of animal manures. J. Anim. Sci. 48:121.

"We and the rest of the world have reached the end of a pleasant era of increasing standards of living and of cheap energy... We now face a period of declining standards of living and the jarring and costly transitions to other energy forms... economic adjustments will affect the purchasing power of consumers and, in turn, their preferences for food and especially for meat."

Philip Abelson

TABLE 10.1. PROPORTION OF DIETARY ENERGY PROVIDED BY CERTAIN KINDS OF FEEDSTUFFS TO VARIOUS KINDS OF LIVESTOCK IN 1972-1973[a]

Feedstuffs	Percentage of dietary energy derived from different feedstuffs by indicated classes of livestock					
	Dairy cattle	Beef cattle	Pigs	Poultry	Other livestock	All livestock
Grains[b]	25.2	21.8	72.0	62.5	74.0	35.6
High-protein feeds[c]	5.3	2.6	11.6	29.5	9.0	7.5
Other by-products[d]	4.7	1.2	2.1	2.1	3.8	2.5
Roughage[e]	64.8	74.4	14.3	2.6	40.2	54.4

[a]J.T. Reid. 1975. CAST Rep. No. 5.
[b]Cereal and sorghum grains.
[c]Includes oilseed meals, slaughter by-products, and grain-protein by-products.
[d]Includes grain-milling by-products.
[e]Includes pasture, range, hay silage and straw.

TABLE 10.2. AVERAGE GROSS ENERGY CONVERSION AND SUBSIDIZED ENERGY UTILIZED TO PRODUCE CARCASS OR PRODUCT FROM VARIOUS ANIMALS[a]

Animal species or product	% gross energy in feed energy converted to food energy for man	Kcal of cultural energy/Kcal of gross energy in product
Pork	17	4.9
Poultry	12	3.8
Rabbit	9	1.6
Eggs	7	2.0
Goat	5	3.0
Lamb	5	2.2
Beef	4	5.2

[a]Cook, 1977, 1979.

TABLE 10.3. ANIMAL EXCRETA PRODUCTION IN THE UNITED STATES IN 1974[a]

Class of animal	Thousands of metric tons of dry waste	
	Production	Collectible
Beef cattle (range)	52,057	1,897
Feeder cattle	16,428	16,000
Dairy cattle	25,210	20,358
Hogs	13,360	5,538
Sheep	3,796	1,700
Laying hens	3,374	3,259
Turkeys	1,251	983
Broilers	2,086	2,434[b]
Total	111,562	52,169

[a]Van Dyne and Gilbertson, 19&8.
[b]Includes litter.

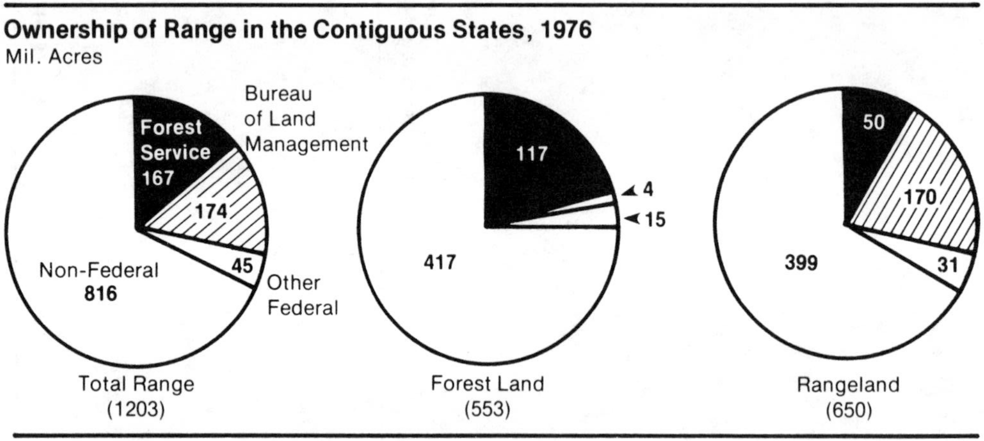

Figure 10.1. Ownership of range land in the United States in 1976.

11

J. Bonnen (Co-Chairman), R. O. Wheeler (Co-Chairman),
L. F. Schrader (Rapporteur)

J. H. Anderson, J. W. Bennett, D. M. Burns,
L. J. Connor, L. R. Ewing, W. D. Hefferman,
J. A. Hoefer, T. B. Kinney, Jr., F. Mulhern, R. P.
Niedermeier, D. Pridgeon, C. Randall, G. W. Salisbury,
D. H. Stroud, J. R. Welser, and S. Wittwer

PUBLIC POLICY

SUMMARY

Government action (public policy) represents the means for society to intervene in markets or to modify directly the action of firms or individuals to advance the common good. Animal agriculture is affected by many policies, laws and regulations, some of which were conceived to influence animal agriculture directly and some of which were instituted for other reasons. Often the unintended impacts of policies are as important as the intended.

The roles of public policy research are to evaluate the impact of policy actions -- existing or proposed -- upon the quality of life of affected groups, and to design optimum strategies or institutions to accomplish the goals of society as perceived by policy makers. Policy research depends upon the physical and biological sciences for the specification of technical relationships and upon the social sciences for knowledge of needed human behavioral relationships. Policy analyses represent a means to bring the results of scientific knowledge to bear on the policy process.

There are serious questions concerning the degree to which certain regulations which impinge upon animal agriculture have accomplished their intended goals and whether the unintended impacts have been adequately accounted for. The process of policy formulation may lead inadvertently to a poor fit of goals and effects and to the generation of unintended effects.

While value conflicts are inherent in the policy area, the process of making policy decisions can be improved. Objective research can provide a solid knowledge base regarding the issues. Such knowledge can refute factually inaccurate claims, can show that certain fears are groundless and can quantify more accurately expected costs and benefits. Thus it helps to narrow the disagreements of opposing interest groups and increases the opportunities for reasonable compromise.

Five priority areas have been identified in which public policy research should be focused to enhance the ability for animal agriculture to satisfy human needs more adequately in the years ahead.

1. Improving the Policy Decision Process
   a. Evaluate the information systems whereby the needs or desires of the body politic which are relevant to animal agriculture, are reflected to public decision makers.
   b. Evaluate the processes whereby programs and regulatory decisions relative to animal agriculture are made and implemented.

2. Improving the Effectiveness of Government Regulations and Programs
   a. Evaluate government programs (e.g., market orders, conservation, etc.) and regulations (e.g., food safety) which affect animal agriculture to determine whether they are meeting their objectives in terms of production, consumption, environmental and other societal goals or are creating unintended consequences.

3. Assessment of the Impact of International Trade
   a. Evaluate the impact on U.S. producers and consumers of animal products of barriers to trade in livestock and livestock products.
   b. Evaluate the resource content of trade in livestock and livestock products in terms of capital, labor and natural resources, including land.

4. Impact of Input Limitations
   Evaluate the supplies of resources pertinent to animal agriculture and the consequences of imbalances and/or policy induced limitations of resources on the capability of animal agriculture to meet human needs.

5. Structure of Animal Agriculture
   a. Evaluate the impact of the structure of the food-animal subsectors on their resource efficiency, distribution and equity of income, stability, environmental impacts and other social goals.
   b. Evaluate the impact of public policies on the structure of the food-animal subsectors.

INTRODUCTION

Animal agriculture is increasingly controlled by laws and regulations. These affect producers, processors, distributors and consumers. The regulations occur at many levels: federal, state, regional and local, and their effect is often unforeseen.

Producers want an economic climate which encourages efficient production and allows them the freedom to make a profit. They would like the cost of their inputs to be stabilized at a reasonable price, and also protection from unreasonable exploitation by supplier, processing and marketing industries.

Consumers want food which is both safe and cheap -- two goals which are sometimes in direct conflict. To produce meat products cheaply, producers strive for greater efficiency, and this frequently means practices which may add an element of risk to the consumption of these products.

Can the regulatory system be improved? Can the process be made more inclusive, so that all elements affected by a policy have an opportunity to participate in formulating that policy? Can we ensure that all participants are adequately heard? Can we ensure that policy-makers have access to the information they need, and in a form they can use.

Policies will never be free of controversy, but the process can be made more orderly, and be better understood. Scientists can do a great deal to improve communication with policy-makers, and they should acknowledge that they have an important role to play in the process. Public understanding of science is critical, and scientists have an indispensable role to play in submitting evidence which will help elevate the public debate on critical issues. Unfortunately, scientific evidence is sometimes ambiguous, and scientists are not always of a single mind on the facts or their interpretation. There must be humility in admitting what we do not know, and candor where we are not certain. Moreover, scientists must also recognize that policy-making cannot always wait until conclusive evidence is gathered. In the meantime, and in the midst of controversy and conflicting claims, scientists must do their best to ensure that the policy process is as rational and as fact-based as possible.

One role of policy research is to evaluate the impact of policy proposals upon level of living and quality of life and the distribution of costs and benefits. Policy research is also directed toward the determination of appropriate policy strategies to correct perceived market failures.

Another dimension of the public policy research is to evaluate the institutions by which resources are organized and income is distributed in society. These institutions may be in the form of specific organizations or they may be in the form of laws, rules and regulations. In any case, they reflect either policy decisions or the end result of policy decisions. More importantly, they determine the way in which individuals as social, economic and political entities interact with each other.

Research which attempts to evaluate these institutions is an important component of social science research. The design of new institutions and the design of institutional innovations is an important output of social science research and is directly analogous to the technological innovation which is produced by biological and physical science.

Policy research cannot be divorced from biological, chemical and mechanical research, or the political process. The former provides the physical and biological performance relationships and the latter provides the goals and the appropriate weights. Behavioral relationships representing consumers, producers and marketing firms must be provided by socio-economic research. Policy research involves and depends on all of these different kinds of research.

With these data and relationships, the policy researcher can proceed to identify the gainers and losers and to quantify these gains and losses, both economic and noneconomic. The need for completeness of the policy model is clear. The unintended consequences of a policy often outweigh the intended consequences, indicating an inadequate scope of the original policy formulation.

POLICY DECISION PROCESSES AND INSTITUTIONS

Statement of Problem

For the first half of this century, the respect earned by and paid to agriculture for its research was enormous and appropriately placed. During those decades most U.S. biological research was produced by the agricultural establishment. The competence of agricultural sciences, as measured by increased productivity in producing food was unequivocally established.

Society and its needs have changed. The research system of agriculture no longer appears to policy-makers either as able or as well equipped to serve today's highly urban society as it served an agrarian society. This problem has been exacerbated by the inward orientation of agricultural research itself. Its leadership has had difficulty in developing a sensitive listening apparatus in order to interpret society's changing needs. Likewise, its leadership has not seemed to understand that an urban society requires a new and different communication process in order to comprehend the importance and vitality of the agricultural research that is benefitting consumers and the entire society.

The flow of information to decision makers from those sensitive to and aware of needed research on animal agriculture is all too often distorted by intervening variables and the lack of an effective transmission system. Of special significance as intervening barriers are middle-level staff in the Executive Office of the President in the White House, in Congress and elsewhere who have little understanding of either agriculture or research systems. Their actions and analysis often act as screens in the flow of information, distorting and losing critical knowledge, as it moves to policy-makers.

Animal agriculture now appears to have lost much of its former respect in the minds of many policy-makers and thus has taken a lower post in the ordering of priorities for allocation of research dollars. If this situation continues, the inefficiencies in resource utilization and loss of nutrients to the American and world publics will be socially unsettling and economically disruptive to animal agriculture and, therefore, to consumers.

The solution lies in the hands of the agricultural research establishment. Some of the problem was created there. During the 1960s and early 1970s agricultural research was not as responsive as desired to public priorities, which included values not well understood by agriculture. In the late 1970's agricultural research has become far more responsive. Animal agricultural research funds have been reallocated from production research to food safety, health and environmental implications and other goals on the newer research agenda. This research response to a new set of societal needs has been at the expense of its capacity to generate greater food output. Society must understand the severe limits of present funding upon the research response to both its new and its traditional needs.

The primary beneficiary of increases in the productivity of animal agriculture is the consumer -- a fact poorly understood by many policy-makers. Productivity gains have resulted in billions in dollar savings to consumers, in foreign exchange needed to pay for petroleum and other imports and in greater farm income. This lack of understanding is now influencing the priorities for animal research. Research that increases productivity, makes foods safer or improves the environment which benefits the entire society.

Critical questions are being asked about animal foods. Although the animal agriculture establishment has bridled at such criticism and even responded forcefully, apparently it has not done so convincingly. It needs to develop a better understanding of the environment from which the challenges are being issued and the persons issuing them; it needs to produce the research that will establish the facts. Among the most prominent groups of such questions are those raised about the healthfulness and safety of meat, dairy and poultry products.

The concept of spending research dollars to expand use of animals to produce nutrients has been emphatically challenged. Advocates hold that the land and other resources would be more wisely, morally and efficiently used to produce the same nutrients, less expensively and more abundantly through a nonanimal medium. They advocate massive transfer of grains currently used to feed livestock to nutrient poor countries. These advocates ignore two major issues. Except in the form of aid, transfers of either plant or animal nutrients to low-income, food-deficit countries is likely to be difficult and dominated by market forces. Moreover, despite great differences in culture, as income has risen around the world, consumers have exhibited a preference for obtaining nutrients from animal sources. Further, the nutrients are "packaged" in highly assimilable form, and their production is far more efficient than critics of the system appear to understand.

Nevertheless, the debates continue. Policy-makers do not have an adequate information base upon which to judge the factors essential to making well-informed decisions. In fact, the greater the conflict on a given issue, the more need for research to provide an accurate data base from which to make judgments. Such research would help reduce purely emotional conflict and eliminate factually inaccurate assertions. It narrows the gap between adversarial views and improves the atmosphere for reasoned resolution of differences.

State of the Arts and Science

However good the information that is provided by scientific research, it is a fact that this information is usually processed by large, complex government organizations. To be useful, then, scientific information must be either tailored to the needs of these organizations or these organizations must be redesigned so that they can use this information. Public policy research can help to realize this goal. Any formally organized area of policy is supported by a configuration of institutions which constitute a decision process. All such social structures must be adaptive to survive. They must change over time as society's goals and values change and as technology and social organization change. Much is known about old, well-organized policy processes and related institutions, especially at the Federal level. Public administration studies of Cabinet, department and agency organization, their function, behavior and performance are extensive. Newer programs and agencies, however, often have not been examined. The impact of the growing size of Congressional staffs, in particular, has received little attention.

Similarly, policy analyses are generally available for the older policy areas but often not for the newer policies and legislation. Moreover, much policy analysis is narrowly focused. As an example, the achievement of legislated goals has been examined without equally important analyses of the unintended consequences of a program or assessment of who are the gainers and losers. For many policies and programs, the political dynamics and conflicting values which drive the political decision process in the Congress and in the Executive Branch are poorly identified and understood. What makes public policy research even more difficult is that many policy areas, especially new ones, are dynamic or changing processes and constitute moving targets so that 3- or 4-year-old pieces of research are already obsolete.

A unique problem of the decision processes of older policy areas, such as agriculture, is an organizational hardening of the arteries that leads to loss of the adaptive capability necessary for survival. Agriculture's stunning success in increasing productivity and industrializing its production processes has led the institutions of agriculture to be complacent about their roles and current social performance. And, the rest of society is equally complacent about a continued flow of productivity increases from the agricultural cornucopia.

The policy research imperatives implied by this problem involve analyses of the social performance and necessary current role of agricultural institutions, objective examination of who gains or loses from various policies, and the development of a solid knowledge base about those policy information processes and institutions which support and influence major areas of policy for agriculture. It is imperative to organize scientific knowledge on animal agriculture in a manner that can be communicated to and understood by politicians, their staffs, interest groups and the voting public. If policy actors understand clearly the preconditions for the continued productivity and production of safe, healthful and abundant animal products, those preconditions are more likely to be met.

The categories of needed policy process research can be made more concrete by examples. The older decision process for establishing agricultural research policy and funding priorities has been subject to considerable criticism in recent years. It is held responsible by some scientists, especially in the nonagricultural community, for the alleged deficiencies of agricultural research. Neither defensive responses nor heated exchanges can resolve the researchable questions of which funding philosophy, decision criteria and institutional arrangements best support excellence in disciplinary research (e.g., gentics or nutrition), subject matter research (e.g., energy availability in agriculture, arid rangeland animal culture), applied problem solving research (e.g., reduction of fat content in fed beef), or some mix in the same research system of these three fundamentally different types of research.

The relative merits of alternative funding methods can be researched. The classical formula-based institutional funding system of agriculture has distinctive characteristics and some long observed social and institutional consequences. Similarly, the NSF peer group review of project by project funding has a very different but equally distinctive character and set of observable consequences. There are other systems of research funding. The design of research funding to achieve a desired performance from each of the three fundamentally different types of research must itself be researched if this important, contentious issue of the appropriate design of research funding institutions is to be settled intelligently and in a manner that serves society through an effective research establishment.

There are other major researchable questions about the performance and management of research institutions that involve the appropriate mix of disciplinary and applied research and the coordination of research where priorities are established or influenced greatly by one organization (e.g., NAS, Congress) while funding strategies and decisions are made by others (e.g., NSF, USDA). Failure to manage or coordinate well disturbs the research enterprise and creates instability and acrimonious institutional relations that erode research capability.

An entirely different research problem involves a large fraction of the regulatory policy-decision process which is fragmented in structure and leadership. Its poor coordination has many unintended and costly consequences. Regulation responsibility and planning is scattered among both regulatory and nonregulatory agencies, including FDA, USDA, EPA and HHS. The problem of fragmentation is compounded by a further, often ambiguous, distribution of responsibilities between State and Federal planning eschelons. This compound fragmentation results in a mosaic of support patterns which tend to create gaps in communication and inefficiency in total use of resources. Policy research on the nature and effect of this fragmentation with a view to the improved coordination of policy could make a major contribution to improved communication and rationality in the policy dialog and to more effective protection of human health and safety, as well as to possible savings in resources.

GOVERNMENT REGULATIONS AND PROGRAMS

Statement of Problem

Over the past 15 years, there has been a great expansion in the number of regulations and programs enacted and proposed for animal agriculture. These regulations have been enacted or proposed for various purposes, including safety and nutritional well-being of consumers, safety of farmers and farm workers, the maintenance of the quality of our environment, the protection of human subjects in research, the enhancement of income levels of producers through price supports, and the welfare of animals. Examples of regulations include OSHA labor safety requirements, pesticide banning and licensing of operators, permits for large-scale livestock feeding operations, water quality standards for streams, human subjects research rules, animal care regulations, livestock feed additive restraints, bans of certain food preservatives, transportation rate structures, commodity price supports, meat and poultry inspection and grading and food labeling. The impetus for these regulations has come from a number of regulatory and nonregulatory agencies, including USDA, EPA, FDA and DOE. In more than a few cases, the responsibility and planning operations have been divided among them, creating difficulties for researchers in complying with the guidelines relative to methodology and laboratory practices of the different agencies.

State of the Arts and Science

Programs and regulations evolve from the perceived needs of individuals or groups in society; some groups perceive that their needs are not being met or that they are being unjustly disadvantaged. They may feel that current market institutions are not dealing satisfactorily or equitably with their problems. In order to correct these deficiencies, regulations are designed to influence behavior and, hence, assure desired outcomes. Regulations accomplish their behavioral functions by providing positive or negative incentives. Positive incentives include such regulations as cost sharing, price supports and various direct subsidies, such as investment tax credits and lower income tax rates on capital gains. Prohibitions with possible fines or imprisonment are examples of negative incentives. In some cases, programs and regulations may augment or work through the market system, as in the case of price supports and certain tax plans. The regulatory approach used in dealing with a perceived problem is important in both achieving the regulatory objectives, and in distributing the benefits and costs among various segments of society. A wide variety of regulatory approaches has been used in animal agriculture including cost sharing, licensing, permits, labeling, grading, setting rate structures, taxing and prohibiting.

The benefits and costs associated with various regulations are borne in varying proportions by different segments of the population and of animal agriculture. Large growers are typically impacted differently than small growers; affluent consumers are impacted differently than poorer consumers. All regulations are designed to enhance the human needs of some group or groups within society, but they may become obstacles or constraints to others.

Many programs and regulations which were developed to respond to specific needs have become more burdensome than necessary. This burden frequently results because the regulations and programs are not altered as changes occur in technology and in society. Thus it is necessary to have a system to monitor governmental programs and regulations that impact animal agriculture from the time they are proposed to the time when they are no longer serving the needs for which they were adopted. For example, the Delaney Clause was passed at a time when technology to detect contaminants was much less sensitive than it is today. Does the clause still meet Congressional intent?

Experience indicates a complex society must have many regulations; but, specific regulations must bear close scrutiny as to their net societal impact. To what extent have individual regulations interacted in unanticipated ways to increase producer costs or consumer prices for specific livestock and poultry products? Have these increased costs been offset by resulting benefits? Have regulations been predicated upon an adequate and generally accepted scientific base of knowledge? There appear to be varying opinions in the scientific community on this issue. Is there undue overlap and duplication of authority among regulatory agencies, with resulting confusion and cost to farmers and ranchers and to marketing and distribution firms? Have the costs of regulation been equitably borne among various consumers, producers and marketing groups? There are no clear cut answers to these questions. The present knowledge base does not provide definitive answers.

The preceding discussion illustrates the dilemma of policy-makers in dealing with programs and regulations. They are continually being confronted with single interest groups pursuing the enactment of specific regulations to deal with their perception of a problem. The responses of producers, commodity groups, processors and others (sometimes including scientists) often have been to resist and combat these proposed programs and regulations. Thus, policy-makers have been thrust into positions where they have had inadequate, incomplete, or conflicting information on which to make decisions. They need objective answers to these questions. Is there a problem? A problem to whom? What criteria are appropriate for defining a problem? If regulation is needed, what regulatory approach should be used? What are the benefits and costs of the regulations, and how are they borne in society? These information needs are crucial for policy-makers in order to make informed rational decisions.

At present the analysis of most regulations and programs involve the multidisciplinary input of scientists. Usually a benefit/cost, risk/benefit and/or cost effectiveness analysis is utilized, but a system has not been established to evaluate the efficacy of the regulations nor the criteria used for establishing them. Thus, the regulations, their enforcement and their more far-reaching consequences often are left unexamined.

The major difficulty in conducting these analyses is in establishing some consensus with respect to tolerance levels or standards. The scientific community often has difficulties in reaching consensus on appropriate levels or standards (e.g., zero tolerance; how many parts per million). Even when there is a good physical and biological knowledge base, the policy-maker must also be provided information on the distribution of risks.

Because regulations are numerous and cover many areas, and are continually surfacing (or resurfacing), it is difficult to provide a list of research priorities. The analysis and development of tolerance levels and standards are particularly relevant to possible emerging regulations in human and animal nutrition, food processing, animal health, feed production and environmental considerations. The use of benefit/cost analysis and the analysis of alternative regulatory methods for a particular problem are relevant in analyzing almost all potential regulatory problems.

INTERNATIONAL TRADE

Statement of Problem

U.S. export policy in agricultural commodities has focused on disposal of surpluses and stimulation of the development and expansion of foreign markets. Because of U.S. agricultural production capability and the great expansion of the world demand for food, the U.S. has become the supplier of much of the food basket for world's trade in food. U.S. agricultural policy is now predicated on a strong export performance. In addition, surplus balance in agricultural trade helps to pay for the imports of petroleum and other raw materials. World dependency on the U.S. agricultural system offers both an opportunity and a challenge for the future. It also suggests that important public interest questions will be raised about the effects of domestic programs and regulation on agricultural trade as a result of this dependency.

U.S. agriculture faces major barriers in international markets. Large market areas such as the European economic community effectively keep out many agricultural exports by means of a variable levy. These barriers are effective against both grain and livestock products. Japan and other countries severely limit the importation of livestock products by means of quotas and other nontariff barriers such as health and sanitation regulations. Finally, the U.S. is undercut in many markets because countries such as Brazil use large spot subsidies to protect their markets.

Although the U.S. is the largest importer of meat products in the world (of the 7.5 million metric tons of meat products flowing through international trade channels, the U.S. imports approximately 1.2 million metric tons), it also imposes sizeable barriers to trade. Protection of the U.S. dairy industry and the dairy price-support program is an outstanding example. Imports of beef are also regulated by import quotas set automatically by law. Although the Executive Office may override these quotas for selected reasons, and in the past has done so with considerable frequency, the restrictions on total imports are considerable.

U.S. barriers to trade are often justified on the grounds that they benefit the long run consumer interest in providing a more stable industry supply from domestic sources. Exclusion of imports of certain livestock products are generally justified on the basis of protecting the nation from undesirable diseases and pests. Protection from certain contaminants that would enter directly into the human food supply is also a justification for this type of trade barrier. These exclusions come under periodic attack as being designed to limit competition. However, they are generally rationalized as being reasonable and necessary as a defense against undesirable consequences.

Major beef imports into the U.S. include frozen beef and primal beef cuts for fabrication by retailers into smaller cuts. Live animals from Mexico and Canada provide another source of beef imports. Preserved and canned meats (e.g., corned beef, hams and other pork products) are imported, primarily from Argentina in the case of beef, and from Europe in the case of pork.

On a worldwide basis, humans directly consume about 60% of the total grain energy produced and animal agriculture consumes about 37% of the total. Among the major animal agriculture components, the relative importance of grain and oilseed energy use in the world system is: swine, 32%; poultry, 29%; dairy products, 17%; beef and veal, 16%; sheep and goats 2%.

While total consumption of grains directly by humans has remained relatively stable over the past decade, the growth in grain demand has come from the animal agriculture sector. On a worldwide basis, poultry production about doubled between 1967 and 1980. Demand for all animal agriculture products has steadily increased in several countries (the OPEC nations for instance) where grain imports are necessary to provide most of the feed energy base.

Since the U.S. is a major supplier of grain products to support expansion of world animal agriculture, the stability of the animal agriculture sector in many countries is directly linked to U.S. supply capability. Few countries have a natural resource base capable of further expanding grain production to meet the growing needs of their animal-agriculture sectors. They are in effect locked into a dependency on world grain supply of which the U.S. is the major supplier.

In considering public policy research in the international trade area, one cannot view animal agriculture independently of the world's grain sector. The interdependency is so strong that any attempt to isolate animal agriculture would not yield meaningful results. Many researchers have recognized the importance to the livestock industry of the grain sector and grain prices, but many times have not fully appreciated that the livestock sector was equally important as a factor affecting price stability in the grain sectors. Between World War II and the late 1960s, when hugh grain surpluses and farm price support programs resulted in relatively stable grain prices, this interdependence was seldom appreciated.

However the price instabilities in grain and animal markets in the 1970s associated with significant variations in world grain harvests bring new research questions to the forefront. Are we entering an era when animal agriculture of the U.S. and many other nations will be battered by fluctating world grain harvests? What are the cost-benefits of attempting to build large world grain reserves to buffer the animal agriculture industries and the consumer? On the other hand, what are the social cost-benefits of allowing major fluctuations to occur in herds and flocks in order to accommodate changes in world grain harvests? These policy questions are complicated by the fact that the current strategies of many nations tend to isolate their grain and animal economics from major price and volume fluctuations and, consequently, transfer their market risks and instability to the open markets of the United States.

Two points have become clear from the experiences of the 1970s. First, international trade in grain transmits the price effects of variations in harvests any place in the world to those few nations that do not protect their markets by high import barriers as the EEC or large national grain stocks. Variations in world feed availability and prices are transferred to the animal agricultures of those countries. The United States as the world's largest grain exporter, and as a major producer and consumer of food-animals experiences major effects, but numerous smaller grain importers around the world also experience effects of great significance to their economics. Second, in a free market, animal numbers vary inversely to grain supplies. This relationship tempers price swings in grains and reduces the vulnerability of those poorer people of the world dependent on grain for most of their nutrients. Thus the large cutback in grain fed to U.S. animals during 1973-1975 tempered what would have been an even more extreme rise in world grain prices.

If the balance between world grain supplies and demands becomes more precarious in the 21st century, this shock-absorbing role of animal agriculture as a regulator and conservor of human food supplies will become ever more crucial. A major trade policy question is how the United States balances various conflicting goals in its role as the world's grainery. Because of state trading, EEC trade barriers, etc., most of the major nations either had or obtained sufficient grain from the open U.S. market to maintain their livestock production in 1973-1975 when there were harvest shortfalls in the U.S.S.R. and elsewhere. Thus, the major burden of adjustment to shortfalls in the 1970s fell upon U.S. animal agriculture because of our open trade policies. The U.S. clearly needs to find policies short of the extremes of grain export embargoes or state trading which enable a reconciliation of various goals of domestic and foreign policy in an uncertain and shock-prone trade environment.

State of the Arts and Science

The question of whether the world stage is set for wider or sharper oscillating prices during the next 2 decades for grain products, which in turn impact adversely on the stability of the animal-agriculture sector, appears worthy of serious policy research efforts. These policy research efforts would focus on evaluating strategies for U.S. reaction to events in the world grain and livestock system to minimize or at least dampen disruptions to the nation's livestock and grain sectors.

As to other future needs in research, it is clear that short and medium range price-quantity-forcasting capability must be strengthened, particularly by making policy formation internal to the model.

Another little understood aspect of foreign grain trade development is the long-range effect on domestic opportunities due to the exporting of raw materials (grain) so that other countries can build their own meat production systems (and by-product production systems). Are there opportunities for the domestic meat industry to capture a share of such markets by exporting our meat instead of grain? Or does the escalation of tariffs on "processed" products by other countries effectively rule out such trade?

The U.S. also has exported livestock production technology to Developing Countries and others that has helped to create industries which compete - at least theoretically - with U.S. meat industries. The transfer of poultry technology is notable for its success in bringing poultry and egg foods to low-income people in many countries and also in expanding income for their respective domestic producers. But it has done little for the U.S. poultry industry's export opportunities and, in fact, may have diminished them. At the same time, it undoubtedly has increased the demand for feed grain from the United States. There are other considerations as well. In the long run, developing countries must become economically viable, if the world food crisis is to be solvable, and this requires increased productive capacity which includes the country's agricultural resources. An additional consideration is that some technologies, such as in poultry and eggs, are so readily transferred that it is fruitless to design policies around attempts to monopolize that technology.

An examination of U.S. barriers to trade and their triggering mechanisms, as well as a study of their economic impacts, is in order. The review should take into account the emotional underpinnings and reactions of the livestock industry and of consumers - in other words the "political" and "public relations" effects of the setting or lifting of barriers -- as well as the actual effects on tonnage movement, prices and domestic production.

Subsidizing of domestic production of livestock or meat for purposes of creating more competitive opportunities in foreign markets is a common practice in many countries, but has not been widely discussed in the U.S. livestock and meat industry. Likewise, increased duties on imported meat products (as in the case of woolen goods) would not be popular with American consumers, but ought to be studied if other protective devices are believed desirable.

Like all research, the quality of public policy research related to international trade policy is directly affected by the quality of the data available. In this case, information over time on the resource base, input-output rates, consumption and trade flows on a global basis are required. The major data source for this type of information is the Foreign Agricultural Service (FAS) of the USDA, the

Food and Agricultural Organization (FAO) of the U.N. and to a lesser extent the World Bank. Even in areas in which one might think the data to be accurate, serious questions of collection methods and inconsistencies exist. For instance, in the trade of meat products the published export figures from one country to a stated destination do not always mesh with the import figures of the designated importing country. Relevant information on pastures and grazing lands is scanty to nonexistent for many areas of the world. In general, the analytical tools necessary for processing and interpreting information for policy research are far beyond the quality of the data base needed to make effective use of the tools. If meaningful and timely public research for policy formulation and implementation is to take place, more attention must be given to the collection and assembly of relevant data.

A top priority research effort is needed to develop and evaluate alternative national strategies for dealing with sudden and important shocks to the world food system. To do this will require an improved analytical capability to predict price impacts in the short, intermediate and long-run time frame. Most sudden shocks to the system, such as climatic conditions, cannot be controlled while others can be. The strategy must also address the issue of protecting national interests in dealing with the centrally planned economics.

Another public policy trade issue, which needs research, involves the question of the relationship of current world market prices (benefits) for agricultural products and the longer run costs associated with depletion of natural resources. The concern is that the cost of natural resource transfers, such as soil erosion, vastly exceeds the current prices reflected in the international markets.

RESOURCE INPUTS

Statement of Problem

Animal agriculture is inextricably linked to other sectors of the economy and many of the inputs purchased are heavily influenced by domestic and foreign governmental programs, rules and regulations. For example, approximately 60% of U.S. feed grain production is consumed by domestic livestock and poultry. The production, prices and utilization of feed grains are affected by U.S. price support, inventory, acreage control and export control policies, and by European Economic Community variable levies. Similarly, many of the nonrenewable resource inputs such as liquid fuels are critically impacted by government policy rules including price floors, price ceilings, taxes, quotas and priority allocation schemes. Water policy conflicts, particularly in the west, are legendary. Water poses special problems as it is a renewable resource in some areas but is exhaustible in others.

Capital markets are influenced by the federal government's monetary and fiscal policy and by tax legislation, particularly capital gains treatment, e.g., taxation rules are one of the most important determinants of the period of time that breeding stock is held by the producer. Labor and welfare legislation, including work rules, has had a major impact on mechanization.

Land use is directly impacted by tax policy and zoning, and, indirectly by cheap energy and by transportation policies that have encouraged geographic dispersion. The result has been indirect impacts on feed costs and direct conflicts over land use, particularly over odors, between farm and nonfarm families. The impacts of environmental policy are becoming ever more important, particularly when policies to reduce air and water pollution conflict with traditional agricultural production practices. Public ownership of range lands and forests is associated with numerous questions of public policy, especially in the west.

The structure of the food-animal industry in the 21st century will be heavily impacted by public policies toward its vital inputs of feedstuffs, chemicals, energy, land, water, labor and capital. Focus must include indirect, as well as direct linkages.

State of Arts and Science

Public policy research must be based upon sound conceptual foundations, a meaningful data base and an understanding of the empirical relationships necessary to predict (albeit imperfectly) the consequences of alternative policies. Subsequent sections outline the policy research perspective for major input categories.

Feedstuffs. Feed grain, food grain and oilseed policies are probably the most important ones influencing the size and structure of the livestock and poultry industry. The primary purpose of growing feed grains and oilseeds is for animal feed. Because feed and food grains compete for cropland, policies influencing food grains cannot be ignored.

The conceptual foundation underlying grain and oilseeds supply and their demand by livestock and poultry are moderately well understood; poorly understood areas include market participants' response to alternative levels of risk and their rates of adjustment to new market signals. The U.S. data base is adequate, but the data bases of major feed importers and of competing exporters are often inadequate. Behavioral relationships, at aggregate levels, have been studied extensively. Less is known about how the pieces fit together, and initial attempts to develop policy models have been inadequate. Nevertheless, initial progress justifies accelerated research, particularly policy research focusing upon integrating knowledge of the international livestock-feed-grain-food grain-oilseed economy. These investigations must be sufficiently flexible to test the impacts of alternative policy rules, including inventories policy, on the food-animal economy.

Energy. Animal agriculture must expect to do its share in conserving scarce energy; it cannot ask for policies to insulate it from these problems. However, research can help to insure that the scarcities are not compounded by ill-informed policy making. Shrinking energy supplies, will lead to major changes in the structure of food-animal agriculture in the 21st century. Energy prices, relative to all major inputs, declined dramatically until 1973. Electricity prices relative to wage rates, for example, declined in 1973 to one-fifth their ratio in 1937.

Cheap energy contributed to major changes such as substitution of anhydrous ammonia for legumes as a nitrogen source in feed grain production, the development of corn drying technology, shifts in cattle feeding from the Midwest to the High Plains and substitution of energy for labor in food processing. These trends are expected to reverse, but energy is such a pervasive input to all sectors of the economy that only a broad-brush perspective is possible without definitive research. Choice of policy instruments by state and federal governments will be critical to agriculture where the biological nature of most processes requires immediate action and, hence, assured fuel supplies; for example, fuel for planting, harvesting and drying corn; heating and/or ventilation of livestock structures; slaughter of market animals and refrigeration of perishable foods.

The subsidization by state and federal governments of ethyl alcohol production from feed grains provides an example of an important policy issue. This policy encourages the use of grain to produce liquid fuel resulting in a potential reduction in grain supplies for livestock feed. The potential of increased use of grain for alcohol, in turn, adds to the medium protein feed supply (distiller's by-products). The net impact of an increase is alcohol production will be a decrease in the size of the livestock industry.

Land. In some types of animal production systems such as feeder-calf-production land is a major input. In other enterprise systems the requirement of land is less direct, but nevertheless important, because it influences the quantity of feed available to animal agriculture. The loss of agricultural land to nonagricultural uses is well documented. The direct and indirect impact it will have on animal agriculture is less well known. The deterioration of some rangeland and the loss of available grazing on public lands will have a direct impact on animal agriculture. In case of range deterioration, researchable questions arise concerning the cost to all of society of allowing the process to continue and the cost to the society of restoring such land to its original productivity.

Research is being conducted by agronomists, engineers and economists on the potential of pollution from land in agricultural crops. States, in large part with federal dollars, are responsible for developing programs to reduce water sedimentation to tolerable levels. Public policies (e.g., subsidies, mandatory limits on use) which influence adoption of erosion reducing tillage practices, mechanical procedures (terraces) and cropping practices will have a substantial impact on animal agriculture because of impacts on the supply and form of feedstuffs. It is essential that the biological, economic and social consequences of water quality policy be carried beyond cropping patterns and through the impacts on animal agriculture.

Water. Water is being recognized increasingly as a scarce natural resource not only in the semi-arid range country, but also throughout most of the country. In some regions the major concern is adequate water for competing uses, while in other areas the question is water quality, both underground and in rivers. Water rights are a major issue in many states. Research information on the consequences of alternative water supplies on the animal agriculture and through it to local communities and the society at large are needed.

Feed additives, pesticides. Science has developed chemicals which increase agricultural output for any given set of inputs. Examples include antibiotics, growth and metabolic stimulants, and DDT. Sometimes, additional research or perhaps public opinion has raised serious questions concerning the impact (e.g., health) on society).

In most cases, it is impossible to design research which "proves" that a substance when used in animal agriculture has "no risk" to safety. But, it is possible to determine many of the known costs and benefits to society and to estimate the risks involved. It will then be the responsibility of policy-makers to determine the trade-off between the costs and benefits to society and the risk.

STRUCTURE OF THE ANIMAL SECTOR

Structure refers to size and number of units as well as ownership concentration at each stage of a commodity production and marketing system. Structure includes the pattern of integration or coordination of stages or functions within the system. For example, the number of beef packing plants and the proportion of packing capacity under the control of the largest firms affects the technical and economic performance of the entire beef commodity system. Likewise, the mechanisms for coordination of successive stages from input through production, processing and distribution affects the efficiency of the system as well as distribution of economic rewards. Concentration of economic power at the processing stage, vertical integration and contracting have been focal points of controversy for many years.

The primary goal conflicts involve efficiency, equity and freedom of choice. Clearly, the cost advantage of large size has been a major factor in increasing the size of units at nearly all levels of the system. But, fewer market outlets or supply sources provide the opportunity for perceived inequitable treatment of producers. Vertical integration and contract coordination have reduced the costs associated with buying and selling products, and increased efficiency in the use of facilities. But these nonprice coordination systems invariably limit scope for decision and action at one or more stages, as well as limiting access to markets by firms not involved in the vertically coordinated system.

Concentration at the processing-distribution stages facilitates the differentiation of products through advertising and physical product variations. Such moves benefit consumers to the extent that information and variety are increased, but may represent waste if variation is illusory and claims unfounded or irrelevant.

Efficiency and competition are inherently in conflict when one considers the geographic dimensions of agricultural markets. The least cost organization of assembly and processing of animal products typically means a single plant serves a given area, but such an arrangement cannot be described as competitive.

An important consequence of vertical integration, contract coordination and product differentiation is the disappearance of open markets and observable market price information. This situation affects the amount of data which is available to decision makers and ultimately, the social efficiency of the decisions they make.

Changes in technology often have an important effect on the structure of the subsector involved. Advances in nutrition and in the ability to control diseases and parasites made large, mechanized production units feasible. This led to both geographic and structural concentration in the poultry subsectors. Similarly, the emergence of boxed beef is coincident with significant concentration in the beef packing industry. In both cases, increased technical efficiency may have resulted in reduced competition.

Cooperatives have been one means which producers have used to counter the power of the concentrated input and processing industries. Cooperative processing has also provided the means to gain the efficiency of a limited number of plants while avoiding the bargaining disadvantage of separate ownership. The dairy sector has made the most use of cooperatives. Concentration of power in the hands of farmer cooperatives has attracted the attention of consumers and competing firms.

External forces also cause changes in the structure of animal agriculture food system. Energy credit, food safety, labor, environmental and other regulations have induced, and will continue to induce, changes in structure and regional distribution of animal production and marketing. Many of these impacts are unintended.

In animal agriculture, different structures are associated with different animal products. The structure of the broiler subsector differs markedly from that of dairy and the latter is quite different from the beef or swine subsectors.

Some years ago there was great public concern about concentration and vertical integration within the poultry subsector. Currently, concern is focused on concentration in the beef packing industry with some evidence that the bulk of this industry will soon be in the hands of a few firms.

In order for policy makers to assess how animal agriculture is meeting human needs, it is necessary for them to have information on the structure and performance of the food system relating to animal agriculture. Information is needed on a number of questions. To what extent are needs or wants in conflict? What factors are causing changes in the structure of animal agriculture? What are the likely emerging structural trends? What would be the impacts on society of changes in structure induced by various policies -- both those designed to affect structure directly and those intended to affect other dimensions of performance?

State of the Arts and Science

The current structure of production and its evolution since 1940 is well documented in two recent compilations, one by the Department of Agriculture (Schertz et al.) and the other for the U.S. Senate Committee on Agriculture. The features of structural change common to all of animal agriculture are increased size and decreased number of production units. The following data indicate the magnitude of the changes.

Table 11.1. NUMBER OF PRODUCTION UNITS

Item	Year 1940	1974
Dairy	4,663,000	404,000
Hogs	3,768,000	470,000
Eggs	5,125,000	304,000

The changes in number and size have been accompanied by a tendency toward both functional and geographic specialization. For example, a broiler chicken subsector has emerged as a recognizable separate enterprise on farms since 1940.

The size of production units in each species varies greatly. One finds a large number of small units in each which contribute a relatively small proportion of the total commercial supply.

A substantial research effort is now devoted to the study of the relationship of costs to size of production units, and the economic viability of "small" or "family" farms. These research projects (at both the State Experiment Stations and Federal level) now in process are expected to provide the information needed for policy decisions related to farm size limitations. It is also expected to provide evidence regarding unintended impacts of other policies on production unit size.

Much of the recent research on the organization and structure of complete commodity systems and the impact of structure on the performance of the food sector has been accomplished as contributions to North Central Regional Research Project 117, Studies in the Organization and Control of U.S. Food System. This research has provided a much needed description and analysis of coordination between levels in commodity systems as well as building the foundation for analysis of the impacts of concentration on pricing, technical efficiency, responsiveness to consumer needs and other aspects of performance of marketing firms.

Much remains to be done to estimate the relationships between the structure of the food sector and its performance with respect to the desires of consumers and the welfare of producers. Industrial organization theory provides hypotheses for the study of the relationship of performance to structure at a single level of the system. But there is no widely accepted theory which explains the pattern of combinations of successive functions (e.g., growing and processing) in a single firm or the impact of such combinations on subsector performance.

Research in the area is impeded by the fear of adverse impact on the firms which are asked to provide the data. A firm that believes there is a chance that its operation will appear in an unfavorable light or that the research will trigger regulatory action is understandably reluctant to provide information. The more concentrated the industry, the more likely one is to encounter this reaction. However, the more concentrated the industry, the more likely there is an urgent public policy need for such information.

RESEARCH PRIORITIES

The following five research program areas encompass the major gaps in knowledge regarding policies affecting animal agriculture which were identified in the preceeding sections. Each is judged as researchable and each has national policy importance.

1. Improving the policy decision process.
   Policy formulation and regulatory processes are inhibited by lack of information, distortion of information and lack of appropriate analysis. Problems also arise from the fragmentation of regulatory authority. Allocation of research funds suffers the same difficulties. Research should be addressed to:
   a. Evaluate the information systems whereby the needs and desires of the body politic which are relevant to animal agriculture are reflected to public decision makers.
   b. Evaluate the process whereby programs and regulatory decisions relative to animal agriculture are made and implemented.

2. Improving the Effectiveness of Government Regulations and Programs.
   An unknown number of programs and regulations impinge upon animal agriculture at various levels. Systematic review of existing and proposed programs and regulation is needed to:
   a. Evaluate government programs (e.g., market orders, conservation, etc.) and regulations (e.g., food safety) which affect animal agriculture to determine whether they are meeting their objectives in terms of production, consumption, environmental and other societal goals or creating unintended consequences.

3. Assessment of the impact of international trade.
   Trade barriers (quotas, tariffs, etc.) to imports into the U.S. and other countries and to exports from the U.S. and other countries of livestock products have important impacts on the U.S. animal agricultural sector and upon consumers. The mix of products in trade is a factor in the resource balance of the U.S. The social and economic impacts of exporting nutrients in animal-product form needs examination. The following research evaluations are urgently needed to:

a. Evaluate the impact on U.S. producers and consumers of animal products of barriers to trade in livestock and livestock products.
b. Evaluate the resource content of trade in livestock and livestock products in terms of capital, labor and natural resources including land.

4. Impact of input resource limitation.
Competition for resources for nonagricultural uses and resource exhaustion will result in major changes in animal agriculture. Land use restriction, energy policy, water allocation and environmental issues are some of the factors which will affect the sector directly and indirectly through other purchased inputs. Research is needed to:
a. Evaluate the supplies of resources pertinent to animal agriculture and the consequences of imbalances and/or policy induced resource limitations on the capability of animal agriculture to meet human needs.

5. Structure of animal agriculture.
The structure of the animal agriculture sector has shifted rapidly during the past 30 years. Some of these changes have a major impact on the competitive conditions in the sector, as well as on its technical efficiency. Optimum policy formulation depends on being able to:
a. Evaluate the impact of the structure of the food-animal subsectors on their resource efficiency, distribution and equity of income, stability, environmental impacts and other social goals.
b. Evaluate the impact of public policies on the structure of the food-animal subsectors.

REFERENCES

Anderson, J. 1975. Public Policy Making. 2nd Ed. Holt, Rinehardt and Winston, New York.

Baumol, W.J. and W.E. Oates. 1979. Economics, Environmental Policy, and the Quality of Life. Prentice Hall, Englewood Cliffs, NJ.

Downs, G.W. Jr. 1976. Bureaucracy Innovations, Public Policy. Heath Publishers, Lexington, ME.

Haverman, R. and J. Margolis. 1977. Public Expenditure and Policy Analysis. Rand McNally, Chicago.

Jones, C.O. 1977. An Introduction to the Study of Public Policy. 2nd Ed. Duxbury Press, North Scituate, MA.

Kneese, A.V. and C.L. Schultz. 1975. Pollution, Prices and Public Policy. Brookings Institution, Washington, DC.

Landsberg, H.H. 1979. Energy: The Next Twenty Years. Resources for the Future, Cambridge, MA.

Lindblom, C. 1968. The Policy Making Process. Prentice Hall, Englewood Cliff, NJ.

McCraw, T.K. 1975. Regulation in America. Business History Review. 49:159.
Noll, R. 1971. Reforming Regulations. Brookings Institution, Washington, DC.
Okrent, D. 1980. Comment on societal risk. Science No. 4442. 208:372.
Posner, R. 1974. Theories of economic regulation. Bell J. Econ. Manage. Sci. 5(No.2):335.
Schertz, L.P., et al. 1979. Another revolution in farming. USDA.
Strobaugh, R., et al. 1979. Energy Future. Random House, New York.
U.S. Senate Committee on Agriculture, Nutrition and Forestry. 1980. A Historical Perspective on Changes in the Number and Size of Farms. Washington, DC.
Wildavsky, A. 1979. The Politics of the Budgetary Process. 3rd Ed. Little Brown, Boston.
Winrock International. 1980. The world livestock, feed and food grain system -- an economic assessment. Morrilton, AR.

"The U.S. is no longer the world's leader in agricultural research. The free world has seriously fallen behind the Soviets, not only in the arms race, but they have now edged us out in research investments relating to the production and stability of our food supplies."

Sylvan Wittwer

"We know that expenditures for agricultural research are good investments...from 29-90% annual returns."

Anson Bertrand

12

KEYNOTE ADDRESSES

PHILIP H. ABELSON

Introduction
John E. Cantlon

It is a pleasure to be asked to perform this pleasant task of introducing your prestigious keynote speaker. Choosing someone to lay the point of departure on a subject that asks what Animal Agriculture - broadly defined - has to offer toward meeting human needs in the 21st Century is indeed a challenge. Man and his immediate ancestors have been feeding on animals for several million years. He has had domesticated animals such as the dog for almost as long as he has had fire. He began to domesticate sheep about the same time he began to domesticate plants, and great strides have occurred in building an Animal Agriculture. The affluent nations of the world as well as the developing nations will face great stress on their food systems over the next century. The complexity of your agenda conveys one measure of the ways we must prepare for this coming strain.

Your planning committee has chosen a Conference keynote speaker who has the breadth to put these complex matters into perspective. While his own area of training was chemistry and nuclear physics, each year the contents of weekly Science, which he has edited since 1962, cover the full sweep of modern science.

It came to me as I paused at the top of the ski hill west of us last night after dinner that more of us have read Phil Abelson's writings than we have read those of any other of the conferees. There are probably few here who have not read one or more of our speaker's editorials or comments in the weekly, Science. Those of you who have read them know he has treated tough issues such as the safety of nuclear energy, contamination with toxic wastes and stresses on the chemical industries. Recently, he gave scientific laboratory and university administrators a going over for biting too deeply into scarce research resources and thereby contributing to the difficulty young scientists are encountering in getting their research careers underway. From such a past track record, we can expect a thoughtful and thought-provoking keynote paper.

He is a westerner by birth, Tacoma, Washington, 1913, to be specific. He did his undergraduate work in physical chemistry at Washington State University and did his Ph.D. in nuclear physics at the University of California, Berkeley, graduating in 1939.

He joined the Carnegie Institution in 1939. He spent the WW II years 1941-1945 at the Naval Research Laboratory as a physicist, and returned to the Carnegie Institution in 1946 where he lead the biophysics section, the Department of Terrestrial Magnetism and the Geophysics Laboratory, and was the Institutions's president from 1971-1978.

He has been the editor of Science since 1962. This weekly journal is the most widely read, most respected and most influential scientific journal in the world. The Russians give it the signal honor of promptly translating every edition into Russian. As one might expect, their censors have a ball trying to delete such material as letters and reports about Sackharov's exile to Gorky, and even getting rid of pictures of Chinese railroad trains from the front cover of the magazine.

Dr. Abelson has served on or headed up numerous advisory groups for the U.S. National Academy of Science, the Atomic Energy Commission and the National Institutes of Health, to mention a few.

He is a member of the U.S. National Academy of Science and has won numerous awards, citations and honorary degrees from a broad range of national and international organizations and institutions.

It gives me great pleasure to introduce Dr. Philip Abelson who will give the first keynote address for the Conference.

"The systems approach permitted representation of a variety of disciplines to abandon their traditional competitive stance and work in earnest cooperation on strategies... for meeting individual human needs in the 21st century."

Norje Jerome

Address

In choosing the theme of this conference, "Animal Agriculture and Human Needs in the 21st Century", the organizers have identified an important and inspiring challenge. It is desirable at any time that those engaged in agricultural research should contemplate the topic with future-oriented vision. The need to attempt to gauge the future is even more pressing now.

We and the rest of the world have reached the end of a pleasant era of increasing standards of living and of cheap energy in the form of petroleum. We now face a period of declining standards of living and the jarring and costly transition to other energy forms. We face major economic and social adjustments that will affect greatly both those who produce and those who consume food. The economic adjustments will affect the purchasing power of consumers and, in turn, their preferences for food and especially for meat.

This morning I will devote the first half of my talk to economic and societal matters, particularly those that will tend to shape the future of agricultural practice. In the second half, I will set forth what I visualize as opportunities for improvement of the efficiency of meat production, particularly with respect to ruminants.

ECONOMIC AND SOCIETAL FACTORS

Changes in standard of living have important consequences for agriculture. During the 1950s and 1960s the standard moved up steadily. This affected consumers' choices at the supermarkets. They bought more beef and more quick food items. But, during the past several years, consumption of beef has dropped more than 10%, and other shifts in preference will surely occur. Costs for housing, heating and transportation, to name only three factors, have risen far more rapidly than have incomes. For many people these items have a higher priority than quality food, and thus food purchasing practices may be in for greater changes than the economy as a whole. That is not to say that people will stop eating, but their menus will change. In fact, they have already changed somewhat. Less beef in the form of steaks is being bought. A larger fraction of what is sold is in the form of hamburger. Purchases of chicken and pork have also increased.

Why have the standards of living decreased and what are the prospects that the trend will be reversed soon? The decline in standards is the result of the additive effects of many factors, including some that have been gathering force for as long as a decade or more. At least six factors have contributed:

1) increasing prices of liquid fuels and natural gas,
2) expanding costs of big government,
3) inefficiencies arising from thousands of government regulations
4) federally-mandated paperwork,
5) environmentalism, and
6) escalating costs of health care.

At this time the major factor is the cost of liquid fuels and natural gas. Between the two of them, they will take at least $300 billion out of the economy this year and redistribute it, with about $90 billion going abroad. The amount and impact of such withdrawals will increase for another decade. The federal government has done very little to insure future supplies of liquid fuels. Within the next 5 years, it is likely that a price of $4 to $5 a gallon for gasoline will be reached. At present we are at the not so tender mercy of OPEC. However, costs will not escalate indefinitely. In the United States, price is already affecting consumption of oil, which during the first 4 months of 1980 was about 10% below that in the same period in 1979. Another brake on runaway prices is competition from other energy sources. At present, the cost of heat energy from coal is about a factor of five less than that of heat energy from oil. Later, other sources such as bio-energy will also substitute for products derived from oil.

Another factor leading to a lower standard of living is the expanding federal and state budgets, which now consume 30% of the gross national product. The income tax money harvested every year is not available when individuals go to the supermarket.

When steel companies must comply with 5,700 different federal regulations to make a ton of steel, the price of steel must increase to cover all the inefficiency that ensues. In turn, products made from steel, including farm machinery, have a higher price tag.

One of the most iniquitous reasons for a lower standard of living is the $100 billion a year cost that corporations incur in dealing with federally-mandated paperwork. Most of the papers are never looked at. The cost is passed on to the consumer.

Environmental clean-up is one of the more defensible government programs. Initially, it was generally assumed that costs involved would be borne by industry. But these have been passed on to consumers. While real benefits arise from cleaner air and water, these benefits are hard to find in the pocketbook. Instead, clean-up costs of hundreds of billions of dollars to be spent during the decade 1975 to 1985 are coming out of the buying power of consumers.

Health care costs have now soared to about $240 billion a year with little apparent benefit to the health of the average person. A substantial fraction of the total is going to an ill-advised Herculean effort to keep alive old crocks who should be allowed to die in peace.

Since we live in one world, our lives and our economy are necessarily affected by developments in other countries. Most nations also face a troublesome era of adjustment to high prices and scarcity of oil. A few, such as Japan and West Germany, have been able to cope very well. In fact, the Japanese have been able to increase their consumption of beef. However, most countries and especially those of the third world have been heavily impacted. Some may be forced into a form of international bankruptcy. Continuing exponential increases in population, if sustained, must eventually lead to famine. But one should not underestimate the ability of humans to adjust and cope with problems. Hence, I do not place much credence in predictions of general catastrophe arising out of shortages of food. Localized misery is probable - even certain. An increase in demand for U.S. food exports is highly likely.

The higher costs for liquid fuels and nitrogenous fertilizers will surely affect the practice of agriculture. Already in the southeastern United States no-till farming is spreading rapidly. This, of course, saves tractor fuel. As costs rise further, the practice will surely spread to other regions. A combination of concern about soil loss and the cost of fertilizer is likely to lead to crop rotation and expanded culture of legumes. The high cost of transportation is already affecting those who operate feedlots, and decrease in trucking of cattle seems certain. Another effect of high prices for liquid fuels will be to increase even more the attractiveness of alcohol as a fuel. When farmers start to produce alcohol from grain, they will also produce stillage, which must be used quickly as feed for animals.

One of the effects of the high costs of petroleum and its products will be to cause various countries and regions to seek and exploit energy sources within their borders. Another effect will be to give advantages to local small-scale technology where transportation costs can be minimized. These effects were evident at a World Congress on Bio-Energy held April 21-24 in Atlanta. Some of the topics treated and some of the equipment shown there have relevance to the future conduct of agriculture, so I will mention a few. First, the alcohol program in Brazil is going well. At present it is based on sugarcane, but use of cassava is also beginning. The Brazilians are equipping autos to burn an 80% alcohol/20% water mixture. This use of less than 100% alcohol saves much energy otherwise required for distillation. The Swedes, who have the greatest per capita consumption of oil, but have no indigenous coal, oil or natural gas, have become concerned about their great dependence on imports. They believe that the solution to their problem is fast-rotation tree culture involving cropping. They would use willow or birch, and harvest every 3-5 years. Their studies indicate that all of their needs could be met while still maintaining Sweden's present large-scale forest products industry.

Obtaining ethyl-alcohol from wood is one of the major goals of those interested in bio-energy. Progress is being achieved in making cellulose and hemicellulose available for fermentation to alcohol. This progress is relevant to ruminants, as will be discussed later. Another mode of obtaining alcohol is the conventional one of fermentation of grain. A piece of mobile equipment costing only $6,000 was displayed that produces distilled alcohol from grain at the rate of 4 gal an hour. With time, the makers of the equipment will doubtless improve it and lower its selling costs. This would make feasible widespread use of such equipment.

Another piece of mobile equipment was designed to obtain energy from crop residues. Instead of taking the residues to the factory, the processing equipment would be taken to the residues. In this case, the residues are pyrolized to off-gases, fuel oil and char. The off-gases are used to power the equipment. The fuel oil and char are useful products. A particularly impressive piece of mobile equipment was a tree-chopper that can handle trees up to 51 cm (20 in) in diameter. Several trees about 17.8 cm (7 in) in diameter were each processed in about 30 seconds. The product was a chip about the size of a silver dollar. Another small piece of equipment could process the chips into a fine flour suitable for further biological or chemical conversion.

I have already mentioned some of the factors likely to affect the price of grain. First, the costs of fuel and nitrogenous fertilizer will increase further. Second, export demand will increase. And third, grain will be diverted to produce alcohol. Increased prices for grain will particularly affect meat producers, who will be forced to increase their prices, thus meeting more consumer resistance. The obvious general solutions are to increase the efficiency of animals and to expand the numbers of ruminants growing on forage.

OPPORTUNITIES FOR IMPROVING MEAT PRODUCTION EFFICIENCY

There is abundant reason for confidence that further efforts in selection of superior animals will lead to gains in feed efficiency. During the past 30 years efficiencies have doubled and progress during the past decade has, if anything, accelerated. This is true of poultry, swine, beef cattle and dairy cattle. Professor Hafs recently provided figures on the efficiency of conversion of feed to human food. The dairy cow producing milk and the hen laying eggs were tied for top with 25%. Broilers had an efficiency of 23% and pigs 14%, while fed cattle trailed with less than 10%. Elsewhere, figures were given for milk production by a champion cow. In less than a year she produced 25,000 kg (35,000 lb) of milk, which in turn contained more than 2,700 kg (6,000 lb) of solids. This truly amazing performance emphasizes again the contrast between the efficiency of dairy cattle producing food and that of beef cattle converting feed to additional flesh. There is a difference in efficiency of a factor of 2.5 to 3. This large factor should serve as a challenge and a goad to research leading to substantial improvements in the efficiency of beef cattle.

The largest single factor in the improvement of performance of dairy cattle has been the use of artificial insemination. About 60% of dairy cows are made pregnant in this way, using semen from superior sires. In practice, it is fairly simple for the dairy farmer to note estrus in the cows, for the behavior of the animals can be observed during milking. However, no such convenient inspection is available with beef cattle. Because of the practical problems of detecting estrus in such animals only 3-5% of beef cattle are artificially inseminated. Thus, herds of superior beef cattle have not been attained at the rate they might have if artificial insemination were widely used.

Recent developments of drugs that control the timing of estrus in cows open up the possibility of faster rates of improvement. Through use of prostaglandin or progesterone, estrus in a group of animals can be made to occur so close to a specified time that detection of estrus is not essential. Artificial insemination is relatively successful at a fixed time after the prostaglandin treatment.

The prostaglandins have made embryo transfer in cows more feasible. A superior cow can be induced to superovulate, releasing about 10 ova. After fertilization by artificial insemination the embryos can be flushed out and, subsequently, transferred to recipient cows. For the transfer to be effective, the donor cow and the recipient cows must be at the same stage of the estrous cycle. This can be brought about with prostaglandin. At present, the cost of obtaining a calf and bringing it to the age of 6 months is $2,000. But George E. Seidel, Jr., who is knowledgeable in such matters, believes that costs can be greatly reduced. Were this to occur, the rate of improvement of both dairy and beef cattle could be substantially increased.

A number of different developments indicate that performance of animals can be improved generally by a variety of techniques. The gains, though not huge, are substantial and worthwhile. One technique that is widely applicable is control of the daily period of light. A combination of 16 hr light and 8 hr darkness appears to be optimum. It leads to earlier maturity of animals and to an improvement of feed efficiency of 5-10%.

The use of growth stimulants to obtain greater efficiency has long been known. Diethylstilbestrol was the chemical first employed. This stimulant has now been effectively outlawed, but new substitutes are available that serve the function while not being objectionable. Two that have been used are Ralgrow and Sinovex. A new product that probably will be approved soon seems particularly desirable . It is based on estradiol, a hormone that occurs naturally in cattle. The dose is only 20-40 ug per day and is furnished by diffusion from an implant in the ear. The implant can be effective for nearly a year. Once the implant is removed, the level of estradiol in the animal drops to normal in only a day.

At one time in my career I studied synthetic pathways in microorganisms and was fascinated with their great competence as chemical engineers. Thus, I am particularly interested in the microbial flora of the rumen, for it is these organisms that help to give ruminants their unique capabilities for converting cellulosic roughage into products that can be assimilated and used by the animals. For the organisms to grow they must be furnished fixed nitrogen and phosphate as well as potassium, sulfate and trace elements. When grass fed, the animals obtain the necessary fixed nitrogen and other constituents from the forage. However, Virtanen (1966) showed that good growth could be obtained from cellulose alone as a carbon source provided a nitrogen source (urea) and the other essential elements were added to the diet. Urea is formed in many animals as a product of detoxification of ammonia that is released by metabolic processes. However, microorganisms convert urea to ammonia plus carbon dioxide. Thus, urea is probably decomposed in the rumen and the resultant ammonia must be to some degree toxic.

The microbial flora of the rumen are very complex. More than 100 different microorganisms have been identified in the contents of the rumen, as well as many species of protozoans. These are to some degree interdependent and to some degree competitive. Hungate has stated that the principal products of rumen activity on 58 moles of hexose are 60 moles of acetic acid, 22 moles of propionic acid, 18 moles of butyric acid and some methane. Some of the feed is, of course, converted into microbial material. The volatile fatty acids - acetic, propionic and butyric - account for about 75% of the carbon in the input hexose. About 15% of the feed energy is converted to methane. This represents a loss each year of about $1 billion worth of feed energy. A recent grant proposal called for the installation on ruminants of experimental equipment designed to catch the methane. The matter was aired on a radio show, and listeners were asked to vote on whether the grant should be funded. The proposal did not get majority approval, but it came close. Some 48% voted to fund it.

Experts in the field of the study of the rumen agree that propionic acid is an especially useful product. In the animal it is glucogenic; that is, it gives rise to needed glucose. For some years scientists speculated about the desirability of controlling the flora of the rumen with the object of improving yields of propionic acid. A coccidiostat, monensin, produced by Streptomyces cinnamonesis affects the production of propionic acid. What is more important, its use increases the efficiency of feed utilization by about 10%. Production of methane is attenuated but not stopped. It appears that further improvements in rumen efficiency could be obtained. I have been told that numerous pharmaceutical and chemical companies are seeking other additives for improvement of rumen flora.

When feed such as grass or grain that contains protein is consumed by ruminants, the usual fate of the protein is destruction of at least some of the ingested amino acids in the rumen. The addition of fixed nitrogen in the form of urea is said to have a sparing action on the amino acids. But, as noted, above, urea is probably not the ideal nitrogen source. Thus, other possible sources should be evaluated. Asparagine and glutamine are likely candidates, for they are known to participate in transamination reactions. Another means of preserving protein is to treat feeds with formaldehyde. This inhibits breakage of the protein in the rumen without interfering greatly with digestion in the lower alimentary tract.

In recent years considerable effort has been devoted to improving methods of handling silage. If the materials are handled skillfully, substantial losses in the feed value can be avoided. Benefits of as much as 15% have been noted. In one method now being used, grass is mowed, cut into 1.7-2.5 cm (1/2- to 1-inch) lengths and carried directly to the silo, where it is blown in. The material is then treated with a solution of formic acid, which largely inhibits growth of molds. For corn silage, success has been reported with use of ammonia. Apparently, the ammonia is rather quickly fixed and converted to another chemical form.

One is led to speculate that control of the bacterial flora in silage might lead to an improved silage that has better nutritive values and is more appetizing to the cattle.

As food from conventional tillage agriculture becomes more expensive and scarce, the advantages of ruminants will loom more important. Currently the full potential of grazing land is not utilized. In the future more forage may become available as crop rotation is more broadly practiced. Another possibility with considerable potential is the use of processed wood as feed. From a number of sources I have heard that processed wood is already being fed successfully to cattle.

CONCLUSION

We have entered an era of great changes. We will use forms of energy that are less available, and more costly. The general standard of living will decline, and the part of income after taxes that is devoted to food may also decline. Fuel for agriculture will be more costly, and this will influence agricultural practice. Increasing population around the world will tend to increase exports of grain, as will diversion of some grain to production of alcohol. Meat production from grain will encounter high feed costs and consumer resistance to the necessarily higher prices for such meat. In contrast, milk and meat production from ruminants will increase the food supply and will not be so much affected by costs of grain. Improved breeding of cattle designed for use of forage, forage crops and wood should have a high priority, since this will increase the overall food supply while holding down or diminishing its cost.

CAROL TUCKER FOREMAN

Introduction
Gilbert A. Leveille

It is a pleasure to introduce the evening's keynote speech from Carol Tucker Foreman, Assistant Secretary, U.S. Department of Agriculture. Ms. Foreman brings to her post a vast experience in politics and in consumer advocacy. She has enthusiastically undertaken major challenges in attempting to improve the efficiency and effectiveness of the programs under her direction. Ms. Foreman has certainly changed the character of the position she fills. One cannot help but be impressed with Ms. Foreman's enthusiasm, energy, dedication and commitment.

We are priveleged to have this message from Ms. Foreman at this Conference. I am pleased to present to you Mr. Jay R. Reich, a White House Fellow from Ms. Foreman's office, who will deliver this keynote address for Ms. Foreman.

"The organizing framework of the Conference - a systems ecological approach - was the key factor in the Conference's success."

Norge Jerome

Address Delivered by Jay A. Reich

Secretary Bergland said last December that the productive resources of American Agriculture were in near balance with domestic and export demands. The Soviet grain embargo has intervened since that time, of course, but it is probably still fair to say that, barring such extraordinary events, the long-existing slack in American agriculture appears to be gone. We are at a point where policy decisions hinge on some harsh new realities. In addition to fuel shortages, tight money and a falling water table, there are disturbingly graphic illustrations of our relatively inelastic land base.

In the lifetime of the present Secretary of Agriculture, this country has paved over the equivalent of all the cropland in Ohio. Before this century is out, the Secretary estimates that we will also pave over an area the size of Indiana.

"Whither we are tending" is harder to determine than where we are. It is a promising sign that many different disciplines are presented here so that scientists from diverse fields can discuss the interdependence of their actions and ideas.

Basic questions about getting enough to eat and distributing it to the people who need it remain problems in some parts of the underdeveloped world. American agriculture productivity and food assistance programs have made significant progress in solving these problems here at home. But there are other questions we must ask ourselves as we grope toward an understanding of human needs in the 21st century.

What about the unintended consequences of food availability - health problems which may be related to overconsumption or inappropriate food choices?

What are the implications for food production of the reduction in American birth rates, resulting in a population whose average age will be more than 40 in the not-too-distant future, with a substantial percentage over the age of 65?

If energy problems continue to escalate, how successfully will the food production sector compete for energy with industrial and transportation uses. The crudest example of food vs energy tradeoff might be posed by the use of foodstuff for gasahol.

Is the series of agricultural and processing revolutions, which have helped bring this nation unprecedented prosperity, sustainable in a shrinking world of scarcities?

To focus on animal agriculture specifically, what constraints - or opportunities - are of priority consideration? It seems to me there are at least four that I would like to address -one on the supply side and three on the demand side. On the supply side, it is true now and should be in the future, that realistic research policies should be directed toward increasing the opportunities for livestock industries to make a reasonable return on investment. This is a business and unless we understand this, nobody wins.

Increased productivity, efficiency of production, processing and marketing are fundamental to that end. The staggering costs of farm inputs and the inflation of land prices make increased production essential. But, also essential is producing products that consumers want. And that clearly implies the need for more market-oriented research by the production factors. That view is not limited to fuzzy-headed consumer advocates and zealous Naderites.

Last December, at a regional hearing on the structure of American agriculture, the president of the Iowa Farm Bureau Federation told Secretary Bergland that "the best hope for a commercial, family-farm type agriculture lies with a market-oriented agriculture."

He said that government's role should be "to help by providing research . . . and education."

To see how well government is doing that job, we need to look at the three demand side constraints I mentioned earlier. They are: food safety, human nutrition and availability of products at reasonable price.

Let's look first at the overall picture. Anyone who takes more than a cursory look at our more successful agricultural production policies will see that they usually have been based on sound economic and policy research -- both federal and private.

For example, USDA's Science and Education Administration spends $375 million a year on in-house agricultural research. That expenditure has resulted in: 1) dairy cows that produce twice as much milk as they did 30 years ago; 2) svelte swine which today have only 1/2 as much fat as they did 20 years ago; 3) broiler chickens which mature in 6 weeks instead of 3 months; and 4) corn yields which have tripled in the last 40 years.

Federal agricultural research is a good buy for the taxpayers. For every $1 spent on agricultural research, the consumer gets 50% return, compared to a 20-30% return from other types of research. But if we consider food safety specifically, the picture is not all that rosy. For years, we have successfully used chemicals at various points in the food system to dramatically increase production, retard spoilage and preserve foods. The chemicals seemed a stunningly successful solution to those problems. But now, new problems are rising, phoenix-like, from the ashes of past successes. We are learning that some of the same chemicals, even those which had such beneficial effects on plants and animals, treat the human body rather less kindly.

In the past year, we have had DES implanted in hundreds of livestock feedlots in clear violation of the law. We had an accidental PCB spill in which contamination from a single transformer at one slaughter and feed production firm quickly spread to 19 states and two foreign countries. It ultimately involved the destruction of several million pounds of chicken, hogs, eggs and turkeys, and had a financial impact estimated quite conservatively at $10 million.

Some chemicals and pesticides are practically indestructible in the environment. In the Fall of 1978, USDA found a flock of turkeys with violative levels of DDT. DDT had been banned in December of 1972 -- 6 years previously. But those turkeys were raised on land which had once been an apple orchard. And the trees had been sprayed with DDT in the spring of 1972.

These and similar problems are things which you must be concerned about in long-term planning. For example, we still individually inspect, each year, some 4 billion chickens, ducks and turkeys, for diseases which the modern poultry industry has all but eliminated. On the other hand, the Food and Drug Administration has licensed well over 100 chemicals for use in agricultural production. Yet less than 50 have tests for the detection of residues.

Improved poultry inspection procedures and swab tests are significant breakthroughs, and the Food Safety and Quality Service is working on the development of other rapid screening tests. But we need to be concerned more about the problems associated with sophisticated technology, mass production and the interdependence of the food chain. Conferences like this one can help do that job. And the American people will be grateful. Last year's American family report, a survey by Yankelovich, Skelly and White, showed that 65% of those surveyed wanted the government to do more to protect consumers from unsafe products.

You can help too, in the increasing concern about human nutrition. There is significant evidence that the American people are concerned about the food they eat and its effect on the public health.

The Yankelovich study revealed that American family members are concerned now about many food products such as cholesterol, fats and food additives, which did not worry them in the past. The majority of the families surveyed believed overweight to be "a serious health hazard." They were far more concerned with, and conscious of, the need for good nutrition than they had been in the past. There are arguments to the contrary. But a consensus seems to have been reached. In 1977, the dietary goals of the U.S. Senate Select Committee on Nutrition were greeted with screams of outrage and the immediate gathering of discrediting forces.

The publication of "healthy people" by the Surgeon General of the U.S. was met with, if not silence, at least with an absence of shrieks and groans. The American Society for Clinical Nutrition has found four correlations between diet and health to have considerable strength.

So it is really not a responsible public posture to insist that consumer movements are always and everywhere unrepresentative of public concerns. They are, I suspect, as representative, on the whole, of consumers, as this conference is of people with an interest in animal agriculture.

The consumer movement, the "natural" food craze, the concern about environmental contaminants - all of these are comparatively recent phenomena. But to assume that they will be less pervasive in the 21st century is to play futuristic ostrich.

Government - notorious for slow action - has reacted to these concerns in a variety of ways. USDA and the Department of Health and Human Resources, the two federal departments most concerned with food and health, in February published dietary guidelines for Americans. USDA has updated its standards for the purchase of commodities for federal feeding programs. It has increased its nutritional research.

Sometimes we are successful, sometimes we are not. A bad example of where we have been is the tomato developed for mechanical harvesting, with a scientific eye on consumer acceptability and ease of market handling. It was a marvel, all right. Its hard exterior and flat bottom was a boon to mechanical harvesting. No one noticed - except after the fact - that the content of vitamins A and C had been significantly diminished.

A good example is the development of the svelte swine, which I mentioned earlier. The point is, of course, that producers and industry can grasp - as they have in the past - new nutrition research findings, and make them into effective marketing strategies. Hormel and Oscar Mayer, for example, have already test-marketed low-fat hotdogs. Or, industry can dig in their heels, stand pat and help broaden a growing breach between an increasingly defensive industry and an ever-more strident consumer movement. The Federal Government certainly doesn't want that. And I don't believe you do, either.

The third demand-side constraint beyond the need for safety and good nutritional products is the food cost to consumers. With today's inflationary pressures, it would probably come as a surprise to many consumers that, by general income standards, Americans still spend less of their disposable income on food than anyone else in the world. But if input fuel and fertilizer costs continue to rise, that percentage may not be sustainable. Is it probable that better-educated consumers and more productive producers will together construct even more than the 20,000 food choices available in many supermarkets today? Is that desirable in a 21st century when a much larger percentage of the population will be retired on fixed and limited income? We do not have those answers. You can help us find them.

Finally, one cannot, or should not, consider cost to consumers and profit constraints on producers and processers without a word about government regulation.

The brouhaha about the impact of health and safety regulation on the food industry is, to some degree, underfocused and overemphasized. A recent study prepared for the Congress by the Center for Policy Alternatives of the Massachusetts Institute of Technology said that the American public saves billions of dollars a year as a result of federal regulation of health, safety and the environment. The study applauded a variety of FSQS actions which, it said, "typifies what is good in the regulatory area." Consumer and producer confidence in the general food supply "would be unlikely to prevail in the absence of vigorous regulation of the products and practices of the industries involved," the report concluded.

That is not to say that federal regulation does not exact a cost. Of course, it does. But it also pays a dividend in consumer confidence and protection for the producer, not to mention direct health and safety benefits.

Policy-makers at USDA are hopeful that, in the next several years, federal initiatives, such as reformed poultry inspection procedures, quality control in meat processing plants, swab tests and their progeny and flock testing, will increase the efficiency of industry productivity at the same time it increases our ability to assure product safety.

The consumer viewpoint is a necessary note in that desirable chord. To be economically effective, a product must be bought. To be nutritionally effective, it must be eaten. Obviously, producers and consumers need each other. They are natural allies, not born antagonists.

In the 21st century, the dialogue between government, producers and consumers will surely have lost some of its strident, "so's your old man" quality. We are now in a preliminary stage, attempting to integrate new knowledge with an understanding of new dangers. It is not the first time we have been in that position. Seventy-four years ago, an idealistic and ideological young writer set out to describe the privations of slaughterhouse workers, and by doing so further the cause of socialism.

His descriptions of what went on in the sausage-making assembly line, and what went into the sausage itself, was the impetus for the first pure food and drug law.

No one was more surprised than Upton Sinclair at what his book, "The Jungle," wrought. He said later that "I aimed at the public's heart and, by accident, I hit it in the stomach." That was in 1906. In the year 2000, perhaps people will look back at this Conference and see it as having a similar seminal effect. Except that your aims - tripley and aptly targeted at the public's mind, stomach and pocketbook - will hopefully have scored a more decisive crystal ball bull's-eye.

I ask you to seize the moment, and to plan for the next century, in the spirit of the words of John Dewey, who wrote that "every great advance in science was issued from a new audacity of imagination."

ANSON R. BERTRAND

Introduction
James H. Anderson

The setting of this Conference and the signs of the times suggest that I comment briefly on the state/federal partnership that has been responsible for the development of the research and education system of this nation and made this Conference possible.

The congressional acts that established the land-grant system, the Department of Agriculture, the State Agricultural Experiment Stations, and the Cooperative Extension Service, have proven some of the wisest legislation of all time. These historic acts set in place the state/federal partnership that has been responsible for the development and dissemination of information upon which modern agriculture in this nation is built. It has resulted in an agricultural research and education system that is the envy of the world, serving as a model for the industrialized nations, the developing nations and also those of the third world. It is a system that transcends ideological boundaries and its success is recognized by friend and foe alike.

In the beginning of the state/federal partnership, expenditures of both state and federal funds were very meager. After the system gained momentum, new technologies developed faster than they could be applied to the farms and ranches of this nation. Consequently, a backlog of science and technology or a reservoir of unused technology was developed. After World War II, the sophistication of agriculture developed to the extent that we could apply this technology very rapidly. Although we continued to develop new technology, we also drew upon the reservoir of unused technology. The ingenuity of the American farmer, operating in a free enterprise system, was the key ingredient in the application of the new technology.

An assessment of agricultural research needs in the 1970s indicated that the reservoir of unused technology had been depleted. Subsequently, the application of technology to agriculture began to slow down and yields in many of our major crops plateaued. Further massaging of existing technology is not likely to produce large additional gains. We are dependent upon new technology which is yet to be developed.

The current research agenda consists of two major items: A mission-oriented thrust designed to serve the present agricultural industry, and a long-range plan designed to develop the technology base upon which agriculture in the decades ahead will depend. It is imperative that both receive adequate funding. However, this occurs for neither at the present time. We now expend about 65-70% of the resources in agricultural research in the mission-oriented phase, and about 30-35% in the long-range dimension. It is absolutely necessary that we increase the long-range expenditures because the problems facing us are extremely complex, they are time dependent, they must be resource sparing and their solution will demand some of the best trained minds of our time. On the other hand, we cannot jeopardize the mission-oriented research by diverting a greater portion of the present funds to the long-range dimension. As we look to the future this is our challenge. Our keynote speaker of the morning is the key leader in meeting that challenge.

Dr. Anson R. Bertrand was appointed Director of Science and Education for the U.S. Department of Agriculture on July 20, 1979. This position is equivalent to that of an Assistant Secretary and Dr. Bertrand is a general officer of the Department of Agriculture advising the Secretary of Agriculture, as well as members of the Department Program and Budget Review Board. As head of the Science and Education Administration, Dr. Bertrand directs the Department's major resources for agricultural research, human nutrition research, extension, higher education, and technical information systems, including the National Agricultural Library. In addition, he is the administrator in charge of a relatively new competitive grants research program as well as the long-established cooperative relationship between the Department and the Land-Grant universities; the "1890 Institutions" and others active in research, extension and teaching.

Dr. Bertrand has a long and distinguished career in research, teaching and university and federal administration. He served as Dean of the College of Agricultural Sciences of Texas Tech University, Lubbock, Texas, 1971-1978. He was Chairman of the Agronomy Division, College of Agriculture, University of Georgia-Athens, 1967-1971. Earlier, he was Director of the Southern Piedmont Soil and Water Conservation Center, USDA, Watsonville, Georgia, and then Chief of the Southern Branch of the Soil and Water Conservation Research Division of the Agricultural Research Service, USDA. He served for 12 years on the faculty of the Agronomy Department of Purdue University and was involved in teaching and research.

Dr. Bertrand received the B.S. degree in Agricultural Education from Texas A&M University, an M.S. degree in Agronomy from the University of Illinois, and a Ph.D. in Soil Physics from Purdue University. He is the author of numerous publications and a member of many professional and honorary societies. He is a past President of the Soil Science Society of America and has served on study panels and committees for the National Research Council, National Academy of Sciences. In 1976 through 1978, he was a member of the Board of International Food and Agricultural Development in the U.S. Department of State. He has been a consultant to the Food and Agriculture Organization and the Indian government, and has had field experience in Bangladesh, Niger, India and Mexico.

Ladies and gentlemen, it is my pleasure to present to you Dr. Anson Bertrand, Director, Science and Education Administration, U.S. Department of Agriculture, our keynote speaker.

Address

I should like to preface my remarks by calling your attention to the theme of this meeting on animal agriculture, as it is stated in the Conference brochure. The title itself is worth keeping in the forefront of all of your deliberations this week: "Human Needs in the 21st Century." These particular words were not chosen casually, you can be sure of that. The emphasis is on the word Human. Human Needs in the 21st Century. I invite you, also, to look at the disciplines represented by the participants. You will find that sociologists are represented along with animal scientists and animal disease specialists. These two facts, in my opinion, already put this group far along in the kind of planning that needs to be done. It will surely be within the framework of human needs that our livestock products will be fashioned over the 19 years that lie between us and the 21st century.

It is also certain that those human needs will be articulated by the broad spectrum of our society, by consumers as well as by scientists, by private foundations as well as by industry, by public interest advocates as well as by public policy-makers. So it is the identification of human needs, human concerns, if you will, and the consideration of how we can address them, that will produce for us a strong, supportable agenda of research priorities for animal agriculture.

Animal agriculture is obviously a critical part of our agricultural resource base. It plays a vital role in providing quality nutrients in proportions needed by humans. Currently, animal products provide U.S. consumers with one-third of their energy, two-thirds of their protein, two-thirds of their fat, four-fifths of their calcium, two-thirds of their phosphorus, 38% of their iron, 42% of their vitamin A, 37% of their thiamine, 79% of their riboflavin, 47% of their niacin, 60% of their vitamin B_6 and virtually all of their vitamin B_{12} (USDA, 1980a).

Equally important, the ruminant is the key to using vast amounts of land that would otherwise go to waste as a food-producing resource, land that is too steep, too arid or otherwise unsuitable for crop production. To the extent that animals are fed from land unsuitable for crop production, or as part of rotations required for soil conservation, or with wastes or by-products of agricultural or industrial production, they represent a net gain to the food supplies for people. Animal production must build on this advantage. It must strive to become increasingly complementary to plant food production. With the increasing emphasis that is being placed on feeding increasing populations, we must learn to combine animal production with plant production in the most efficient manner possible.

We know also that diets high in animal products do require more resources to produce, and we must produce them as efficiently as possible. For example, we must learn how to get the feed to our animals at less energy cost, how to get the animals to market at less energy cost, how to get the animal products to the consumer at less energy cost. More important, probably, is the real need to decrease the energy used by the animal to convert feed to meat, milk and eggs.

My specific assignment is to discuss identifiable constraints in a motivational manner. This is a fairly formidable task! The list of constraints that we face today is a long one indeed, and I do not intend to provide you with a long tabulation. I have elected instead to discuss three main topics. I propose to discuss these topics under the headings of fat, antibiotics and frozen food; land and water resources; and productivity. In selecting these topics I have omitted many interesting and essential areas, i.e., manpower and technology transfer. SEA exists to foster increased articulation in Research, Extension and Higher Education. It is essential that we not forget needs in all areas. I have selected these topics not simply to illustrate constraints but to set them within the framework of human concerns and research imperatives. I am presenting to you, if you will, a smorgasbord rather than a grocery list.

FAT, ANTIBIOTICS AND FROZEN FOOD

I have selected the fairly specific topics of fat, antibiotics and frozen food in order to make several points. First, fat symbolized the dietary transition taking place in this country. Any deliberations on future research in animal agriculture must take this into account. Consumers are much more concerned than they have ever been about the relationship between the food they eat and their long-term well-being. That interest will continue and increase. People are making it their business to be well-informed on diet/disease relationships, and on the content of the foods they buy and eat. They want at least to be informed of the choices they can make and the level of risk they take when they choose one food product over another. Fat in the human diet, particularly fat of animal origin, is the subject of considerable controversy about what is known and what is not.

The issuance of the joint Department of Agriculture-Department of Health, and Human Resources Dietary Guidelines, has resulted in an increasing and useful dialogue on the whole subject of nutrition. These guidelines were the result of almost 2 years of literature reviews and diet studies. However, reaching consensus among scientists in the two departments and other authoritative scientific groups was not easy, and acceptance of the guidelines has obviously not been unanimous.

One area of agreement, however, is that more research is needed on the diet/disease relationship for fat, saturated fat and cholesterol. In the meantime, it is important to those engaged in animal agriculture to recognize that the guidelines do not advise the public to eat less red

red meat. To the contrary, they recommend more lean red meat. They do not say eliminate saturated fat. They say avoid too much. They advise a balanced diet, moderation and variety. Obviously, animal products hold an important place in moderate, balanced, varied diets.

A second point to be made is that we know that research can accomodate food production to meet nutritional needs. For example, rather great modifications can be made in fat to protein ratios in meat animals while still retaining the taste and tenderness desired by consumers. The meat-type hog is a notable example of our success in managing fat. That research effort reduced lard content per hog carcass from 13.6 kg (30 lb) in 1960 to about 6 kg (13 lb) today, with a corresponding increase in lean meat. It is a dramatic success story in livestock research and development. Recent research in beef also has shown us that genetically fast-growing animals can reach desired carcass weights at youthful ages with a low fat content and desired palatability.

There is also much more that can be done by going down to the cellular and molecular level and building our understanding from there. We need to know the rates, amounts and efficiencies of lean and fat tissue deposition, in terms of muscle fiber number and size, and in terms of fat cell number and size. With this basic understanding we might be able to improve the rate of growth and efficiency of growth of our meat animals as well as optimize the lean to fat ratio. We have a long way to go in terms of gaining this basic understanding because at this time we cannot identify the precursor cell of the fat cell.

Thus far I have discussed fat largely in terms of human nutrition needs. Fat is also an economic problem. Use of animal fat is a difficult problem in much of the world. For example, recent efforts to negotiate a science and technology agreement with the Japanese ran into difficulty with regard to joint work on animal fat utilization. The abdominal fat in broiler chickens, for example, accounts today for 3.5% of the chicken's live weight. This is fat that is largely discarded. If poultry breeders could cut abdominal fat to 1% of the chicken's live weight, we could save an estimated $177 million a year. The money could be saved all along the line, by the processor, the retailer and the consumer. Money, by the way, is another human need.

If consumers are increasingly interested in the nutritional aspects of their food, they are doubly concerned about the safety and quality of their food supply. Research in this area often has an urgency attached to it, while at the same time it can be enormously frustrating because of the complexity of the problems and the many and often conflicting interests involved. The use of antibiotics in animal feeds is a case in point. The recent review of this issue in Science (Marshall, 1980) illustrates the fact that there really are no simple answers to complex problems. To illustrate how inter-related and inter-dependent are the interests of all who are concerned about animal agriculture, what one area of activity does can have enormous implications for all the rest of us. That is why it is so essential that conferences such as this one take place and why such importance will be attached to the outcome of your deliberations.

The list of researchable problems in food safety is a long one. Certainly we need fast, positive methodology to detect undesirable materials in animal products, whether by natural toxicants or by production practices or inadvertently introduced by packaging materials or other environmental factors. We need improved monitoring technology that will make it possible to identify the source of the problem and to track the success of measures taken to alleviate the problem. The success we are having with sulfonamides in pork is an example. We can profitably aim research at helping improve regulatory actions or make them unnecessary by eliminating the problems they are designed to control.

Innovation is needed not only in the regulatory area but also throughout our food system. Much of the current technology was developed in an era of cheap energy. For example, it has been 50 years since frozen foods were introduced to the American people in retail packages. Eighteen cuts of frozen meats were among the 26 items that went on display in 10 grocery stores in Springfield, Massachusetts in March of 1930 (The Washington Star, 1980). No truly new preservation method has been developed and commercialized since that time. We need new methods that will increase shelf-life and storage stability without requiring such large energy inputs as are required in refrigeration and freezing.

One key to new technology may be the control of microorganisms present on meat and in meat products through the use of methods that are known to hold promise but are not now in practice. For example, the surface pasteurization of meat is feasible by the use of ionizing radiation. Even though seemingly expensive, the scaling-up of this procedure might possibly prove to be cost-saving in the long run. Certainly if sterilization could be achieved through the use of this type of radiation or by other new technology, and the dependence on chilling or freezing reduced or eliminated, world trade of meat could be greatly increased. The problem presented to researchers is how to accomplish this without compromising all the attractive features of meat as a food - its aroma, flavor, texture and nutritional value.

LAND AND WATER RESOURCES

When we talk about land and water, we are talking about severe and growing problems. Our annual soil loss is now estimated as averaging 9 tons per acre from U.S. farms. And water may soon outrank land as a major constraint to U.S. food production. The finiteness of our land and water resources; the impact of overuse and erosion of our lands on the environment; the increasing pressures of people, not only as more mouths to feed but also in their shifting geographic distribution and in their growing demands for, and use of, land and water resources - all of these things and more will affect our agricultural production capacity.

We know we are not going to be able to always produce vast acreages of irrigated forages, as we have in the past. Irrigation itself consumes about 20% of the total energy used in agricultural production and irrigation costs are increasing dramatically. We must have more drought-resistant crop varieties; we must make better use of water of inferior quality; we must minimize evaporation and seepage losses and make more efficient use of water harvesting.

We must continue our work to solve pollution problems as they relate to land and water resources. The problems of run-off from animal-holding areas and of waste handling, for example. We are making good progress in these areas, I believe, but we cannot afford to let up on these efforts.

It is water - or the lack of it - that may spell the most dramatic changes for animal agriculture. Those of us who are from the Western and Southwestern parts of this country know first-hand that a crisis is already upon us. The combination of increasing human needs and decreasing water supplies has set us on a collision course.

Will we still be able to support 25% of the cow herd and the majority of the sheep in the arid and semi-arid parts of the country? Or will we see continuing movement of livestock production to the humid regions of the Southeast, where water is not such a limiting factor? We know that improved Bermuda and other tropical grasses such as Star Grass, with their great forage yields, are making a tremendous impact on the livestock industry in the South and in Puerto Rico. But these high yielding improved grasses require large amounts of nitrogen fertilizers. Will energy shortages and costs of fertilizer result in reversion to low stocking rates? Or will we see a continuation of the increased use in the last few years of range forages in livestock production? If that is the case, what are the environmental consequences? Recent estimates indicate that more than one-half of the range in the contiguous 48 states is in unsatisfactory condition. Some areas are in high risk of desertification, with soils so fragile they cannot sustain more intensive use.

And what of the interest of the nontraditional users of rangelands - the environmentalists, the recreationists and others, who believe that the range is for their use? There is also some interest in using these arid lands to grow special use crops such as jojoba and guayule. There will be growing conflicts between traditional and unconventional land use on our rangelands.

I have posed questions for you here by way of suggesting that we need answers. Lots of them.

It should be abundantly clear to us that a rapidly changing society and economy is upon us. We need to know, for example, how the increasing demands on land and water resources are likely to affect animal agriculture. What research is needed so that we will be assured of a voice in helping to manage the changes that are coming?

PRODUCTIVITY

We have achieved many research breakthroughs in animal agriculture - the eradication of hog cholera, the development and application of artificial insemination, the genetic evaluation of dairy bulls and the improved methods of feeding to maximize production. We have seen the production of milk per dairy cow double in the last 25 years or so. The amount of beef marketed per breeding female increased over 1 1/2 times on a live weight basis in the same period. Pork production per breeding female increased about 25% in the same time span, while the amount of feed needed to produce a pound of broiler chicken has been cut almost in half (USDA, 1980b).

Now we are seeing a leveling off in our rates of increase all across agriculture. We do not see the great increases that we saw in the 1940s, 1950s and 1960s. As all of you know, for many years in the United States we have been able to point with pride to a very steep curve of productivity when plotted against time. But that has changed, and for many commodities we appear to be on a plateau of that curve.

Total productivity on farms and ranches (expressed as an index of farm output divided by farm input) trended upward in the 1970s, This trend can be attributed to external inputs of chemicals, machinery, fossil fuels and research results. It does not take a very smart person to realize that all of these externally generated inputs are in jeopardy.

Time will not permit me to deal in depth with the contributions made by the first three of these inputs. Suffice it to say that the development and availability of agricultural chemicals is not ensured in the face of increasing costs of manufacture, use-clearance requirements and environmental concerns. Machinery costs are soaring. Fossil fuels to operate that machinery is scarce and costly. However, I do wish to develop the fourth input I mentioned - the research component - a little more fully.

A number of our problems in animal reproduction, genetics, nutrition and disease control are particularly complex and will be difficult to solve. At the same time there are areas that seem especially promising and that could have important impacts on production efficiency. For example, we are learning how to obtain and transfer embryos on a routine basis in cattle, and the process now needs to be improved and extended to other animals.

Control of the sex ratio and the number of offspring born may be possible within this decade if research progress continues. If each beef cow could be induced to give birth to two male calves, productivity and profitability in the cow-calf business could be more than doubled. This situation would enable a rancher to respond to changes in supply and demand.

We need to continue to improve our ability to make early evaluation of the genetic worth of animals and to maximize the use of superior germ plasm in animal breeding. We need to better understand the effects of genetic interactions. For example, new work with dairy animals indicates that the fetal placental complex that develops during pregnancy has an effect on the cow's subsequent milk production. This effect varies depending upon the sire of the fetus, and is thought to be related to hormonal secretions from the placental complex. Findings of this nature indicate that the complexity of the genetic influence on milk production may be greater than has been supposed in the past.

So far, we have made great strides in animal nutrition and we know a great deal about the digestibility of different feeds and the physiology of metabolism. But we need to probe deeper into the inter-relationships among nutrients, how they are transported and how they enter and participate in the function of cells. We need new vaccines and methods without environmental side effects. We must apply the genetic engineering of microorganisms, gene-splicing, mutation and selection in ways to attack livestock problems more effectively. The work underway now in developing a safe, effective and inexpensive vaccine for foot and mouth disease through recombinant DNA techniques is an example. Many other possibilities lie ahead, genetic control of the animal's immune response to protect it from diseases and parasites and genetic sterility for the pests, to name just two.

Techniques such as fiber optics, a spin-off of space-age technology, are allowing studies of the internal workings of organs such as the mammary gland. These and other studies of the physiology and immunology of the mammary gland's defense mechanisms may lead to ways to prevent mastitis.

Through continued research and application, it will be possible to increase our rate of productivity considerably by the year 2000. Examples of progress could be described for all our livestock, i.e., beef cattle, dairy cattle, sheep, poultry, but let me use swine as an illustration. It could be commonplace in the year 2000, for a hog producer to select a superior young gilt with exactly the right ratios of preferred cuts and of lean to fat tissue, and by means of superovulation and sex selection techniques have her produce 30-50 fertilized eggs instead of the usual 14-17 of today. Surrogate mothers would be used for embryo transfer, and almost all of the fertilized eggs would be brought to full term. The baby pigs would be born with greatly enhanced opportunities for survival and rapid growth. First, they would be in an optimum weight range as a result of new knowledge obtained from nutrition studies during gestation. Second, new techniques would control and prevent diseases such as transmissible gastroenteritis and baby-pig pneumonia. Third, many of the components making up their feed nutrients would come from sources not now used in swine diets.

Pigs born to these superior sows would go to market at 120 days of age, weighing 127 kg (280 lb) tailor-made for the consumer and the processor. In the meantime, their mothers would have been bred back within 20 days of farrowing, on their way to producing nearly 3 litters of pigs per year compared to the present 1.7.

As in the scenario I have just described for swine, a similar one could be developed for other domestic animals. Thus, it is possible to increase our productivity greatly, but it will not be done without a strong, undergirding basic research effort. Unless we begin now a long-term commitment to restock our storehouse of fundamental knowledge, it will have been picked bare long before the year 2000.

As we look to the future, we must be aware of the critical need for basic research information of animals. More investigation is needed at the cellular and molecular level, and we need to get started on attempts to find ways of utilizing genetic engineering techniques in livestock. Areas such as nuclear transplantation and cloning are promising, and more basic information on cytogentics is needed if we are to be able to manipulate chromosomes and even genes. The cell and its genetic machinery is vital to every process important in animal production. We must probe it deeper than we have in the past.

While your efforts have been directed to identify research priorities in terms of human needs, in terms of problems that must be addressed, in terms of pay-off, in terms of proper climate for obtaining funds for one project or another, all of which are important, I urge you to keep in mind that the base of fundamental knowledge from which most of us in science are working today is not very great. It is, in fact, a very thin foundation.

I do not believe that the general public, whose human needs are in the forefront in setting our priorities, has the slightest idea of how precarious our position is.

Some other basic research areas worthy of mention are: 1) physiological and biochemical factors and their relation to diet and nutrient absorption; 2) effects of nutrient availability on regulation of cell differentiation, growth and function; 3) hormone/enzyme effects in reproduction and growth; 4) micronutrient effects on gene expression; 5) relationships of nutrients to membrane structure and function; 6) role of cell mediated immune responses; 7) genetic control of fat distribution; and 8) a host of genetic factor relationships.

Time will not permit further development of these important areas in this paper.

A good illustration of the application of basic research can be seen in the accomplishments that are now being reported on work with recombinant DNA. From many of the press accounts, it would seem that these accomplishments were pulled from a hat just yesterday and that their promise today will be reality tomorrow.

But as Lewis Thomas put it so well in one of his recent columns:

> "These things had their beginnings in lines of work so far removed from today's enterprises that it is already difficult to track back through the network of experiments to find the beginnings. Recombinant DNA techniques could not have evolved without the 30-year background of research in virology and molecular genetics, almost all of it done without the faintest inkling that anything like recombinant DNA lay ahead." (Thomas, 1980).

Thomas (1980) also points out the stake that private industry has in basic research and the need for more stable funding of such long-term work. The comments are pertinent for federal and university researchers who are operating in a climate of extremely tight budgets and accountability, where basic work must compete with quick pay-offs and defensive research.

Despite current fiscal constraints that are necessary to fight inflation, I believe we are beginning to see an encouraging turnaround in the funding for agricultural research. We are seeing heightened concern for science and education within the USDA and a tremendously increased level of concern from the President's Office of Science and Technology Policy. Research and science are clearly recognized as fundamental to increasing productivity. We know that expenditures for agricultural research are good investments. They pay annual dividends far in excess of average rates to both producers and consumers. Estimates appearing in the literature range from 29-90% annual returns.

Research that you are considering must be identified as having potential to meet human needs if it is to compete in this era of scarce public funds. Well-thought-out programs of research on animal agriculture must define the needs and consider the users if they are to find a listening ear.

It is now abundantly clear that the food and agricultural industry of the United States, including the research and education component, faces challenges which will tap its resources and imagination to the utmost. As we move through the last decades of the 20th century and into the 21st, we must learn to make creative decisions that balance our priorities in fairness to all. We will have to act in a context of a rapidly changing society and economy. We will have to get used to operating in a world of vastly lowered degrees of certainty. This will call for more flexibility in the system and in individuals. Federal, state and industry researchers must make whole-hearted efforts to reach agreement on research goals that are of highest priority. Agricultural researchers need not be reminded that number one on the list of human needs is a dependable food supply. In the United States that means a food supply that is nutritious, abundant, varied, safe, economical and good tasting.

Americans have had that kind of food supply to a greater degree and for a longer period of time than any other people at any other time in the history of this planet. If we are to have it in the 21st century, the science of animal agriculture must be strengthened so that it can continue to play a major role in filling this human need not only for the people of this country but also for those in much of the rest of the world.

REFERENCES

Marshall, E. 1980. Antibiotics in the Barnyard. Science 208:18.

The Washington Star. 1980. 1930-1980: 50 Years of Frozen Foods. March 5.

Thomas, L. 1980. Notes of a Biology Watcher on Science Business. New England J. Med. 302:151.

USDA. 1980a. Human Nutrition Center, Consumer and Food Economics Institute, Science and Education Administration, Washington, DC.

USDA. 1980b. National Program Staff, Livestock and Veterinary Sciences, Science and Education Administration, Washington, DC.

GLENN W. SALISBURY

Introduction
Harold D. Hafs

Relative to the major thrust of this Conference, we in the United States have a problem! Notwithstanding the subject of our working group on Human Nutrition, most people like those of us at this Conference do not comprehend hunger, because we have enjoyed an uninterrupted and unprecedented 40-year period of food affluence.

Most thoughtful people were shocked when only about 5 years ago drought in two areas of the world produced massive starvation, so delicate is the balance of our population with our food supply. But with return of better weather, most of us choose to overlook facts such as the following. Dr. Norge Jerome informs me that it was estimated last year, during the International Year of the Child, one-third of the children born would die before age five due directly or indirectly to malnutrition. We do not like to acknowledge that demographers predict twice as many people to feed within as little as 35 years. This is particularly discomforting in the light of our shrinking land resources to produce food, referred to earlier in this conference.

The business of this Conference is to decide how to maximize the potential usefulness of animal agriculture for human needs. To my knowledge, never before has such a diverse assembly of talent and technical expertise been brought together, as you are related in this Conference. It is already clear that you are creating approaches to solving some of our problems. To the extent you succeed, we can buy time to balance our ability to produce people with our ability to feed people.

Our fourth keynote speaker has devoted his entire professional life to such a goal. For the past year especially, he has made an opportunity to stand back and look at some of our public institutions. Before that, he had very broad experience as Director of the Agricultural Experiment Station for 10 years and as Chairman of the Department of Dairy Science for 22 years at the University of Illinois. As a researcher he contributed over 255 scientific publications, many of which formed the foundation for our present-day highly successful cattle artificial insemination organizations. Previous keynote speakers have referred to the enormous genetic impact of artificial insemination. Still he found time to write a major scholarly textbook on Reproduction and Artificial Insemination of Cattle, initiate a livestock rehabilitation program in

Europe after World War II, serve as an advisor to governments in five other countries, and participate in President Kennedy's Science Advisory Committee. He has numerous awards, and is a member of the National Academy of Sciences.

Dr. Glenn Wade Salisbury, a native of Ohio, obtained a B.S. degree from Ohio State University and his Ph.D. from Cornell University, where he served on the faculty for 13 years before he accepted appointment at the University of Illinois. I know you will agree this man has given much more than most. I take great pleasure in introducing Glenn Salisbury.

"This conference has provided a unique opportunity to think, plan and exchange ideas and understanding about present and future problems in animal agriculture..."

Maryln Cordle

"...the documents produced will be used, perhaps in ways we have not yet recognized... many persons... will refer to the documents as evidence of the sincerity and good faith of the industry to prepare properly to deal rationally with the... problems facing animal agriculture."

Dick Sechrist

Address

The first 80 years of the 20th century which we are now using to estimate the parameters of the 21st century for a very limited segment of the human future ought to have taught us a great deal. Many things happened in the 20th century that had never happened to man before. Many of these happenings and their potential impacts on the future of American animal agriculture have been mentioned already by others and careful attempts are being made to forecast potential impacts of events only unclearly perceptible now.

Among those happenings was the production of clear cut evidence of a fruitful outcome of a dream originating in the 19th century with the creation by Congress of the Land Grant University system in 1862 and of the USDA in 1863. As a product of the Land Grant University System and having been associated with it for a lifetime, I am convinced that the initiating partnership of the Land Grant system and the USDA, along with cooperation from the state agricultural experiment stations, is one of the great social inventions of all time. For animal agriculture and many other segments of agriculture it has been the world leader. The future progress of the human race may rest on the performance of the totality of its function.

I am glad that Dr. Phillip Abelson, in his opening address of this Conference, made assertions of the competence of agricultural science and particularly of its accomplishments in research on domestic food animals which I can only echo. To reiterate his comments would serve no useful purpose. There have been other comments made about slippages in the system, about matters we are not attending to in sufficient degree and about problems of all sorts looming on the horizon.

The author acknowledges with appreciation the help of his colleague, Professor J. Robert Lodge, for his aid in this research and with the manuscript.

The substance and data for this talk have been taken from a manuscript of a Special Publication of the University of Illinois Agricultural Experiment Station Bulletin series to be published in the early Fall, 1980.

In seeking to aid in the development of our research priorities for Animal Agriculture and Human Needs for the 21st century, I shall concentrate on one part of the national research establishment which I consider to be critical to the generation of information necessary for a favorable outlook, the research of the State Agricultural Experiment Stations. During the period of my life devoted to research management, I soon learned that an agricultural experiment station has to concentrate on the sciences basic to its endeavor and that it is something like a livestock feeding operation; it has a maintenance requirement just to keep it alive. Only when an information generating organism, which it is, has sustenance beyond its maintenance requirement can it be productive.

It also has to exist, like livestock, in an environment conducive to productivity. That environment is University oriented and has permitted the development of agricultural scientists educated in modern basic science. When young agricultural scientists enter university employment they sometimes hear two drums sounded in different cadence. One is that of the internal promotion and rewards system which sometimes relates only randomly to the cadence of service to agriculture. The other, the discipline necessary to the function of demonstrable service to people through agriculture, is asserted by the administrators of college and station and sometimes at variance with the disinterested critique of a large segment of the university community.

I hear nothing in this Conference suggesting a lesser future role for the agricultural experiment stations than that of the past. Certainly research on the productivity and efficiency of domestic animals as producers of disease-free, contaminant-free, nutritious human food must be conducted with those animals and in a setting where they can be maintained.

Years ago I became interested in the question of research productivity of state agricultural experiment stations for obvious reasons. I needed the comparative evidence to justify our station's existence at the University of Illinois. Next I needed evidence on the costs of its own and other stations' research in relation to their research productivity. Since publication of original research in peer-edited scientific journals of national stature was the university standard of excellence, I accepted that standard as my measure of productivity. I accepted the evidence presented over the years by the Old Office of Experiment Stations and more frequently the Cooperative State Research Service of the USDA for my evidence on station funding.

First, we should recognize that funding agricultural research by the states bears not a very strong relationship to the agricultural output and gross farm income for the various states; Illinois and Iowa, for example, vying each year for the third or fourth position of the states in gross farm income, stood 18th and 16th, respectively, in national rank in their state funding of agricultural research. The states with the highest state funding do so presumably on the basis of the density of each state's population and the diversity of their cropping system. The Federal contribution through institutional funding to the state agricultural experiment stations bears neither a relationship to agricultural productivity nor to research productivity of the stations. It is funding which bears the closest relationship with research output.

In the 20th century our country has learned, as has the rest of the industralized world, and the Third World is being taught, that the great time-compressor for productive change is scientific research as the scouting party of what ought to be done next. I shall use two examples from agriculture, corn production and milk production, to illustrate changes over time in national productivity and in individual yield of these commodities. Most of the yield changes can be shown to be correlated with the results of recognized scientific research in a time series. The actual production figures indicate how well farming practice and national policy have put together the essential inputs and the necessary incentives to achieve the potentials. Let us examine how we as a nation have provided for our agricultural research scouts, those in animal agriculture in particular, and determine if we can, how we ought to think of them in the future.

First, we must recognize that all of the benefits of agricultural research do not, in fact, accrue primarily to the producer (the owners and operators of farms). Most research which deals with production agriculture comes down to two items, savings of land and of labor or, if you prefer, yield and labor. The benefits of most research dealing with saving labor inputs in farming accrue to farm operators. Except for the first adapters of research beneficially affecting yield (the saving of land), the ultimate beneficiary is the consuming public based on the inverse supply-demand influence on pricing.

Figure 12.1 shows the historical trend for corn production in the U.S. from the Department of Agriculture record keeping in 1866. It ends with 1976, but all of us know that 1979 ended with the highest average per acre yield and the largest national production of corn on record. There was no marked change in yield per acre until well into the 20th century and then mostly after World War II when the greatly increased yield capacity of hybrid varieties met with the availability of relatively cheap nitrogen from petroleum products. That, in combination with good soils, the right weather and great farm management resulted in relatively cheap feed grains.

That change resulted in a great expansion in the American livestock industry and an increase to record levels in human consumption of red meat, primarily beef, expanding more than 2.2 times per capita after World War II, from 27 kg (59.4 lb) in 1945 to 58.5 kg (128.8 lb) in 1976.

Milk production through market pricing agreements has remained static since World War II at 54.6 kg (120 billion lb). Human per capita consumption of cow's milk decreased after the wartime peak at a rate almost exactly counterbalancing the increase in the human population numbers. As shown in figure 12.2, dairy cow numbers decreased by one-half from the peak numbers of 25,650,000 in July of 1944-1970 at a rate of 518,000 cows per year. The rate was much less in the 1970s. The average annual yield per milking cow has increased 2.3-2.4 times since 1945, due basically to the application on farms of the scientific findings about genetics, statistics, biochemistry, nutrition, suction pumps, the health sciences and the availability of cheap feed! It has been a revolution in dairy farming, a triumph of applied research that has not yet exhausted its genetic variance for improvement and productive gain. These two figures represent something about research priorities for animal agriculture of the future, however, which I think we must consider. There are tradeoffs to national advantage which many of us in specialities of science do not recognize. If we had had to make a choice in the past, should we have put our bets on corn or cows?

Now, we ought to examine the productivity of the research organizations established by the American public to create such change, the USDA and the state agricultural experiment stations of the Land Grant Colleges of Agriculture.

Actual state dollar inputs have gone up about 19-20 times in those 30 years; federal dollars (about 22-25% of the total over those years) have gone up between 18 and 19 times. In fact, the increase in total since 1945 has been at a compound annual interest rate of 10.4%, while the Federal contribution has been lower, 9.7%. This means that if extended at the same rate to the year 2000, the combined state and federal dollars would be over $3 billion ($3,143,691,000), a highly unlikely occurrence. Discounting this by the inflation rate which has persisted all during the 30-year period, the total in constant value dollars has doubled a bit more than two times in the period. In general, research productivity of these state stations in terms of the number of technical journal articles they have published has followed closely the trend line in constant value dollars (R^2 = .9463). Those trend lines are for the nation as a whole, 1945-1975, inclusive. Clearly the input into and the output from the stations of the various states must vary tremendously. They do vary in both input and output. Table 12.1 presents, for the 1 year 1975, the first 10 of these state stations in rank order of the number of technical journal articles each published that year. The second part of that table shows the rank order for a top 10 if the staff size and the total dollars invested statistically held constant for the stations in each state. Much of the variance among years and among stations is due to highly correlated differences in inputs of dollars and science-power.

The OES and CSRS records were of no value in assaying the question as to whether or not such a close relationship existed between measurable inputs and scientific output among the several subject matters of agricultural research. Clearly an appropriate hypothesis is that the same facts apply to the segments as to the whole.

What about the broad subject area of domestic animal research? These studies started more than 20 years ago. The first of these consisted of enumerating all communications (full-length, peer-edited articles, abstracts and letters) appearing in five selected journals in the three years, 1958-1960. The journals were the Amer. J. Vet. Res., J. Anim. Sci., J. Dairy Sci., Poul. Sci. and Vet. Med. The subject matter fields were domestic animal production and animal health and protection.

A similar study, which included only four of the above five journals at the two time periods 1958-1960 and 1965-1967, contained an opinion survey (for 1965-1967) of 105 animal scientists (85 replied) of the ranking of all states as to the quantity and quality of their research on large domestic animals. There was a correlation of opinions with actual rank order in the numbers of publications of .8307.

In the last year of the 1960s my responsibilities for knowledge about research competence expanded to include all of the field of agriculture. My colleagues and I then expanded the literature studies to include all subject matters in agriculture under 11 headings. In that undertaking, 66 peer-edited journals for review were included, and all full-length publications for the calendar years of 1967-1969 were recorded to the credit of the employers of approximately 75,000 scientists contributing 31,911 full-length papers. Credit was given for multiple authors to the employer of each on a proportionate basis regardless of the order of authors.

Of the total of 31,911 papers, state agricultural experiment station employees contributed 34.2% and other departments in land grant universities 6.3%, for a total of 40.5% from that source. The USDA 12.3% and other federal agencies 4.8% for a U.S. government total of 17.1%, other U.S. universities 13.2%, U.S. private research institutions 6.1% and foreign country scientists, 23.1%.

This kind of study is tremendously costly. Some day library information retrieval systems may be ready to record employers of authors and provide such information on publication output cheaply. That day is not yet with us, however. In this study, we had corollary data from USDA-CSRS reports of the available funding and scientist years of all the state stations for fiscal year 1967, 1968 and 1969, but not for the separate subject matter fields. The station to station statistical results were entirely similar for the 3-year period as we had found earlier for the technical journal articles for the 1945-1975 period and for the 1 year, 1975. I have been unable to acquire detailed data about subject matter funding and research output on those subject matters for all of the stations or even those of the first 10 in a subject matter rank order.

Thus, for the final study of the allocation of funding and of professorial staff in relation to research output, our data are for the Illinois station only. The period covered was 1949-1978, inclusive. In this presentation I shall deal only with animal production subjects, not animal pathologies and disease control. The data are for the two departments under animal production research as organized at Illinois. From my many discussions with scientists in the field and with research directors I am confident that the findings at Illinois reflect the general national situation. The first thing which strikes one on making analyses of this nature is the changes that have occurred in the intrainstitutional funding of various subject matter fields. At Illinois the fastest growing field is Food Science. Financial support for the animal production field has shifted from 30.8% of the Illinois station total expenditures for the 1948-1950 bienium to 22.3% in 1976-1978. I expect that trend to continue.

The next two figures (figures 12.3 and 12.4) illustrate publication and financing trends of the Illinois station overall for the 15 biennia and those of animal production research by itself. In the first figure, total funding in actual dollars and constant value dollars and output of peer-edited publications are shown from an index of 1 for 1948-1950 for the station as a whole during the 15 biennia, 30-year period. Funding in actual dollars for the entire station increased 6.5 times from the 1948-1950 base. In constant value dollars, purchasing power went up only 2.22 times. Peer-edited publication output seems to have reached a peak in 1970-1972 at about 4.9 times its base level and has decreased since.

The illustration for animal agriculture (figure 12.4) is not based on an index. It is based on actual figures, but fortunately the actual dollars expended by the two departments at the base biennium of 1948-1950 were almost exactly $1 million, so the scale is interpretable on the same basis as for the station as a whole, including animal production. In this case the actual dollars expended over time for animal production research have increased only 4.70 times, the constant value dollars increased to a maximum ratio of 2.34 in the 1966-1968 biennium and had dropped to only 1.61 times the initial level in 1976-1968. Publication of peer-edited papers, however, had increased to the highest value of 3.04 times the base of 1948-1950 in the 1974-1976 biennium but had decreased slightly in 1976-1978. Starting in the first biennium with 32.9% of the station's output of publications in peer-edited journals, the period ended with only about 20-25% of the station total. Public research funding for domestic animal agriculture has not kept up with that in the other agricultural subject matter fields for at least a decade and the relative research output is showing it.

Lastly, the efficiency of research (that is, the output of peer-edited publications per million dollars of constant value) has its cycles, both for all subject matters in the Illinois station and for animal agriculture as well. These curves require some interpretation, but in the final analysis they show that it is not the individual scientists either of the station at large or of the domestic animal departments in particular who are to be held responsible for the apparent decreasing output of the state agricultural experiment station.

Animal agriculture at Illinois has required the largest investment per scientist of any agricultural subject matter in the Illinois station. However, its scientists have produced peer-edited scientific publications at higher rates per full-time equivalent professorial staff than has the rest of the station's staff sufficient to equalize the dollar efficiency of the average of the rest of the experiment station scientists.

I interpret the recent fall-off in productivity of the station as a failure to provide all of the necessary nutrients to meet both its maintenance requirement and nutrients for production. Clearly if the state stations are to produce the knowledge base for animal agriculture research required to meet the multivariate and interacting human needs of the 21st century, effective means must be found to finance the research.

TABLE 12.1. THE 10 LEADING STATE AGRICULTURAL EXPERIMENT STATIONS BASED ON TECHNICAL JOURNAL ARTICLES REPORTED TO USDA COOPERATIVE STATE RESEARCH SERVICES[a]

	No. of technical journal articles	Adjusted for		
		Staff size	Federal dollars	State dollars
1 California	1,720	California	California	Wisconsin
2 Wisconsin	1,023	Wisconsin	Wisconsin	Tennessee
3 New York	911	New York	New York	California
4 Texas	731	Tennessee	Oregon	New York
5 Tennessee	625	Texas	Kansas	Illinois
6 Illinois	601	Illinois	Minnesota	Oregon
7 Georgia	513	Georgia	Michigan	Michigan
8 Oregon	499	Michigan	Nebraska	Indiana
9 Indiana	481	Minnesota	Illinois	Texas
10 Michigan	471	Indiana	Tennessee	Utah

[a]Tennessee is a chance arrival this year (1975). It published an average of 124 technical journal articles in the 9 prior years of the 1966-1975 decade.

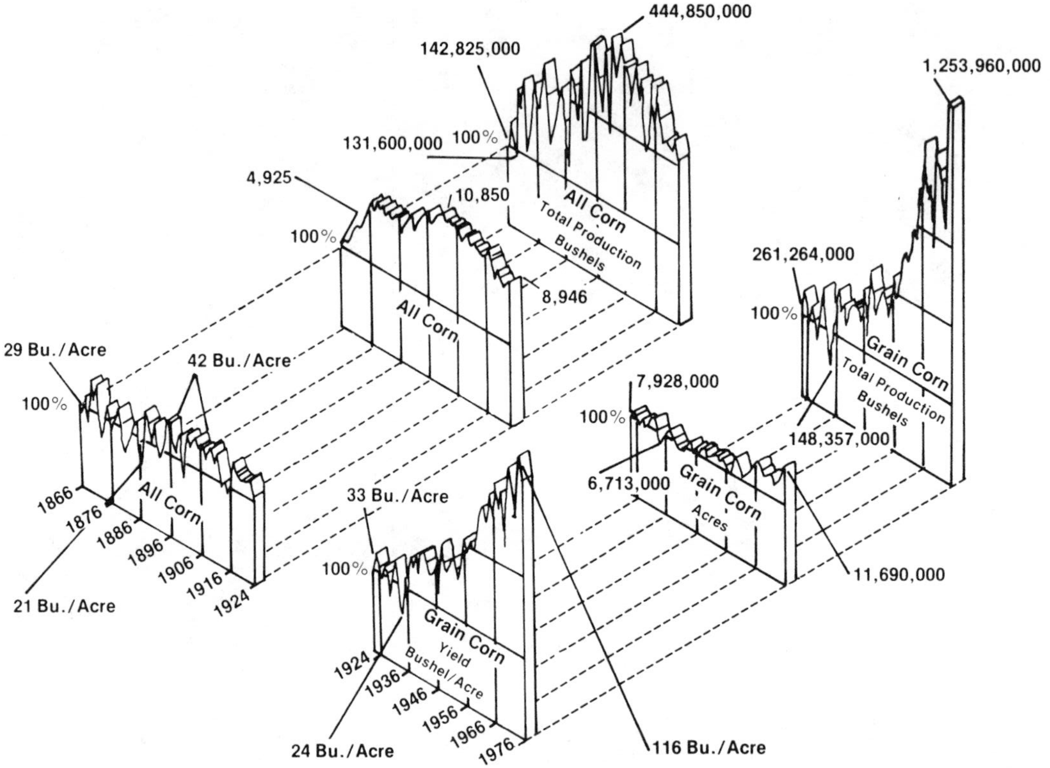

Figure 12.1. Historical trend for corn production in the United States, 1866-1976.

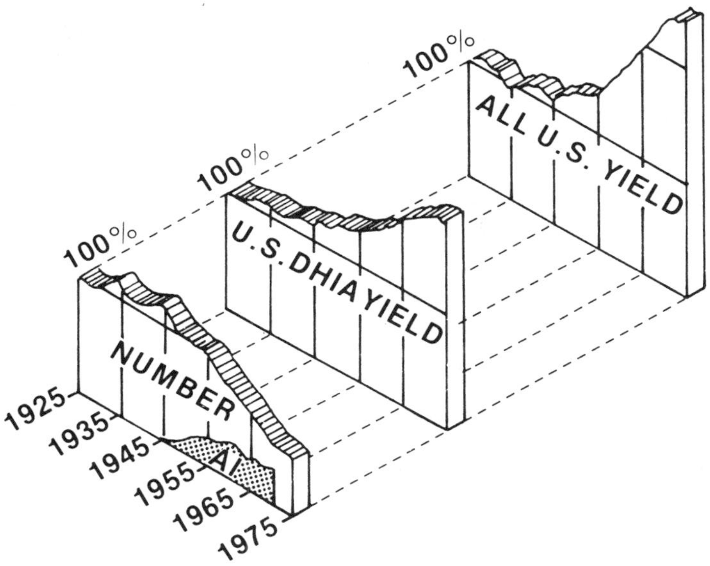

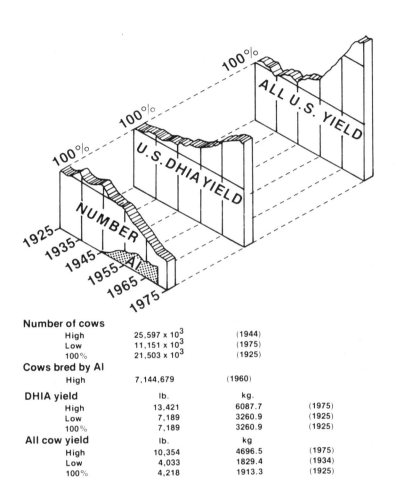

Number of cows			
High	$25{,}597 \times 10^3$	(1944)	
Low	$11{,}151 \times 10^3$	(1975)	
100%	$21{,}503 \times 10^3$	(1925)	
Cows bred by AI			
High	7,144,679	(1960)	
DHIA yield	lb.	kg.	
High	13,421	6087.7	(1975)
Low	7,189	3260.9	(1925)
100%	7,189	3260.9	(1925)
All cow yield	lb.	kg	
High	10,354	4696.5	(1975)
Low	4,033	1829.4	(1934)
100%	4,218	1913.3	(1925)

Figure 12.2. Dairy cow numbers and milk yields in the United States, 1925-1975.

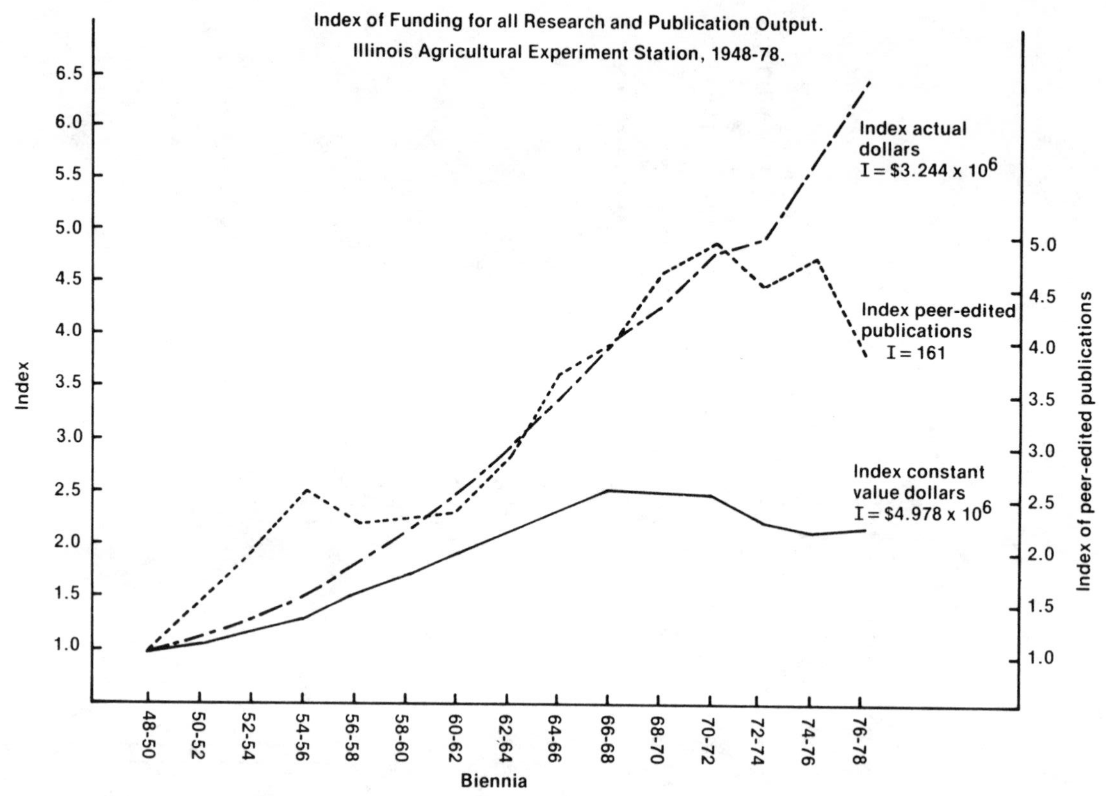

Figure 12.3. Index of funding for all research and publication output, Illinois Agricultural Experiment Station, 1948-1978.

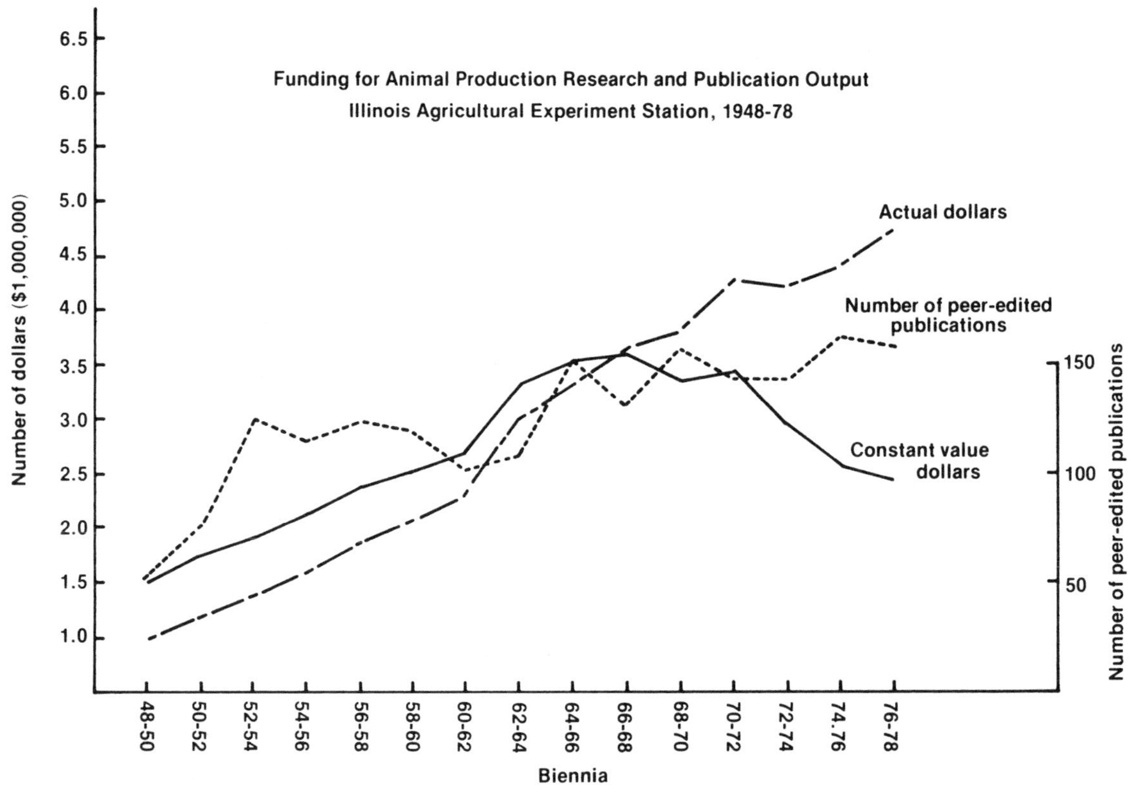

Figure 12.4. Funding for animal production research and publication output, Illinois Agricultural Experiment Station, 1948-1978.

SYLVAN H. WITTWER

Introduction
John R. Welser

I am pleased to participate in this Animal Agriculture Conference, especially to have the opportunity to serve on the Steering Committee.

The Conference is an attempt to define a systems approach to animal agriculture. In the past, many of us individually or collectively have been so busy pursuing our own disiplines or interests that we occasionally have lost sight of the goal of all our efforts - producing safe, wholesome food.

A systems approach which considers all factors influencing animal productivity not only will improve animal agriculture, but will serve the consumers of our products by enhancing human nutrition. I am confident that the long-term beneficiaries of this conference will be all humanity.

The persons who conceived and planned this conference are to be thanked and complimented for designing a conference to develop a comprehensive approach to this critical issue.

The Conference planners also are to be commended for recognizing that any meeting on this subject would not be complete without a major presentation from Sylvan Wittwer, director of Michigan State University's Agricultural Experiment Station. Wittwer deserves a place on the program not because the Experiment Station is a major supporter of the Conference but because he is an internationally-known scientist, having published over 600 papers and scientific reports and three books in the food and agriculture field. Sylvan truly is an international scientist, having travelled and lectured extensively in Europe, Russia, Asia, Africa and South America.

Born in Hurricane, Utah, Wittwer was educated at Utah State University and the University of Missouri. He has been on the faculty at Michigan State since 1946.

The receipient of numerous honors and awards, Wittwer recently was the second U.S. scientist elected to the prestigious Lenin All Union Academy of Agriculture Sciences of the U.S.S.R.

Sylvan Wittwer was a principal driving force in establishing and implementing this Conference. He gave the Steering Committee the courage to develop the broad base and national scope of this meeting.

On behalf of the Steering Committee, the Conference participants, the animal agriculture industry, and the consuming public, I thank him for his efforts and his expertise.

I am sure Sylvan Wittwer's remarks today will be forthright, insightful and most helpful toward meeting the goals of this Conference.

"...this conference probably brought together the largest gathering of multidisciplinary scientists and industry ever assembled for [a consideration of animal agricultural research priorities]."

Burt Hawkins

"Consumers and farmers will reap benefits from this historic event before the 20th century ends."

Norje Jerome

Address

Agriculture is the world's most basic enterprise. Food is first among the needs of people, and it is our most important renewable resource. Plants provide, directly or indirectly, up to 95% of the world's food supply. Nevertheless, there are more than 3 billion livestock in the United States to supply the animal protein consumed. The livestock of the nation consume 10 times as much grain as the entire U.S. human population. The enormous quantities of grain the U.S. exports abroad now approaches $35 billion annually. Most of it goes for feeding livestock and poultry with little being used directly for human food. A total of 5.4 million metric tons of protein is produced annually which supplies an average of 70 g of animal protein daily per capita in the U.S. (Pimentel et al., 1980). Mayer (1976) stated, "It appears to be historically inevitable that as people or societies become wealthier, their consumption of animal products increases".

Technological Inputs into U.S. Agriculture

Projected annual increases in food production between now and the 21st century to meet human needs are 3-4%. Increases in demand will come from both growing populations and increases in consumer incomes. Expanded crop areas or higher production per unit land area per unit time are the only sources for the 3-4% increases of production. It is projected that these increases could come 25% from an expansion of arable land and 75% from intensification of land use through higher yields and increasing the number of crops produced per year.

To do this, one must consider the resource inputs (land, water, energy, fertilizer, pesticides, human labor,machinery) their costs, availability and renewability. In the U.S., the development of labor-saving technology has been a major goal. One farm worker today produces enough food for himself and 56 other people. This compares with 29 people in 1970, and with only 11 just 40 years ago. There is no such parallel in all of history. Never have so few people produced so much.

In addition to the massive resource inputs of land, water, energy, fertilizer and pesticides, the increased productivity per unit of labor input may be attributed, in part, to better management, more timely operations, and more efficient and productive equipment. Mechanization in the U.S. has enabled farmers to carry out their field work on a timely basis and at the same time to allow for management activities. It also has been forced because of the unavailabilities, uncertainties and rising costs of human labor. In Japan and some other industrialized nations quite the reverse has occurred where resources (land, energy) are limited and labor is more plentiful. Here yields are higher, but output per farm worker is much less.

Thus, there are two general types of food production technologies for the future: mechanical-labor saving and land resource intensive on the one hand, and biological-chemical or land resource sparing on the other. There will be a worldwide shift from less natural resource-based to more science-based agriculture. The emphasis will be to raise output per unit of resource input and to release constraints imposed by relatively inelastic supplies of land, water, fertilizer, pesticides and energy. Ruttan (1977) pointed out that this has occurred already during the first part of the 20th century in Japan and certain European countries. Whereas the U.S. has followed a mechanical-resource intensive technology pathway, Japan has followed a biological-chemical resource sparing one. Incentives for increasing yield technologies in the U.S. have lagged until recently compared with Japan and some European countries because of the abundance and low cost of resources. It is projected now, however, that almost all future increases in food production will be a result of increases in yield (output per unit land area per unit time) and from growing additional crops during a given year on the same land. There are really no other viable options. This underscores in a dramatic way the importance of science and new technolgies for meeting future national and world food needs. Environmental issues will mount as more land, water, fertilizers and pesticides are diverted to food production and there is forcing of higher productivity. New technologies will call for the use cf more, not less, chemicals (fertilizer, pesticides) to increase food production. This will occur as populations increase and demands rise.

Degradations of environment arising from agricultural production are present in many forms. There are, also, innumerable examples of farming practices which can improve rather than threaten environments. Agricultural food production is the chief user of our land and water resources. Toxic chemicals in the environment, many of them pesticides and fertilizers used for food production, have been declared environmental threats and hazards to human health and well-being. Debates will continue on issues of food safety, deleterious effects on fish and wildlife, endangered species and carcinogenicity. It has not yet been clarified by anyone as to what an environmentally sustainable set of agricultural production technologies might be.

One of the most debilitating myths in setting the agricultural research agenda for the future is that all we have to do is put to use the technology we now have, and most problems of food production and distribution in our nation and those of developing countries soon would vanish. This implies we do not need more research but better dissemination of the results of the research we already have done.

Nothing could be further from the truth. The American agricultural research agenda, both publicly supported and private, have focused on large-scale single crops or livestock operations, and labor saving technologies that are capital, management and resource intensive. The economics as well as the technology of development are changing so rapidly that older assumptions--yesterday's assumptions --are no longer valid. We only now are beginning to realize how much the changing cost of energy is undermining most all our previous assumptions on costs and feasibility of increasing agricultural production at home as well as in the third world. We no longer can plan research programs patterned after the conventional and those of the past.

Meanwhile, many societal global problems have emerged relating to agricultural production and the integrity of food supply. These include poverty, malnutrition, inflation, unemployment, soil erosion, deforestation, desertification, firewood shortages, toxic chemicals in our environment, increased climatic impacts and agricultural production stability (table 12.2). These problems strongly suggest that for assisting the developing world and ourselves we should seek agricultural technologies that will result in stable production at high levels, that are scale neutral, sparing of resources and management, and are more labor than capital intensive, that will increase the demand for underutilized labor resources and put people to work more days per year and more productively, that are nonpolluting and literally will add to the resources of the world. Such technologies exist. We have not addressed ourselves seriously to their development. An example is hybrid cotton and the current "White Revolution" of milk production in the Gujarat State of India.

Over a million farmers with the water buffalo as the milk animal in India have formed cooperatives for marketing and distribution of milk to the almost unlimited markets in Bombay, New Delhi and other large cities. It is known as "Operation Flood". Milk production at the farm is enhanced by genetic improvement through artificial insemination, by better feeding with the highly productive hybrid Napier grass and by disease control. The result is a dependable daily cash income, better utilization of wastes and by-products, improvement of the nutritional status of poor people, an opportunity for landless laborers, and an enterprise which is labor intensive at the production level.

During the decades of the 1950s and the 1960s, we witnessed the golden age of American agricultural productivity. We rediscovered our great production potential such that during the seventies we increased by several-fold our agricultural exports and reigned supreme as a nation exporting abroad up to one-third of the U.S. total production. Never in the history of mankind had any nation produced or exported so much for the welfare of others. Until recently, the aim was to "produce two blades of grass or two ears of corn where only one grew before". We were eminently successful.

Times now have changed (table 12.3). Stable high production still remains as a goal, but it must be with a minimum of resource inputs. We must move from a resource (energy, land, water) based agriculture to one which emphasizes more science, chemistry and biological opportunities. This means low inputs to agriculture of fossil energy and one which emphasizes efficiency of water use and enhancement of productivity per unit of land from both higher yields and more crops per year (table 12.4).

Agricultural research is currently at a crossroads. No more than 3% of the world's total expenditures for research is allocated to agriculture (table 12.5). For the United States it is only 2% or significantly less than the world's average (table 12.6). Ninety percent of the world's total expenditure on agricultural research is devoted primarily to needs of large-scale commericial agriculture in the northern temperate zones (Presidential Commission on World Hunger, Preliminary Report, 1979). The U.S. is no longer the world's leader in agricultural research (Wittwer, 1980).

Resource Constraints and Inputs

Volumes have been published on the importance of land, water and energy resources for food production. To assess a degree of priority to their relative importance would be foolhardy. All are essential. Soil erosion is a global problem and a threat to the world's capacity to feed itself in the future (Eckholm, 1976). We continue to lose enormous amounts (9 tons per hectare per year, for a total of 2.8 billion tons) of topsoil from our best lands.

Water may become the most seriously limiting of all resources for further advances in food production. Agricultural production, mostly for irrigation, consumes 80-85% of the total diverted fresh water resource in the United States (Water Resources Council, 1978) (table 12.7). The consumption of water by agriculture in the total U.S. water budget far exceeds that of energy (3%) in the energy budget. The future value of water for food production will not be limited to its use in desert areas. Water added to achieve optimum moisture could be the next major means for increasing crop productivity in what traditionally have been subhumid areas.

Food production is energy dependent. The most vulnerable input is nitrogen fertilizer followed by irrigation. Energy inputs for livestock are reduced by about 60% if livestock are grazed on pastures or forest ranges compared to grain feeding (Pimentel et al., 1980). There is a rising interest in production of biomass for energy purposes. Energy crops will compete not only for land and water but also for investment capital, fertilizer, farm management skills, agricultural credit and technical advisory services (Brown, 1980). A major new variable is being introduced into the food-population equation and a food-feed-energy conflict relative to resource inputs (National Research Council, 1979). In any event, new technological pathways must be appropriate to emerging resource scarcities and to rising environmental pressures (Crosson, 1980).

Agricultural Sciences and Human Resources

A series of editorials in Science and elsewhere warn of over-management and regulation in agricultural research. There is strong evidence that the competitive grant and contract system is burdensome, time consuming, detracting from research and loaded with paperwork (Leopold, 1979; table 12.8, 12.9). Academic research, in particular, is being impaired by unrealistic accountability which spawns bureaucracy, destroys morale and hinders progress (Abelson, 1980). Recent attention to minor details and over planning and coordination have reached the point of stifling scientific productivity among the very clientele it is designed to assist (table 12.10). Schultz (1979) emphasized that we now are over organizing and programming agricultural research. Funding is too far away from those that do research. Research scientists and their funds are being held in hostage by regulators and programmers.

It took 40 years for hybrid corn to be 95% adopted by American farmers (table 12.11). For Iowa farmers, it was 7 years (NSF, 1973). In India's Punjab, grain production has been increased threefold in 10 years. The time between discovery of new technology and its adoption must be shortened. There is a need to speed the flow of information from scientists in research centers to farmers. New breeds of technologies and technologists are called for. Provision is needed for communication among scientists constituting a blend extending from the applied field worker to the molecular biologist. The time between a basic research discovery and its first application averages 13 years. It takes at least 10 years for a new agricultural technology to graduate from the research stage to field application. The half-life of a new disease resistant cereal variety is 5 years. The time from introduction of a new technology until its adoption reaches an expected ceiling is 35 years. It now takes 6 to 10 years to train scientists to do food research. We must force the pace of agricultural research and development. One of the problems we now face is the result of meager research efforts through the 1970s which already have set the stage for slower growth and reduced farm efficiency for the decade of the 1980s. One of the most remarkable events in scientific history was the introduction of the vaccine for control of Marek's disease in poultry. The rapidity of acceptance was such that within months the nation and much of the world had adopted its use. Never before in the history of agricultural research had there been a parallel achievement. We should seek to identify the elements of acceptance.

Animal agriculture scientists fail to think big enough. During the week of this Conference May 9, 1980, an editorial appeared in Science, "Beef Production and Consumption", outlined by the editor, Philip H. Abelson, the keynote speaker for this Conference. A critique of the editorial pointing out errors in the editorial was distributed. Why did not someone of this Animal Science Conference write the editorial? Letters to the editor seldom get published, or if they do, they have little impact. It is the original that is important. We tend to react rather than act. February 26, 1980, a feature article appeared in Science 207:843. "The Potential for Grass Fed Livestock: Resource Constraints" by an entomologist and his collegues. Why not animal scientists as authors? The American Association for the Advancement of Science is helping us convene this Conference. We already have an entry to Science, why not use it. Why not a special edition of Science on "Animal Agriculture Meeting Human Needs for the 21st Century"? Science magazine is the most prestigious of scientific journals and worldwide has the greatest distribution.

The role of science for future increases of food production will focus on biology. Early in this century we went through an industrial revolution. The next revolution will be a biological one. The 21st century can be projected as a biological opportunity century with renewed emphasis during the decades of the eighties and nineties not only in the production, but also the use of renewable resources.

Control of the basic biological processes that regulate and limit crop and livestock productivity will become increasingly important. This means greater emphasis on genetic improvements, reproductive efficiency, improved animal health, emphasis on protein synthesis and less fat, alleviation of stress and resource conservation.

Finally, serious attention must be given to support of food production research. No more than 3% of the world's expenditure for research is allocated to agriculture. Most of this is for large-scale operations in temperate zone agriculture in developed countries. As for the United States, support of agriculture research is approaching a disastrous crisis stage. Only 2% of our total research and development is now going to agriculture and food research. This is considerably below the global percentage investment. Reputable and credible reports in the mid-1970s suggested that one-third of the world's agricultural research was conducted in the United States. That is not true for 1980. The free world has fallen seriously behind the Soviets not only in the arms race, but also they have now edged us out in research investments relating to production and stability of our food supplies.

This is a most unusual Conference. The assemblage consists of some 200 invited experts, mostly working scientists, not bureaucrats or administrators, gathered together in a remote area in Michigan for a week to determine national research priorities for the 1980s and the 21st century. The intent is to look at human needs and what can be done about them. Those, the actors, that will be involved in implementation of the recommendations are purposely a part of the process. This is a genuine cooperative effort of state agricultural experiment stations and the federal agricultural research system. This is an interdisciplinary effort. Not just animal scientists are involved. Representatives from academia, government and the private sector are here. By choice, all are the best minds to be recruited. Some of those that determine food, agriculture and nutrition policy have participated as keynote speakers. This is not just another group of self-serving professionals with only their provincial interests to espouse.

A special working group in this Conference is addressing research imperatives relating to environmental and climatic stresses on animals. It is projected that the rising atmospheric carbon dioxide (CO_2) will change climate. Research seeking technologies to alleviate these stresses induced by climate change far transcends the

CO_2 issue and is important now to cope with interannual variations which already exist. Here, and in the overall sponsorship of this Conference, the American Association for the Advancement of Science is playing a key role. Assistance in the area of environmental stress is provided through the CO_2 research program of the Department of Energy.

Topics in this Conference range from health and nutrition issues to environment concerns, resource inputs, economics, food safety, biological efficiencies and public policy, to the socio-political aspects. Food acceptability is more than good nutrition. There are issues of appearance, texture, flavor and color. Also, food is consumed not only to sustain life but also because it tastes good. There is joy in eating.

Finally, this Conference is following the model of its counterpart of 5 years ago, "Crop Productivity--Research Imperatives" (Brown et al.,, 1975) (table 12.12). The product of that Conference, jointly sponsored by the Michigan Agricultural Experiment Station, the Charles F. Kettering Foundation and four federal agencies, has had a major impact for increased funding of the plant sciences through a competitive grant program now in its third year in the Department of Agriculture. We look forward to an equally significant impact of this Conference and its proceedings on corrective measures for the currently low research support for animal agriculture, the kinds of research to be done and how such research should be managed.

REFERENCES

Abelson, P.H. 1980. Diversion of funds from research. Science (Editorial) 208, April 25, 1980.

Brown, L.R. 1980. Food or Fuel: New Competition for the World's Cropland. Worldwatch Paper 35. Washington, DC.

Brown, A.W.A., T.C. Byerly, M. Gibbs and A. San Pietro. 1975. Crop Productivity-Research Imperatives. Michigan State Univ. Agr. Exp. Sta. and Charles F. Kettering Foundation.

Crosson, P. 1980. Draft Report to the Food and Agriculture Program of the International Institute of Applied Systems Analysis on Limits and Consequences of Food Production Technologies. Commission on International Relations, National Research Council. Washington, DC.

Eckholm, E.P. 1976. Losing Ground, Environmental Stress and World Food Prospects. W.W. Norton and Co., Inc. NY. 223 p.

Horwitz, A. 1977. Worldwide perspective on animal disease losses and their impact on nutrition, health and development. Pan American Health Organization. Washington DC.

Leopold, A.C. 1979. The burden of competitive grants. Science (Editorial) 203, February 16, 1979.

Mayer, J. 1976. The dimensions of human hunger. Sci. Amer. 235(3)46:462.

National Academy of Sciences. 1979. Energy in Transition 1985-2000. National Academy of Science-National Research Council, Washington, DC.

National Science Foundation. 1973. Interactions of science and technology in the innovative process: some case studies. Battelle Columbus Laboratories Contract NSF-C 667.

Pimentel, D., P.A. Oltenacu, M.C. Nesheim, J. Krummel, M.S. Allen and S. Chick. 1980. The potential for grass-fed livestock: resource constraints. Science 207:843.

Presidential Commission on World Hunger. 1979. Preliminary Report. Dec. Washington, DC.

Ruttan, V.W. 1977. Induced innovation and agricultural development. Food Policy 2(3):196.

Schultz, T.W. 1979. The Economics of Research and Agricultural Productivity. Paper presented at the Seminar on Socio-Economic Aspects of Agricultural Research in Developing Countries May 7-11, 1979, Santiago, Chile, and published as an occasional paper of the International Agricultural Development Service.

Water Resources Council. 1978. The Nation's Water Resources 1975-2000. Washington, DC.

Wittwer, S.H. 1980. U.S. and Soviet agriculural research agendas. Science (Editorial) 208, April 18, 1980.

TABLE 12.2. GLOBAL PROBLEMS AND AGRICULTURAL PRODUCTIVITY

Poverty
Population increase
Inflation
Shortage of firewood
Malnutrition
Water-logging and salinization
Underemployment
Uncertainties of energy supplies
Deforestation
Toxic chemicals in the environment
Soil erosion
Improving production and yield stability
Changing climate
Grain--food/energy conflicts
Communication gap between agriculturists and policy makers

TABLE 12.3. AGRICULTURAL AND FOOD EVENTS, 1971-1978

Item	Year		
	1971	1974	1978
Set aside land, 10^6 acres	37	0	14
Grain stocks, 10^6 metric tons	51	31	74
Agricultural exports, 10^6 \$	8	22	29
Milk production per cow, 10^3 lb	10	10.3	11.2
Yields of corn, bu/acre	88	71	109 (1979)
Nitrogen fertilizer, ¢/lb	4.8	15	10
Corn, \$/bu	1.08	3.03	2.20
Wheat, \$/bu	1.34	4.09	2.94
Soybeans, \$/bu	3.03	6.64	6.75
Sugar, ¢/lb	8.5	29.5	11.0
Fuels, 1967=100	114	208	322
Income for food, %	16.5	17.1	16.5
Agricultural income, 10^9 \$	61	100	126

TABLE 12.4. CAUSES OF PLATEAUING IN AGRICULTURAL PRODUCTIVITY

Soil erosion--loss of topsoil
Loss of organic matter--soil compaction
Chemical soil residues--air pollution
More less productive land under cultivation
Increased pressures on productive land base
Fewer options for water, fertilizer, pesticide uses
Climate and weather fluctuations
Increased regulatory constraints
Decreased support for agricultural research

TABLE 12.5. GLOBAL RESEARCH AND DEVELOPMENT BUDGET (1978)

Program	Share, %
Military	24
Basic Research	15
Space	8
Energy	8
Health	7
Information Processing	5
Transportation	5
Pollution Control	5
Agriculture	3
All Others	20
Total	100

TABLE 12.6. U.S. RESEARCH AND DEVELOPMENT BUDGET[a]

Program	1980	1981 Proposed	% change
		Billions of dollars	
Defense	13.8	16.6	20.3
Energy	4.9	5.1	4.1
Space	5.1	5.6	9.8
Health	3.8	4.0	5.3
NSF	0.9	1.1	22.2
Agriculture	.74	.78	5.4
EPA	.415	.445	12.5
All Other	---	2.5	---
Total		36.1	

[a]The Agricultural Research Budget is 2.2% of the total.

TABLE 12.7. ESTIMATED AND PROJECTED CONSUMPTIVE WATER USE IN U.S.

Use	Year			
	1965	1980	2000	2020
	(Billions of gallons consumed daily)			
Agriculture - Irrigation	65	82	90	96.9
Agriculture - Livestock	1.6	2.2	3.1	4.2
Total - All uses	78	104	128	157

TABLE 12.8. RANK OF STATES IN 1978 USDA COMPETITIVE GRANTS PROGRAM

State	Millions of dollars
California	1.291
Massachusetts	1.147
New York	1.007
Illinois	.913
Michigan	.775
Ohio	.727
Minnesota	.615
Oregon	.599
Wisconsin	.534
Maryland	.526
Texas	.490

TABLE 12.9. USDA GRANTS AWARDED-FY 1979

Rank	State	Special and Competitive
		Millions of Dollars
1	California	2.663
2	New York	2.162
3	Michigan	1.628
4	Iowa	1.572
5	Illinois	1.430
6	Wisconsin	1.277
7	Ohio	1.169
8	Texas	1.107
9	Minnesota	1.106
10	Missouri	.969

TABLE 12.10. RECENTLY IMPOSED REGULATIONS ON AGRICULTURAL RESEARCH AND PRODUCTIVITY

Occupational safety and health
Human subjects requirements
Personnel relations
Union/management relationships
Affirmative action
Waste handling and disposition
Handling and disposition of chemicals
Laboratory animals
Building codes, ordinances, and standards
Accountability
Budget management
Freedom of information
Acts--Clean Water; Federal Insecticide, Fungicide, and Rodenticide (FIFRA); Toxic Substance; Clean Air; Rebuttable Presumption Against Registration (RPAR)

TABLE 12.11. TIME FOR ADOPTION (95% acceptance) OF NEW TECHNOLOGY IN U.S.

Examples	Interval	No. Years
Hybrid corn	1933-1969	36
Hybrid corn (Iowa)	1933-1940	7
Hybrid sorghum	1955-1970	15
Monogum sugar beet seed	1956-1965	9
Mechanical harvesting of grapes	1968-1971	3
Vaccine for Marek's Disease	1971-1973	2

TABLE 12.12. NEXT GENERATION OF AGRICULTURAL RESEARCH

Greater photosynthetic activity
Improved biological fixation
Genetic improvement with new cell fusion technologies
More efficiency in nutrient uptake and utilization
Greater resistance to competing biological systems
Greater resistance to environmental stress
Identification of hormonal systems and mechanisms

CONFERENCE PARTICIPANTS

Philip Abelson
Science Magazine
1515 Massachusetts Ave., N.W.
Washington, DC 20005

C. Eugene Allen
Dept. of Animal Science
University of Minnesota
St. Paul, MN 55108

David R. Ames
Dept. of Animal Science & Industry
Kansas State University
Manhattan, KS 66506

Clarence B. Ammerman
Dept. of Animal Science
University of Florida
Gainesville, FL 32611

David P. Anderson
College of Veterinary Medicine
University of Georgia
Athens, GA 30601

James H. Anderson
College of Agriculture & Natural Resources
Michigan State University
East Lansing, MI 48824

Steven Aust
Dept. of Biochemistry
Michigan State University
East Lansing, MI 48824

Daniel Badger
Dept. of Agricultural Economics
Oklahoma State University
Stillwater, OK 74074

C. A. Baile
Dept. of Clinical Studies
School of Veterinary Medicine
University of Pennsylvania
New Bolton Center, R. D. #1
Kennett Square, PA 19348

Frank Baker
Dept. of Animal Science
Oklahoma State University
Stillwater, OK 74078

Jon F. Bartholic
Agricultural Experiment Station
Michigan State University
East Lansing, MI 48824

Julius F. Bauerman
Horace W. Longacre, Inc.
Box 8
Franconia, PA 18924

Fuller Bazer
Animal Science Dept.
University of Florida
Gainesville, FL 32611

John W. Bennett
Dept. of Anthropology
Washington University
St. Louis, MO 63130

W. G. Bergen
Dept. of Animal Husbandry
Michigan State University
East Lansing, MI 48824

Anson R. Bertrand
Science & Education Administration
Rm. 302-A Administration Bldg.
U.S. Dept. of Agriculture
Washington, DC 20250

William G. Bickert
Dept. of Agricultural Engineering
Michigan State University
East Lansing, MI 48824

John Birdsall
American Meat Institute
PO Box 3556
Washington, DC 20007

J. R. Black
Farm Management
Dept. of Agricultural Economics
Michigan State University
East Lansing, MI 48824

Timothy H. Blosser
Program Coordinator-Dairy
Beltsville Agr. Research
Ctr.-West
Beltsville, MD 20705

Larry Bohl
Dept. of Agricultural Economics
Purdue University
W. Lafayette, IN 47907

Jenny Bond
Dept. of Food Science &
Nutrition
Michigan State University
East Lansing, MI 48824

James Bonnen
Dept. of Agricultural Economics
Michigan State University
East Lansing, MI 48824

G. E. Bradford
Animal Science Dept.
University of California-Davis
Davis, CA 95616

Robert Bray
College of Agr. and Life Sciences
University of Wisconsin
Madison, WI 53706

K. I. Brown
Dept. of Poultry Science
Ohio Agr. Research & Dev. Center
Wooster, OH 44591

Myrtle L. Brown
Food & Nutrition Board
National Academy of Sciences
National Research Council
Washington, DC 20418

William B. Buck
Dept. of Physiology
College of Veterinary Medicine
University of Illinois
Urbana, IL 61801

J. Bruce Bullock
Dept. of Agricultural Economics
Oklahoma State University
Stillwater, OK 74078

Clark R. Burbee
National Economics Div., ESCS/USDA
Rm. 260, 500 12th St., S.W.
Washington, DC 20250

David M. Burns
American Assoc. for the
Advancement of Science
1776 Massachusetts Ave., NW
Washington, DC 20036

Jerry Callis
Plum Island Animal Disease Center
PO Box 848
Greenport, NY 11994

John E. Cantlon
VP for Research & Graduate Studies
Michigan State University
East Lansing, MI 48824

Zerle L. Carpenter
Animal Science Dept.
Texas A & M University
College Station, TX 77843

A. Barry Carr
Food & Agriculture
Congressional Research Service
Library of Congress
Washington, DC 20540

T. C. Cartwright
Dept. of Animal Science
Texas A & M University
College Station, TX 77843

Robert G. Cassens
Dept. of Meat & Animal Science
University of Wisconsin
Madison, WI 53706

Ivy Celender
Director of Nutrition
General Mills, Inc.
9200 Wayzata Blvd.
Minneapolis, MN 55426

Larry Connor
Agricultural Economics Dept.
Michigan State University
East Lansing, MI 48824

Maryln K. Cordle
FDA/Bureau of Foods
HFF-350
200 C. Street, SW
Washington, DC 20204

G. L. Cromwell
Animal Science Dept.
University of Kentucky
Lexington, KY 40546

Audrey Tittle Cross
Coordinator for Human Nutrition Policy
Office of the Secretary, USDA
Washington, DC 20250

Raymond H. Cypess
Dept. of Preventive Medicine
College of Veterinary Medicine
PO Box 786
Cornell University
Ithaca, NY 14853

Jack Dendel
Michigan Animal Breeders Coop.
3655 Forest Road
East Lansing, MI 48823

C. W. Deyoe
Grain Science and Industry
Kansas State University
Manhattan, KS 66506

Gordon E. Dickerson
Res. Animal Geneticist
SEA/AR/USDA
225 MBH
University of Nebraska
Lincoln, NE 68583

Richard Dierks
College of Veterinary Medicine
University of Illinois
Urbana, IL 61801

Theron W. Downes
School of Packaging
Michigan State University
East Lansing, MI 48824

Walter L. Dunkley
Dept. of Food Science & Technology
University of California
Davis, CA 95616

Robert Dunlop
College of Veterinary Medicine
University of Minnesota
St. Paul, MN 55108

Johanna Dwyer
Frances Stern Nutrition Center
New England Medical Center
Boston, MA 02111

James Egan
National Dairy Herd Improvement Coop.
Rt. 3
Malone, NY 12953

Roy S. Emery
Dept. of Dairy Science
Michigan State University
East Lansing, MI 48824

Larry R. Ewing
Michigan Farm Bureau
PO Box 30960
Lansing, MI 48909

Theodore M. Farber
Food Animal Additives Eval.
(HFF-154)
FDA, 200 C St., SW
Washington, DC 20204

Daniel F. Farkas
Dept. of Food Science & Human Nutrition
University of Delaware
Newark, DE 19711

D. E. Farris
Agricultural Economics Dept.
Texas A & M University
College Station, TX 77843

Ray A. Field
Animal Science Dept.
University of Wyoming
Laramie, WY 82071

Neal First
Meat & Animal Science Dept.
University of Wisconsin
Madison, WI 53706

Hank Fitzhugh
Winrock International Livestock Research & Training Center
Morrilton, AR 72110

Joseph P. Fontenot
Dept. of Animal Science
VPI & State University
Blacksburg, VA 24061

R. H. Foote
Dept. of Animal Science
Cornell University
Ithaca, NY 14853

E. M. Foster
Food Research Institute & Dept. of Food Microbiology & Toxicology
University of Wisconsin
Madison, WI 53706

Eldon E. Fredericks
Dept. of Information Services
Michigan State University
East Lansing, MI 48824

K. J. Frey
Agronomy Dept.
Iowa State University
Ames, IA 50011

Seymour L. Friess
Drill, Friess, Hays, Loomis & Shaffer, Inc.
1901 N. Fort Myers Dr., Suite 204
Arlington, VA 22209

R. J. Gerrits
USDA/SEA/AR/NPS
Rm. 209, Bldg. 005, BARC-West
Beltsville, MD 20705

Darrel E. Goll
Nutrition & Food Science Dept.
University of Arizona
Tucson, AZ 85721

Jack Gorski
Biochemistry Dept.
University of Wisconsin
Madison, WI 53706

Truman Graf
Agricultural Economics Dept.
University of Wisconsin
Madison, WI 53706

Donald Gustafson
School of Veterinary Medicine
Purdue University
W. Lafayette, IN 47907

Helen A. Guthrie
Dept. of Nutrition
Pennsylvania State University
University Park, PA 16802

Gordon E. Guyer
Cooperative Extension Service
Michigan State University
East Lansing, MI 48824

Harold D. Hafs
Dept. of Dairy Science
Michigan State University
East Lansing, MI 48824

Richard Hagen
National Food Processors
1133 20th St., NW
Washington, DC 20036

LeRoy Hahn
Livestock Environment Research
USDA/SEA/AR
Clay Center, NE 68933

Diane E. Halverson
Animal Welfare Institute
Rt. 1
Nerstrand, MN 55053

Alfred E. Harper
Dept. of Nutritional Sciences
University of Wisconsin
Madison, WI 53706

Robert Harris
US Forest Service
32660 Lake Point Court
Wilsonville, OR 97070

Bert Hawkins
U.S. Animal Health Assn.
Route 1, Box 355
Ontario, OR 97914

Virgil Hays
Dept. of Animal Sciences
University of Kentucky
Lexington, KY 40506

H. F. Heady
Agricultural & University Services
University of California
Berkeley, CA 94720

W. D. Heffernan
Dept. of Rural Sociology
University of Missouri
Columbia, MO 65211

Mark Hegsted
USDA/SEA, Rm. 426A Admin. Bldg.
Washington, DC 20250

Zane Helsel
Dept. of Crop & Soil Sciences
Michigan State University
East Lansing, MI 48824

L. M. Henderson
Dept. of Biochemistry
University of Minnesota
St. Paul, MN 55108

C. H. Herbel
USDA/SEA Agr. Research Western Region
Jornada Experimental Range
PO Box 698
Las Cruces, NM 88001

J. S. Hillman
Dept. of Agricultural Economics
University of Arizona
Tucson, AZ 85721

H. J. Hodgson
Wisconsin Agricultural Experiment Station
University of Wisconsin
Madison, WI 53706

Jacob A. Hoefer,
Agricultural Experiment Station
Michigan State University
East Lansing, MI 48824

Ann Holt, Director
Division of Drugs for Swine & Minor Species
DHEW/FDA/BVM/DDSMS
5600 Fishers Lane, Rm. 6B-24
Rockville, MD 20857

R. E. Hungate
Dept. of Agricultural Bacteriology
Professor Emeritus - Microbiological-Ruminant Studies
University of California
Davis, CA 95616

N. L. Jacobson
201 Beardshear Hall
Iowa State University
Ames, IA 50011

L. S. Jensen
Poultry Science Dept.
University of Georgia
Athens, GA 30602

Norge Jerome
Dept. of Community Health
University of Kansas Medical Center
Kansas City, KS 66103

Neal A. Jorgensen
Dairy Science Dept.
University of Wisconsin
Madison, WI 53706

John W. Judy, Jr.
College of Veterinary Medicine
Michigan State University
East Lansing, MI 48824

Colin Kaltenbach
Dept. of Animal Science
University of Wyoming
Laramie, WY 82071

Max Kellough
Route 1
Friend, NE 68359

Charles A. Kiddy
USDA/SEA/AR
Rm. 206, Bldg. 005, BARC-West
Beltsville, MD 20705

Paul Kifer
Dept. of Food Science & Technology
Oregon State University
Corvallis, OR 97331

T. B. Kinney, Jr.
SEA/AR/USDA
Rm. 340 Administration Bldg.
Washington, DC 20250

J. R. Kirk
Dept. of Food Science and Human Nutrition
University of Florida
Gainesville, FL 32611

T. J. Klopfenstein
Animal Science Dept.
University of Nebraska
Lincoln, NE 68503

L. J. Koong
Animal Science Dept.
University of California
Davis, CA 95616

R. B. Land
Animal Breeding Research Organization
W. Mains Rd.
Edinburg, Scotland EH9 3JQ
United Kingdom

W. E. Larson
Soil & Water Management Research Unit
USDA/SEA/AR, NC Region
201 Soil Science Bldg..
University of Minnesota
St. Paul, MN 55108

Earl Lasley
Farmers Hybrid Co., Inc.
Box 4528
Des Moines, IA 50306

R. V. Lechowich
Food Science & Technology Dept.
Virginia Polytechnic Institute & State Univ.
Blacksburg, VA 24061

V. L. Lechtenberg
Dept. of Agronomy
Purdue University
W. Lafayette, IN 47907

A. D. Leman
College of Veterinary Medicine
University of Minnesota
1988 Fitch Ave.
St. Paul, MN 55108

Gilbert Leveille
Food Science & Human Nutrition Dept.
Michigan State University
East Lansing, MI 48824

Patrick J. Luby
Oscar Mayer & Co.
Madison, WI 53704

Daryl Lund
Dept. of Food Science
University of Wisconsin
Madison, WI 53706

Chester J. Mackson
School of Packaging
Michigan State University
East Lansing, MI 48824

Lon McGilliard
Dept. of Dairy Science
Michigan State University
East Lansing, MI 48824

W. W. Marion
Food Technology Dept.
Iowa State University
Ames, IA 50011

G. C. Marten
Plant Science Research Unit
USDA/SEA/AR
Agronomy & Plant Genetics Dept.
University of Minnesota
St. Paul, MN 55108

Stanley W. Martin
Assoc. Prof., Epidemiology
Ontario Vet. College
Univ. of Guelph
Guelph, ONT, N1G 2W1

Melinda S. Meade
Dept. of Geography
University of North Carolina
Chapel Hill, NC 27514

David Meisinger
National Pork Producers Council
Box 10383
Des Moines, IA 50312

Robert Merkel
Dept. of Animal Husbandry
Michigan State University
East Lansing, MI 48824

J. R. Miner
Agricultural Engineering Dept.
Oregon State University
Corvallis, OR 97331

Paul W. Moe
Ruminant Nutrition Laboratory
Animal Science Institute
USDA/SEA/AR
Beltsville Agricultural Research
Beltsville, MD 20705

J. G. Morris
Animal Science Dept.
University of California-Davis
Davis, CA 95616

Frank Mulhern
Animal Health, Inter-America
Institute for Agricultural
Science
Apartado 55-Coranado
San Jose, COSTA RICA

Ronald H. Nelson
Dept. of Animal Husbandry
Michigan State University
East Lansing, MI 48824

M. C. Nesheim
Div. of Nutritional Sciences
Savage Hall
Cornell University
Ithaca, NY 14853

James E. Newman
Dept. of Agronomy
Purdue University
W. Lafayette, IN 47907

R. P. Niedermeier
Dept. of Dairy Science
University of Wisconsin
Madison, WI 53706

Thelma D. Njaka
State-Federal Cooperative Animal
Health Diagnostic Lab.
4720 Brenda Lane, Bldg. #2
Charleston, WV 25312

J. E. Oldfield
Dept. of Animal Science
Oregon State University
Corvallis, OR 97331

R. R. Oltjen
Hruska U.S. Meat Animal Research Center
Clay Center, NE 68933

I. T. Omtvedt
Animal Science Dept.
University of Nebraska
Lincoln, NE 68506

Bennie I. Osburn
School of Veterinary Medicine
Dean's Office
University of California
Davis, CA 95616

F. N. Owens
Animal Science Dept.
Oklahoma State University
Stillwater, OK 74074

W. D. Oxender
College of Veterinary Medicine
Michigan State University
East Lansing, MI 48824

C. F. Parker
Dept. of Animal Science
Ohio Agr. Research & Dev. Center
Wooster, OH 44691

Allen Paul
USDA/ESCS
500 12th St.
Washington, DC 20250

A. M. Pearson
Dept. of Food Science & Human Nutrition
Michigan State University
East Lansing, MI 48824

Donald Polin
Dept. of Poultry Science
Michigan State University
East Lansing, MI 48824

Wilson Pond
Hruska US Meat Animal Res. Center
Clay Center, NE 68933

R. L. Preston
Animal Sciences Dept.
Washington State University
Pullman, WA 99164

Dean M. Pridgeon
Michigan Dept. of Agriculture
Lansing, MI 48909

William R. Pritchard
School of Veterinary Medicine
University of California
Davis, CA 95616

J. C. Purcell
IR-6, Rm. 15, Bldg. 005, BARC-West
Beltsville, MD 20705

Graham Purchase
Livestock & Veterinary Services
USDA/SEA/AR/NPS-LVS
Rm. 213, Bldg. 005, BARC-West
Beltsville, MD 20705

Chesley Randall
Livestock & Poultry Div.
Agriculture Canada
Sir John Carling Bldg.
Ottawa K1A 0C5 CANADA

Susan Reardon
College of Veterinary Medicine
Michigan State University
East Lansing, MI 48824

Naomi Revzin
Dept. of Dairy Science
Michigan State University
East Lansing, MI 48824

V. James Rhodes
Dept. of Agricultural Economics
University of Missouri
Columbia, MO 65211

Stan Richert
Protein Research, Ralston Purina Co.
Checkerboard Square
St. Louis, MO 63188

Lynne C. Rienner
Westview Press
5500 Central Ave.
Boulder, CO 80301

Larry Rittenhouse
Dept. of Range Science
Colorado State University
Ft. Collins, CO 80523

W. M. Roberts
Food Science Dept.
North Carolina State University
Raleigh, NC 27650

Lawson M. Safley, Jr.
Dept. of Ag. Engineering
Box 1071
University of Tennessee
Knoxville, TN 37901

Glenn W. Salisbury
Agricultural Experiment Station
University of Illinois
Urbana, IL 61801

L. F. Schrader
Dept. of Agricultural Economics
Purdue University
W. Lafayette, IN 47907

Hilman Schroeder
National Pork Producers Council
Box 10383
Des Moines, IA 50306

Edward Schuh
Dept. of Agricultural Economics
University of Minnesota
St. Paul, MN 55018

R. S. Sechrist
National DHIA, Inc.
3021 E. Dublin-Granville Rd.
Columbus, OH 43229

George Seidel, Jr.
Dept. of Physiology & Biophysics
Colorado State University
Ft. Collins, CO 80523

C. Richard Shumway
Dept. of Agricultural Economics
Texas A & M University
College Station, TX 77843

Artemis P. Simopoulos
Nutritional Coordinating
Committee, NIH
Office of the Director
Bldg. 31, Rm. 4B59
Bethesda, MD 20205

Lewis W. Smith
USDA/SEA/AR, Animal Science Inst.
Bldg. 200, Rm. 217, BARC-East
Beltsville, MD 20705

Elwood Speckmann
National Dairy Council
6300 N. River Rd.
Rosemont, IL 60018

William Stadelman
Dept.of Animal Science
Purdue University
W. Lafayette, IN 47907

J. J. Stockton
School of Veterinary Science
& Medicine
Purdue University
W. Lafayette, IN 47907

J. C. Street
Dept. of Animal, Dairy &
Veterinary Science
Utah State University
Logan, UT 84322

David H. Stroud
PO Box 11227
Chicago, IL 60611

W. Bert Sundquist
Dept. of Agricultural & Applied
Economics
University of Minnesota
St. Paul, MN 55108

J. M. Sweeten
Agricultural Engineering Dept.
303 Ag. Eng. Bldg.
Texas A & M University
College Station, TX 77840

Howard Teague
USDA/SEA/CR
Washington, DC 20250

Mary E. Templeton
Division of Resource Management
West Virginia University
Morgantown, WV 26505

W. C. Templeton, Jr.
Director, US Regional Pasture Research Lab.
USDA/SEA/AR
University Park, PA 16802

William W. Thatcher
Dept. of Dairy Science
University of Florida
Gainesville, FL 32601

G. W. Thomas
Office of the President
New Mexico State University
Box 3Z
Las Cruces, NM 88003

D. G. Topel
Animal and Dairy Science Dept.
Auburn University
Auburn, AL 36830

R. W. Touchberry
Dept. of Animal Science
University of Minnesota
St. Paul, MN 55108

H. F. Troutt
College of Veterinary Science
Virginia Polytechnic Institute
Blacksburg, VA 24061

H. A. Tucker
Dairy Science Dept.
Michigan State University
East Lansing, MI 48824

D. E. Ullrey
Dept. of Animal Husbandry
Michigan State University
East Lansing, MI 48824

Roy N. Van Arsdall
Ag. Economist, ESCS/USDA
305 Mumford Hall
Urbana, IL 61801

P. J. Van Soest
Dept. of Animal Science
Cornell University
Ithaca, NY 14853

R. L. Vetter
A O Smith Harvestore Products, Inc.
PO Box 395
Arlington Heights, IL 60006

W. J. Visek
College of Medicine
Rm. 190, Med. Sci. Bldg.
University of Illinois
Urbana, IL 61801

W. J. Waldrip
Spade Ranches
Box 2763
Lubbock, TX 79408

J. N. Walker
Dept. of Agricultural Engineering
University of Kentucky
Lexington, KY 40546

Aaron Wasserman
USDA/Eastern Regional Research Lab.
USDA/AR/SEA
600 E. Mermaid Lane
Philadelphia, PA 19118

W. F. Wedin
Dept. of Agronomy
Iowa State University
Ames, IA 50011

John R. Welser
College of Veterinary Medicine
Michigan State University
East Lansing, MI 48824

R. O. Wheeler
Winrock International
Route 3
Morrilton, AR 72110

R. K. White
Dept. of Agricultural Engineering
Ohio State University
Columbus, OH 43210

Jeffrey Williams
College of Veterinary Medicine
Michigan State University
East Lansing, MI 48824

J. Karl Wise
American Veterinary Medical Staff
930 N. Meacham Rd.
Schaumburg, IL 60196

Milton B. Wise
College of Agriculture & Life Sciences
Virginia Polytechnic Institute
Blacksburg, VA 27061

Sylvan Wittwer
Agricultural Experiment Station
Michigan State University
East Lansing, MI 48824

Virgil O. Wodicka
Food Technologist
1307 Norman Place
Fullerton, CA 92631

Catherine Woteki
USDA/SEA/HN/CFEI
Rm. 324, 6505 Belcrest Rd.
Hyattsville, MD 20782

Jane Wyatt
Dept. of Food Science
Oregon State University
Corvallis, OR 97331

Robert G. Zimbelman
Agricultural Division
The Upjohn Company
Kalamazoo, MI 49001